Analytical Sedimentology

Analytical Sedimentology

Douglas W. Lewis
David McConchie

CHAPMAN & HALL
New York • London

An Apteryx Book

This edition published by
Chapman & Hall
One Penn Plaza
New York, NY 10119

Published in Great Britain by
Chapman & Hall
2-6 Boundary Row
London SE1 8HN

Printed in the United States of America

Library of Congress Cataloging in Publication Data

Lewis, D.W. (Douglas W.), 1937-

    Analytical sedimentology / D.W. Lewis and David M. McConchie.
     p. cm.

    Includes bibliographical references and index.
    ISBN 0-442-01216-0
    1. Sediments (Geology)—Analysis. I. McConchie, David. II. Title.
QE471.2.L49 1993
552' .5—dc20   93-15894
                                      CIP

British Library Cataloguing in Publication Data available

Please send your order for this or any Chapman & Hall book to **Chapman &
Hall, 29 West 35th Street, New York, NY 10001, Attn: Customer Service
Department.** You may also call our Order Department at 1-212-244-3336 or fax
your purchase order to 1-800-248-4724.

For a complete listing of Chapman & Hall's titles, send your requests to
Chapman & Hall, Dept. BC, One Penn Plaza, New York, NY 10119.

CONTENTS

PREFACE

The first edition of *Practical Sedimentology* contained discussions of principles and techniques that could be applied to the analysis of sediments in the field and in laboratories supplied with inexpensive and commonly available equipment. When considering a revised edition, we felt that it was inappropriate to restrict consideration to the simple and common techniques because so many modern analyses of sediments use sophisticated and often expensive equipment to examine sediments and sedimentary rocks. A review of the wide range of available techniques and equipment was not feasible in the same volume as a review of principles. The original intent to produce a concise summary of practical sediment studies in an inexpensive format was maintained, but now in the form of two volumes: one on principles and one on analytical procedures. Despite the existence of a few excellent books on sediment analysis (e.g., R. E. Carver's *Procedures in Sedimentary Petrology* [Wiley-Interscience, New York, 1971] and M. Tucker's *Techniques in Sedimentology* [Blackwell Scientific Publications, Oxford, 1988]) none was as wide-ranging as we thought necessary, nor did any attempt the equivalent treatment of the now-outdated Krumbein and Pettijohn book (*Manual of Sedimentary Petrography* [Appleton-Century-Crofts, 1938]). Our aim for *Analytical Sedimentology* is similar to that of the 1938 book—to describe details of simple and common analytical procedures, and to survey principles, advantages, and limitations of the wider array of techniques that can be applied to sediments. In addition, it appeared appropriate to extend consideration to field mapping procedures.

Although the procedures and advice presented in this book have been developed with operator safety as an important consideration, the authors accept no responsibility for accidents that may arise through the application of any of the methods described.

ACKNOWLEDGMENTS

Colleagues at the University of Canterbury and the University of New England, Lismore, have helped with practical advice on their experiences with various methodologies discussed in this volume. At the University of Canterbury, we are particularly grateful to K. Swanson for advice on preparing materials for scanning electron microscopy and paleontological specimens; to G. Coates (working at the university at the time of the first edition of *Practical Sedimentology*) for compilation of, and additions to, the procedures for textural analysis and some tables and sketches; to Ted Montague for the bulk of the chapter on borehole sedimentology; to Dr. J. Newman for her contributions concerning coal petrologic techniques; to Dr. D. Wise (University of Massachusetts) for his permission to use unpublished material on the preparation and analysis of rose diagrams for directional and orientational data on paleocurrents. At the University of New England, Fiona Davies-McConchie provided practical suggestions from her experience with laboratory techniques and executed several library searches for relevant publications, as well as typing drafts of several chapters. Dr. C. Pailles provided the section on diatom extraction and analysis, G. Lancaster assisted with developing some of the laboratory procedures, and M. Clark helped develop and test several selective extraction techniques described. Mrs. Julie-Ann Higgins and Mrs. Joan McTurk (both of the University of Canterbury) kindly typed some of the more complicated tables. Marjorie Spencer, of Van Nostrand Reinhold Publishing Company, initially put much time and effort into the planning of the books. Bernice Pettinato (Beehive Production Services) capably and professionally handled the manuscript editing and production of the final presentation.

1
Introduction

Analytical Sedimentology (henceforth **AS**) is a companion volume of *Practical Sedimentology,* second edition (henceforth **PS**), and cross-references between the volumes are commonplace because the techniques reviewed in the former are applied to principles and interpretations reviewed in the latter. Nonetheless, we hope that users will find each volume to be of practical value on its own, or in conjunction with their preferred book covering the other aspect of sedimentology. In this volume, we describe analytical methodology applied to sediment studies, point out options available to accomplish the various analytical purposes, and provide practical advice on procedures with which we or some of our colleagues have had experience. Our coverage ranges from consideration of mapping sediments in the field, through sieve/pipette techniques for analyzing grain size and petrographical procedures for mineral identification of components, to instrumental analytical procedures such as X-ray fluorescence, scanning electron microscopy, and atomic absorption analysis.

For some methods we provide "cookbook recipes" for part or all of the preparation and analytical procedure; in others we only describe the techniques in general terms because they can vary depending on the specific kind of equipment available or the sample characteristics. We outline the principles on which techniques are based, but have not attempted to go into depth on most: the book would have quadrupled in size without any change in scope, and we have attempted to focus on the practical requirements of most investigators. Most techniques and reasons for their application are generally described so that a competent scientist or technician with laboratory experience can apply them with an understanding of their purpose and the likely causes of imprecision or error; nonetheless, some will be useful only to those with some specialist technical or geological training (e.g., those requiring

knowledge of the petrographic microscope and characteristics of minerals in thin section). We have provided selected bibliographies on the various methodologies, so that workers desirous of undertaking any of the inadequately described analytical methods can go to authoritative and comprehensive descriptions of the relevant instrumental principles and procedures. Although this book provides useful guidelines for the selection of analytical techniques suited to particular studies, we advise readers intending to use any of the methods, particularly the more complex instrumental techniques, to consult the relevant references listed in each chapter, the operation and procedures manuals for their particular instrument, and any colleagues with experience in the application of the procedure or use of the instrument. Most analysts have found out the hard way that taking the time to learn to operate an instrument or carry out a procedure correctly will be rewarded by not having to repeat work they bungled because they overlooked some small but important detail or tried to take an unsuccessful shortcut.

We stress that, despite the cautions expressed in the text and in the special section below on safety, only personnel trained in field and laboratory safety procedures should perform many of the procedures discussed in this book.

SAFETY CONSIDERATIONS

Both field and laboratory work are potentially hazardous and it is a mistake to underestimate the risks; for example, there are geologists who have lost an eye to a hammer or rock fragment while not wearing glasses in the field and others who have received severe burns or lost fingers to laboratory reagents such as hydrofluoric acid. Everyone planning field or laboratory work should make their own safety and that of others an important part of the planning process. Proper planning and familiarity with appropriate safety precautions are the best ways to avoid accidents. Do not assume that because a procedure has always worked before it will work well this time.

Laboratory Work

The laboratory is potentially a very hazardous environment, but the hazards can be minimized by careful planning and by discussing proposed activities with colleagues who are experienced in the procedures to be used. Planning laboratory work carefully and anticipating what could go wrong are the best foundations for laboratory safety; numerous books are available to assist with this planning (e.g., Manufacturing Chemists Association 1972; Sax and Lewis 1986; Bretherick 1986; Furr 1990). Sedimentologists commonly use a wide variety of potentially hazardous laboratory reagents including bromoform and tetrabromoethane for heavy mineral separation, concentrated corrosive reagents and oxidizing agents for rock or mineral digestions (be particularly wary of hydrofluoric acid, perchloric acid, or concentrated hydrogen peroxide and nitric, hydrochloric, or sulfuric acids), strong reducing agents (e.g., sodium borohydride) for selective extractions and hydride generation, toxic reagents (e.g., cyanide gold leach solutions), narcotic or intoxicating reagents (e.g., ethyl acetate or chloroform) for elemental separations, and highly flammable chemicals. Particularly when developing new procedures there will always be some risk involved that should never be underestimated. If formal safety procedures have not been developed for the laboratory or need to be refined because laboratory use has changed, develop and adopt an appropriate safety program (e.g., Furr 1990). In most countries there are formal legal requirements that must be met in the design and operation of laboratories, and these requirements will also need to be identified and addressed as part of the development of a safety program.

As a general rule, keep the work area clean and well organized not only because of safety considerations but also to minimize the risk of sample contamination. The following are particularly important points to consider before starting work in the laboratory:

1. Familiarize yourself with the location and operation of all safety facilities in the laboratory (e.g., fire-fighting equipment, safety showers, first aid supplies, eyewash bottles, emergency telephone numbers, etc.) and follow the safety procedures instituted in your laboratory.

2. Familiarize yourself with all hazards associated with the specific reagents or procedures likely to be used by talking to experienced colleagues and consulting appropriate literature. Ensure that first aid materials (e.g., antidotes, neutralizing solutions, etc.) appropriate to these reagents or procedures are available. Safety data sheets should be supplied with most laboratory reagents; if they are not, see sources such as Bretherick 1986 or Lenga 1988, where reagents are listed alphabetically and data are provided on selected properties, toxicity, health hazards, first aid procedures, handling and storage considerations, waste disposal requirements, etc. (Many nonspecialist doctors are unfamiliar with procedures for treating some chemical burns or poisonings.)

3. Wear safety clothing, footwear, gloves, and glasses appropriate to the procedures and reagents used. For particularly hazardous reagents (e.g., hydrofluoric acid) that require the use of safety gloves, it is safer to use disposable gloves because previously used gloves may have been improperly cleaned, punctured, or contaminated.

4. Where alternative reagents are available for a particular task, use the least hazardous alternative.

5. Ensure that all reagents are correctly stored when not in use (consult references such as Lenga 1988, Furr 1990, and the legal requirements that apply in the state or country where the laboratory is located).

6. Check fume hoods regularly to ensure that they are operating efficiently and ensure that all electrical equipment in the laboratory is well maintained and subjected to regular safety checks. Many laboratory fires are started by faulty electrical equipment; even low concentrations of corrosive vapors, which commonly exist in a laboratory, can accelerate the corrosion of insulation and electrical contacts.

7. When carrying out any laboratory work that involves hazardous reagents or procedures, do not work alone.

8. Be aware of all work being conducted in the laboratory by others. Not all workers may be as safety conscious as you are and their carelessness could put you at risk.

9. If using radioactive materials or equipment that produces ionizing radiation (e.g., X-ray or Mössbauer equipment), make sure that you are familiar with safe handling and operating procedures (e.g., see Martin and Harbison 1986) and that you hold any operator's licenses that may be required by radiation safety regulations in your area.

10. Ensure that the correct procedures are used for disposal of waste materials or reagents (e.g., see Armour, Browne, and Weir 1984). Clean up any chemical spills promptly; a colleague may not know what was spilled and will thus be unaware of the nature and extent of the hazard associated with the spill.

Fieldwork

Many accidents can be avoided by wearing appropriately robust field clothing and boots; eye injuries can be avoided by wearing safety glasses when using rock hammers. Carry spare warm clothing in case the weather becomes unexpectedly bad or you get caught out at night—hypothermia (or exposure) is a common cause of fatalities. Carry a comprehensive first aid kit and know how to apply it for cuts, broken bones, animal bites, and allergic reactions: take a short course in first aid! The use of explosives—for sampling or in seismic programs—is particularly hazardous and in most places this work can be carried out only by an experienced and licensed shotfirer; all other persons working at the site must follow the shotfirer's instructions.

In the field, most safety incidents are associated with road vehicles or boats. There will always be some risk of accident, particularly when a vehicle is being used in adverse conditions or near its limits, but a few simple rules will help minimize this risk:

1. Be aware of your own capabilities and experience at operating the vehicle; take a training course with an experienced operator if you can.
2. Know the capabilities of your vehicle and do not try to push it further; it may be quicker to walk that last kilometer to the outcrop than to walk back to the highway and locate a tow truck!
3. When using public roads or waterways, obey all relevant laws, use a defensive driving approach, and ensure that you hold the relevant licenses. In many places, if the driver does not hold the relevant license or breaks certain laws, some or all of the insurance policies may be void.
4. Alcohol and driving do not mix. The drink will taste much better back at base than in a hospital.

Further, in particularly remote areas:

5. Never work alone and whenever possible use two or more vehicles in case one is immobilized.
6. Advise someone where you are going and when you expect to return; an efficient two-way radio may be a wise precaution.
7. Ensure that your vehicle is in good mechanical condition and carry a safety margin of fuel. Because minor problems such as blocked fuel lines, electrical faults, flat tires, and broken fan belts are inevitable, carry appropriate spare parts and tools.
8. For the breakdown you cannot repair, carry sufficient food and water to ensure that you can survive until help arrives. Familiarize yourself with the accepted survival strategies for the area you plan to work in.

See American Geological Institute (1992) for a comprehensive discussion of field safety.

UTILIZING PERSONAL COMPUTERS

Computer-based analysis spans all aspects of data collection and manipulation. Barely over a decade ago, the major applications of computers in earth science focused on the manipulation of very large databases (such as geophysical information) via number-crunching programs requiring professional programmers and expensive mainframe computers. Today, a microcomputer (or workstation that behaves like one) sits on virtually every earth scientist's desk and multiple

programs are used by people who could not write a computer program for love or money. Specialist programs and programmers still exist, and mainframe or minicomputers still manipulate massive databases and create simulations, but the focus of most individuals is on the utilization of commercial software written without any thought of its application to the earth sciences. Utilization of microcomputers for scientific purposes ranges from acquisition of data via instrument logging and importation from external databases, to data manipulation and analysis, to word processing and preparation of maps and diagrams for reports—computers are now used in every aspect of analysis. There are a growing variety of inexpensive (and expensive!) specialist earth science programs that have been created specifically for microcomputers by individuals (who either publish the program in their selected language, offer to provide a working copy upon request, or provide it via public domain or shareware routes) and a very few commercial organizations. Even detailed contour maps (e.g., Isdale 1991) can now can be constructed by inexpensive commercial microcomputer programs.

It is difficult to review microcomputer applications to sedimentology not only because of the wide range of applications but also because of the exponential rate of change in the character of the equipment and programs available: any review such as this is bound to be antiquated by the time it goes to press. New programs and updates of old programs are so numerous in the various branches of earth science, both on a local as well as an international scale, that it is even impossible to collate and list available programs, let alone provide an evaluation of their quality. Mainframe earth science programs are not available to us, and in general are tailored by professional programmers to specific demands of individual organizations; hence discussion of them is not attempted. We attempt here to merely introduce some of the salient considerations for microcomputer hardware and some potential applications of software packages, based on our limited experience as users: we regret that we have not provided a well-balanced overview of equipment and software available, and those items specifically mentioned are merely examples and are not necessarily recommended. Fortuitously, one of us is a Mac-fanatic and the other a True Blue (clone) user; hence treatment will not be biased as to operating system and program types. ("Mac" is henceforth used as an abbreviation for the Macintosh™ computer systems and "PC" for the IBM™-style microcomputing systems).*

Hardware

Although the statement will appear heretical to existing computer owners, the kind of computer (PC or Mac) purchased is not nearly as important as the software that is available, affordable, and necessary for future applications! Decide on what work you require the computer for (e.g., numerical ma-

---

*All brand and product names are trademarks of their respective owners. All trademarks mentioned and used in this chapter are acknowledged.

nipulation, instrument logging, or graphic constructions), then settle on the most appropriate software, and finally choose a computer to run it. Differences between the Mac and PC platforms are closing rapidly and the next generation of both will probably be fully compatible. Software (such as Mac Link PC™) facilitates data transfer between Macs and PCs, and many modern computers (such as the Mac Power-Book 180™) can run either Mac or PC software.

In both the PC and Mac worlds, specific hardware configuration generally depends on how much money you want to invest. The most desirable microcomputers for all-round use are those with the biggest RAM memories and the fastest speeds; a hard disk (120 megabytes probably the very minimum) is another vital component, and a high-resolution monitor is an important consideration for those likely to work with a computer for many hours per day. Programs are getting larger and more complex every year, and what was a good bottom-of-the-line computer a few years ago (e.g., an XT-type PC or the original Mac) is no longer worth having for serious scientific work. The industry standard at present is approximately as follows: in the PC line—an 80486, 33- or 50-megahertz machine with 4–8 megabytes of RAM (minimum for satisfactory multitasking operations), a 120-megabyte (preferably larger) hard disk, and a (Super-)VGA color monitor; in the Mac line—a Mac IIsi or better with 4–10-Mb RAM, 80+-Mb drive, and 25-MHZ or faster clock speed. A math coprocessor is desirable with either type of computer if much number crunching is intended (the PC 486 machines and Mac Quadras™ have the equivalent built in). High-capacity disk drives (taking1.4-Mb 3.5-in. disks) are most desirable, and an internal or external CD-ROM player is becoming a standard accessory. Once you achieve these components, start adding megabytes of RAM, a high-resolution 17-in. color monitor, a larger-capacity hard disk, and a tape or catridge hard-disk backup system.

When choosing a computer system it is also worth considering the merits of portable or laptop machines (e.g., the Mac PowerBook™ model 180 and up, or one of a wide range of PC machines), particularly if a high proportion of your time is spent working away from the office. Many modern portable machines have 4- (or more) Mb RAM, 40–160-Mb hard disks, and can be used to send faxes or communicate via a modem with the central office computer. In the field, a portable will enable you to download data from a variety of field instruments, plot and review field data as they are obtained, and start writing up the work in the evenings or when adverse weather prevents regular fieldwork. In many areas, cellular or mobile phones can now be combined with your portable computer to transfer information direct to the office while you are still in the field. Cellular phones can also be linked to data loggers so that data can be regularly downloaded to your office computer without having to make regular visits to the field to collect it personally.

Laser printers are effectively the standard these days, but if the cost is too high for initial individual purchase, ensure that the printer you select is acceptably fast on NLQ (near letter quality) and will print in graphics mode (or else the range of fonts available to you, as well as many attributes of graphics images, will be severely restricted). DPI (dots per inch) is one of the critical features—the larger the DPI, the higher the quality of the output (and commonly the slower the rate of printing).

A modem is essential for communicating with other computer systems via the telephone; commercial information sources and public-domain bulletin boards become accessible. When purchasing a modem, ensure that the communication software is fully standard, that there is good error checking both in transmission and reception, and that information is transmitted at a fast baud rate (9600 is a relatively slow standard these days). Local area networks (LANs) established by your organization may provide access to these sources without the necessary use of a modem, but purchase of the correct network (e.g., EtherNet™) card must be made. With a modem you can also access bulletin boards that provide a wide range of geological and computing information; information on other bulletin boards and a vast array of other information can be accessed at no charge through COGSnet™ (sponsored by the Computer Oriented Geological Society and the Society of Mining, Metallurgy and Engineering). With the appropriate communications software, your modem or local area network will allow you access to InterNet and expose you to the world of e-mail (a wonderfully efficient means of communication, but also a source of voluminous "junk mail"). InterNet will also allow you to do on-line searches of many library catalogues around the world. If your organization has, or has licensed access to, an electronic abstract or reference database (e.g., GeoRef™, Current Contents™), your modem or network will also enable you to carry out interactive literature searches. Modems provide many benefits, but be cautious when using them to communicate with other computers because they are effective means for transmitting viruses.

Software

There are three "layers" of software: public-domain and shareware programs (inexpensive and commonly good for specific applications, but without all the "bells and whistles" or else not terribly robust); commercial programs (the vast majority in use); and highly expensive specialist programs designed for very particular and demanding purposes. For most of us, ingenuity in adapting shareware and commercial software to sedimentological projects is the most important key to successful computer utilization (see Green 1991). Many programs can be made to do things the programmers never imagined! It is important to check the capability of the program to accomplish the goals of all projects you are likely to undertake: it is not easy or cheap to change programs after you learn how to utilize one properly! Ease of relearning is important because there are some programs that may be used only once every couple of months, and if you have to spend hours relearning its operation each time, you will waste time and grow annoyed (Mac software has an edge here over software for PCs because the operational architecture and many features and commands are common to all Mac packages; in addition, it is relatively easy to transfer text, data, graphics, etc. between different packages). Word-of-mouth from colleagues is commonly the best guide to the benefits and drawbacks of particular software packages; for most commercial programs, descriptions even by reviewers in computer magazines seldom adequately review the advantages/disadvantages for earth science capabilities.

Multitasking software to permit simultaneous operation of several programs (e.g., word processor, bibliographic database, and spreadsheet) is generally desirable if your computer operates sufficiently fast and has sufficient accessible RAM or virtual memory. Common PC multitasking systems are Microsoft's WINDOWS™ (full graphics environment), International Business Machine's PS2™ and Quarterdeck's DESQVIEW™ (text-based). Memory-management software is provided by these same companies (Quarterdeck provides a separate product called QEMM™ for DESQVIEW), and modern DOS operating systems (such as Digital Research's DR DOS6™, IBM's OS2/2™, and Microsoft's MSDOS 6™) also provide systems that may be compatible with these multitasking systems. The Macs with sufficient memory at present achieve multitasking through MULTIFINDER™ or SYSTEM 7. (At least 4-Mb RAM is needed for useful multitasking in both Mac and PC environments.)

Warning

It is very important to develop a habit of routinely backing up all your software (and new or changed files more often) onto floppy disks, tape-backup systems, or external hard disks; power failures or spikes at inopportune times, or the arrival of a virus, may otherwise destroy weeks of work. Good commercial programs for rapid backing up are reasonably priced (e.g., for PCs, FASTBACK™ from Fifth Generation Systems Inc.™, THE NORTON BACKUP™ from Symantec Corp.™, Peter Norton Group, PCBACKUP™ from Central Point Software™; for Macs, backing up individual files, applications, etc., is simply a matter of dragging appropriate icons to the icon for the backup disk or tape, although there are programs such as AUTOSAVE II™ from Magic Software Inc. It is also vital to acquire a good, regularly updated virus-detection program (e.g., for PCs Symantec's ANTIVIRUS™ or McAfee's SCAN™ and CLEAN™ from McAfee Associates, 4423 Cheeney Street, Santa Clara, CA 95054-0253 or F-PROT from Fizidrik Skulason, Postholf 7180, 15–127, Reykjavik, Iceland [free for individual use]; for Macs, a recent version of DISINFECTANT™, Northwestern University [Evanston, IL], is a widely used antiviral package) and to routinely check all floppy disks and executable (or overlay) files before you add them to your hard disk (e.g., via modem). Virus-protection software that *prevents* unauthorized writing to system files and areas (e.g., available in or from the same sources as mentioned above) is also desirable to avoid the extensive damage and time-consuming replacement operation necessitated by acquiring one of these products of deranged minds.

Integrated Software
There are many advantages in utilizing programs that can easily and rapidly exchange information (e.g., graphics and text), and some thought should be given to this capability when selecting programs. This point is particularly important for PC users, because the major strength of the Mac system is that each program *must* meet the same basic communication standard to operate. Some companies sell separate software for each purpose and ensure easy exchange between their own products (e.g., WordPerfect™, DrawPerfect™, PlanPerfect™, DataPerfect™); others sell packages that integrate several capabilities (e.g., graphics, spreadsheet, word processing such as Symphony™, Framework™, Q&A™) in a single product that effectively carries out the exchanges automatically (note that each component is usually less powerful than the better stand-alone products). Commonly, exchange of information between products of different major companies is easy enough (e.g., via ASCII, DIF, or DXF formats, which are available in most programs), and there are commercial translation programs that take care of the more difficult cases. However, there continue to be major difficulties in importing, exporting, or exchanging graphics images in PC systems and between PCs and Macs at the time of writing.

Word-Processing Software
Word-processing software should be easy to use and intuitive, but in most packages it is reasonably easy to learn the basics and consequently the most important consideration is the range of capabilities that are available: even if you are unlikely to use some of them more than a few times a year, the time saved by executing them may be substantial (e.g., compiling an index or table of contents automatically). Range of printers supported is important, as is the range of fonts available for each printer. Capabilities of the spell-checker and size of the thesaurus should be examined, together with the speed at which these subroutines are operated. The program should have the ability to easily preserve style sheets and varied page formats. An important consideration is what word processor is being used by your immediate colleagues (in the company, department, or whatever)—uniformity facilitates exchange of files and getting help! ASCII and many different word-processing files can be readily exchanged between Mac and PC computer systems using specialist software packages (e.g., MacLinkPlus/PC™ by DataViz, 35 Corporate Drive, Trumbull, CT 06611), and some word-processing packages (e.g., WordPerfect 6™; MacWrite II™ from the Claris Corporation) can translate text prepared in most other Mac or PC packages and save it as if it were prepared using another package.

Ensure that the program you select has a what-you-see-is-what-you-get method of constructing formulae; it is difficult to format your work if you cannot see what the finished product will look like until you print it. The capability of viewing several documents at the same time is important, and particularly the ability to transfer blocks of words both within and between documents (how many times do we revise our reports?). Easy importation of spreadsheet data is a necessary feature, and easy importation of graphics images is desirable. Operating a word processor in a multitasking environment is particularly desirable because of a common need to consult databases (e.g., bibliographic) and spreadsheets for data; some word processors operate better in such environments than others. The main word-processing programs have the capability for creating *macros,* which can be specifically designed by the organization or the individual to implant commonly used technical information (e.g., formulae, Greek letters, phrases, figure boxes whose numbers will automatically change as new ones are inserted) or to manipulate data as it is typed in (e.g., "take this set of phrases, insert them into a table, with each phrase alphabetically sorted on the first word"). Use of specialist abbreviations can speed typing for most of us; the powerful "find and replace" function will replace them with the desired expansion throughout the document when it is finalized. The "find and replace" function in the better word-processing packages also will allow the user to find words or segments of text with particular attributes and change those attributes; this feature is particularly useful if, for example, you have only sometimes italicized a word and want to ensure that your format is consistent throughout the text.

Spreadsheets, Statistical and Graphing Programs

Spreadsheets are primarily for the storage and manipulation of numeric data (go to a specialist database program for storing text, although there are some spreadsheets with increasing capabilities of dealing with alphabetic data); the intersections of rows and columns define cells in which data or complex formulae can be inserted. Most spreadsheets operate similarly and even look alike in their menu selections. Macro capabilities are more important than the range of formulae that are built in—one can always create a formula that can be executed with a macro, so long as the macro capabilities are sufficiently powerful! It is easy to construct a file that consists of (unlocked) cells in which data are to be entered plus (locked) cells that comprise headings, formulae that manipulate the data, and results. Exporting of data (from data or results cells) to other files of the same program is also easy. Hence, separate (or integrated) files can be constructed for sieve and pipette or hydrometer analyses (e.g., Chapter 7), then results exported to another file that combines the results and performs statistical analyses of multiple data sets. Geochemical data are particularly amenable to spreadsheet manipulations, and a variety of earth science problems can be resolved with judicious constructions (e.g., Koch 1990).

A feature to investigate when selecting a spreadsheet is its graphing capability; virtually all will construct simple business-type graphs, but few will permit insertion of error bars or logarithmic scales, permit construction of ternary dia-

grams, and support other scientific requirements. It is usually necessary to acquire a specialist scientific graphing program. Many graphing programs readily accept numeric data from spreadsheets, but all are not equal in the ease of manipulating that data, the range of capabilities, and the quality (and control) of the output. Seek one that will permit statistical analysis of the graphical data, multiple choice of scales (e.g., arithmetic plus logarithmic), and the construction of ternary diagrams. Also ensure that graph images can be easily imported into your other graphics/drafting programs—commonly you will wish to combine graphs and other kinds of graphical image, or add text to the graph.

Virtually all statistical programs must have graphing capabilities, but the variety and quality of possible graphs vary greatly. For both statistical and graphing programs, select only those capable of handling a larger number of data points than you think your maximum will be! The range of statistical manipulations that can be performed varies extensively between programs and although most will satisfy routine needs, your needs may require a specialist earth science statistical package (e.g., MGAP™ for the Mac, or G-STAT™ for the PC).

Database and Bibliographic Programs

Databases are programs that permit easy storage and retrieval of any kind of data, particularly data involving a high proportion of text (e.g., field notes and descriptive analytical details). Some are purely archival—for storage of data unlikely to be utilized (e.g., seismic data once collected and interpreted)—but most are active in the sense that they are online, maintained (updated and integrated), and available to many users. Fundamentally, a database file consists of *records* composed of a variety of *fields*. Each field has a name given to it by the user, is of any size set by the user, and is placed on the screen in a position selected by the user. Hence they can show any collection of information in any pattern. User-specified fields are indexed for easy and rapid retrieval of the records containing words (or other data) stipulated by the worker. *Flat* databases deal with single files only for a given operation; *relational* databases permit concurrent activity on multiple files (e.g., when a record is changed in one, there can be concurrent modification of other files). Databases are best suited to systematic cataloguing or indexing of samples and any kind of field or laboratory data. (When dominantly numeric data are involved, use a spreadsheet instead.) Most earth science usage of databases is for data retrieval alone and CD-ROMs are becoming an ever more popular means of storing large data compilations, but most database programs have powerful abilities to manipulate the data as well.

For most scientists a comprehensive bibliographic database (e.g., Procite™, Papyrus™, or Endnote™) is an essential tool for the work environment. Key capabilities to look for are author and key-word indexing, with display of available key words; output formatting to the varied specifications of different journals; easy importing of downloaded data from standard mainframe citation index systems (e.g., Georef™,

Oracle™). Current Contents on Diskette™ is a valuable database (for PCs or Macs) that provides weekly lists of articles in most journals covering physics, chemistry, and earth sciences. A particularly desirable feature of some bibliographic databases is automatic perusal of a document and compilation of the references cited therein. Nonbibliographic commercial databases can easily generate satisfactory files for storage and retrieval purposes, but it will usually be necessary to modify their output with a word processor, and even with automated find-and-replace functions, this task can be onerous.

The most complex and powerful databases are those that combine data and graphical information, such as the geographic information systems (GIS—e.g., see Chapter 2; Rhind, Raper, and Mounsey 1992), whereby operators can obtain a wide variety of data about any selected locality within a wide map area. Addition of project-specific information (such as geochemical or mineral data) effectively permits these programs to be utilized for exploration and exploitation, as well as depiction of selectable data combinations (e.g., see Raper 1990; Bonham-Carter and Agterberg in Hanley and Merriam 1990; Haines-Young, Green, and Cousins 1992). The time required for data input is well worthwhile because of the multiple ways in which varied data can be retrieved and manipulated.

Graphics Programs

Graphics programs are primarily designed to create or edit pictorial images. These images may include graphs, but the best graphing programs are specialized and deal essentially with graphs, not other types of images. There are two types of general graphics images: vector-based (information is stored as commands to draw from point to point) and pixel-based (information is stored for each pixel of the screen—the pixel can be on or off); for scientific purposes, only the vector graphics programs can produce publication-quality diagrams. With vector programs, seek one that has the ability to *modify* curves—it is painful to have to redraw complex curved lines!

A variety of vector-based programs are available: in the PC world a fast and easy-to-learn program is WordPerfect Presentations™, which is fully compatible with the common word-processor WordPerfect™; a high-quality program with abundant capabilities is CorelDraw™, which operates only under Windows 3™. The scientific standard is AutoCAD™, which has been customized for many earth science operations (optional costly extras) and has the extra capability of precise three-dimensional drawings (the only drawbacks to AutoCAD™ are the cost and steep learning curve; little-brother AutoSketch™ is cheaper, easier to operate, and is the way to "taste" the capabilities of AutoCAD™; GenericCAD™ and a few others also provide many of the capabilities at lower cost). In the Mac world Adobe Illustrator™ or Aldus Freehand™, Canvas™· and Superpaint™ all provide good vector-based results. Some of these also provide a pixel-based graphics layer (for easier picture—*bitmap*—editing), and

some have the capacity to transfer images between layers; modern versions of these packages all provide full gray-scale and color images.

Editing of images created by the scanning of drawings is a common function of graphics programs. Flat-bed scanners are generally necessary to produce undistorted images—few of us can move a hand scanner sufficiently consistently. Ensure that the graphics program will import most types of scanned image file, that it has the capability of displaying the total image as well as selected enlargements of it, and that the speed of movement around the image is sufficiently fast so that you are not constantly waiting for redraws. Unfortunately, images are scanned in as pixel images (bitmaps), and currently only a few programs can adequately transform those into vector images (e.g., CorelDraw™ and Canvas™). A way around the problem of conversion is to use the bit-mapped image as a template, overlay the desired vector lines, then delete the bit map. Some packages (such as Adobe Photoshop™) manipulate digitized color images, obtained using a video camera or color scanner, and can provide a high-resolution color output. If you do not have a scanner but do have a modem, you can still get a scanned image into your computer by faxing the relevant text or diagrams to your computer. Modern fax software for your computer should have an OCR (optical character recognition) capability so that text received by fax can be converted into normal editable text for transfer to your word processor.

There is a growing use of computer images as sources for publication and talk illustrations; check for the capability in the graphics software you are contemplating for making color slides (although it is quite feasible to photograph the monitor screen using a tripod for your camera; just ensure slow shutter speed to avoid capturing scan lines) and overhead transparencies. Flat monitor screens are particularly suited to photographing without distortion. Digital color images can be manipulated using packages such as Adobe Photoshop™, and used to combine text with photographs, or to combine two or more photographic images (e.g., an outcrop photograph with a close-up image or a stratigraphic column). The possibilities are limited only by your imagination. Computer-generated color slides are now achieving excellent quality and can considerably aid presentations; the days of slides of black-and-white graphs are fast becoming history. Many photographic shops can also transfer slide or print images onto CD-ROM.

Modeling: Simulation and Expert Programs

Simulation programs depict (mathematically, graphically, or by other ingenious methods) the results of situations established by the user. For example, they may predict what should happen were the chemistry of a solid-aqueous system to change in a specified way (e.g., see Butler in Hanley and Merriam 1990) or they may attempt to predict the behavior of sediments in response to changing environmental parameters (e.g., Tetzlaff and Harbaugh 1989; Tetzlaff 1990). Models can be established for such complex systems as paleo-climates, where by varying some parameter such as rainfall patterns, one can evaluate the likely result in terms of hydro-geologic consequences (e.g., Robers, Craig, and Stamm in Hanley and Merriam 1990). Most models are established by mathematical relationships and thus on a foundation of a spreadsheet or spreadsheet-like program; simple "what if" simulations are easily constructed by relative novices on standard spreadsheets if the data and results can be expressed numerically.

Expert programs provide best-bet solutions to questions involving complex data combinations progressively selected by users from multiple branches of an entwined array of variables. Rules are constructed that relate data to interpretation by a team of experts in the field, then are integrated in a progressively more complex system of choice-dependent questions. For example, interpretation of the origin of porosity in carbonate rocks can be predicted from the pore and cement types present (e.g., Watney, Anderson, and Wong in Hanley and Merriam 1990) or from well-log correlation and interpretation procedures established via an expert system (e.g., see papers in Simaan and Aminzadeh 1989). One important consequence of using expert programs is that multiple workers with different backgrounds can be led to reach the same conclusion from a set of objective observations. Access to basic expertise outside the user's ability is provided by these programs, and in remote areas or when the necessary experts are unavailable, these programs can be vital to reach rapid decisions.

Major drawbacks to simulation and expert programs are the long time necessary to establish useful and reliable ones, and the nagging worry that one may not have included all the influential factors in the construction of the framework (definitions in "gray" or subtle areas depend on the "expert" used!). These limitations are largely acceptable in systems where all variables and cause-and-effect relationships can be identified and quantified (e.g., mechanical systems), but there are many problems with the application of simulation programs in the earth and environmental sciences, where significant variables may not be identified, covariant relationships and feedback loops may be unrecognized or poorly defined, and many aspects of the model may be shrouded in uncertainty.

Specialist Sedimentological Programs

Programs specifically designed for sedimentological use of the microcomputer are few, but programs adaptable with little effort are many. Potential applications include hydrological and geochemical mapping (e.g., a number of Rock-Ware products), contour mapping (e.g., Jones, Hamilton, and Johnson 1986; Jones 1989; Moretti and Larrere 1989), creation of graphs, and multiple kinds of statistical applications. Relevant specialist programs are particularly generated for the petroleum industry (e.g., to evaluate pore-geometry attributes, see Habesch in Hanley and Merriam 1990) and in hydrology (e.g., Bellotti in Hanley and Merriam 1990; there are many programs from all over the world for hydrology). Well logs can be prepared with programs that read standard text files (e.g., Viewlog™, or the cheaper Logger™ and Mac-

Section™ by RockWare); porosity, permeability, lithology, and/or other parameters can be plotted in a variety of ways from logged data. A selection of algorithms is available in some programs (e.g., RockWare's Gridzo™—both PC and Mac versions; Golden's Surfer™) for contouring of data points to produce plan-view or three-dimensional (e.g., net) maps. And of course there are many specialist programs designed for the hard mineral exploitation industries, where for example the calculation of potential ore volumes is desired (e.g., Kimberley in Hanley and Merriam 1990 for a program that computes solid or fluid volumes for stratigraphic units from geologic cross-sections). For the student, RockWare™ can supply useful mineral (thin section and hand specimen) databases with search forms that enable users to enter those properties they can identify, and the computer then selects the mineral or minerals that match the identified properties. Also see discussion and references in Wadge (1993).

Experiments and analyses are commonly designed for concurrent data entry to computer programs via direct linkages to the equipment utilized (at the simplest level, to record grain size distributions in settling tubes). Macs are less amenable than PCs to receiving data directly from sensors (e.g., electrochemical probes) or analytical instruments, but some of the necessary software and hardware is available from third-party suppliers and the range is beginning to expand. Rose diagrams for depiction of directional data have been of particular interest to many programmers (see discussion in Chapter 6). Cumulative curves for size analyses are difficult for most commercial programs to construct accurately, and some attempts have been made to write programs for this purpose (e.g., Middleton 1990). A wide variety of graphical dissection and statistical manipulation techniques (e.g., factorial and multivariate analysis) are available in many commercial scientific and business programs, and there are a few programs specifically written to deal with earth science variograms (e.g., Hohn and Fontana in Hanley and Merriam 1990). (Standard commercial statistical programs commonly have difficulty in dealing with earth science data spatially related to dependent data that are not evenly distributed, or to data imprecisely qualified as "greater than" or "less than.") Simulation and modeling programs (e.g., of basin subsidence rates or of flow-regime conditions and results) are more readily available (e.g., Franseen et al. 1991).

We do not know of a good way at present to become aware of the variety of programs and their advantages/disadvantages. A few sources are the Geosoftware section in the journal *Geotimes* and the journals *Computers and Geosciences* (Pergamon Press, D. F. Merriam, editor) and *Geobyte* (published by the American Association of Petroleum Geologists), which publish articles on the use of computers in earth sciences. The Computer Oriented Geological Society (COGS, P.O. Box 1317, Denver, CO 80201-1317) provides a catalogue of public domain and shareware geological programs available at nominal cost. The National Geophysical Data Center (National Oceanographic and Atmospheric Association, E/GC Dept. 806, 325 Broadway,

Boulder, CO 80303-3328) has large data collections (e.g., Deep Sea Drilling Project, digital cartography, atlases) available at reasonable prices on CD-ROM. RockWare™ (4251 Kipling St., Suite 595, Wheat Ridge, CO 80033) is probably the largest commercial supplier of geological software.

PROJECT DESIGN

Prior to discussing specific procedures and techniques for analyzing sediments, it is desirable to give some consideration to the overall design of projects requiring analytical work. In the initial enthusiasm to initiate any project, be it for research, consultancy, or exploration, the thought allocated to comprehensive project planning is often minimized (unless it has been required by the funding organization). However, experience reveals that time spent on planning will save time over the life of the project and reduce costs. All projects will differ in their requirements, but there are some general considerations that should be made before starting manual work.

Clear definition of the objective(s) of the project is the obvious starting point—i.e., what questions are the data required to answer? The answer is commonly not as obvious as initially thought, and there is often a broad gap between the ideal and the practicable, especially when constraints of time and cost are considered. The planner must decide on what kind of data are needed, then what data are practicable to collect (commonly widely divergent concepts in earth science programs). What accuracy (approach to the "true answer," generally requiring application of multiple techniques to acquire the same data—e.g., Griffiths 1967) and precision (the level of reproducibility of the techniques applied) are required? How will we know when we have answered the initial questions to the required level of accuracy or precision? Are all variables (or alternative hypotheses) that may affect the expected link between data and the original questions adequately identified and allowed for? Are control or baseline reference sites needed (e.g., to evaluate pollution loads in estuarine sediments), and if so, what criteria will determine the suitability of such sites?

The sampling program is a particularly vital aspect of project planning (hence is treated in detail in Chapter 3). Although few scientists admit it, many either consciously or unconsciously expect that if they collect and examine enough samples the information they seek will materialize from the resulting plethora of data. Such an approach may succeed, but it does little to optimize success and it may add unnecessarily to the cost of a project. Scientists usually also pay a lot of attention to ensuring that their analyses are precise and accurate, but the ultimate usefulness of the data is always constrained by the original sampling strategy: no amount of analytical accuracy or elaborate statistical manipulation of the data can compensate for poor sample site selection or poor sampling procedures. Some of the general questions that should be asked before and/or during the project are: Precisely what should be sampled (or observed) to answer

the original questions, and why? Are there any externally imposed constraints on the sampling program (such as accessibility of sample sites or statistical or engineering requirements as might apply in ore reserve estimation)? What will determine when we need to rethink the sampling program (i.e., how will we know when continued work is taking us no closer to answering the original questions because the wrong type of samples have been collected or the wrong type of observations have been made and the whole approach needs rethinking)?

Once the ideal requirements for sampling and observation are decided, practical considerations may dictate further steps prior to actual collection. For example: Are suitable base maps and air photos or satellite images (see Chapter 2) available? Is permission required for access to any land in the proposed field area and are sampling permits required (e.g., for parks or reserves as well as private land)? What logistical arrangements are necessary for the transport of personnel, equipment, and samples in the field? Will weather conditions impose any constraints on the timing of fieldwork or on the nature of the work that can be undertaken (e.g., limited access due to floods, snow, etc.)? Is all necessary field equipment available and has it all been checked to ensure that it is in good working order (there are few things more annoying than discovering that an important piece of equipment is defective or has flat batteries after one is in a remote field area)? Is the laboratory adequately stocked with analytical equipment and consumables? Will some analyses have to be carried out elsewhere, in which case can the external organization prepare results within the time constraints of the project? Will field data be recorded in an electronic database for processing by computer, and if so, will this impose any constraints on the way in which data should be recorded (e.g., the format for recording the attitude of bedding, the orientation of paleocurrent data, fossil record data, sample sites, photograph localities, etc.)? Is a sample numbering system established that is informative, easy to use, and compatible with later analytical treatment (it is often confusing and time consuming to use a different sample numbering system in the field from that to be used for final cataloguing of samples)? Are special types of sample containers required, or are there special requirements for the preparation of sample containers (e.g., all containers used in many environmental geochemical studies involving trace elements must be washed and acid rinsed before use; this cleaning is much easier to do before going into the field)? Are there any special requirements for the transport of samples (e.g., sediments to be analyzed as part of nutrient concentration and dispersion studies may need to be frozen to prevent bacterial modification during transport to the laboratory)?

Analyses Carried Out by Others

When planning projects, consideration must be given to farming out work to external contractors in the interests of cost effectiveness. However, considerations must be given to factors other than the cost alone. When scientists carry out their own analyses they take a personal interest in the quality of the analytical data and usually have some additional knowledge to judge when results are anomalous and warrant further investigation (either an unexpected relationship is revealed or possibly because of analytical error). However, many sedimentologists send their chemical analytical work to a separate laboratory where both systematic and non-systematic errors may occur and where the analysts do not have the benefit of that additional knowledge. In a reliable laboratory, the analysts will do the best they can and they will run internal quality assurance checks on the data they produce. Confidence in the accuracy and precision of the data that are provided can be raised by applying the following simple approaches and controls:

1. Provide the analyst with a clear idea of the goals of the investigation, with particular emphasis on the use to which the analytical data will be put. Also provide any relevant data about the particular samples, particularly if there are foreseen problems (e.g., if the samples contain components that are likely to oxidize if left exposed to air or that may decompose if the sample is heated above a particular temperature).

2. Specify the analytical sensitivity required and indicate the likely concentration range for each component.

3. Send a few duplicates of some samples (labeled with different code numbers) and check the reproducibility of the results. In addition, include a few standards labeled as if they were part of the sample batch and check how well the results compare with the accepted data for the standards (e.g., Abbey 1983; Govindaraju 1984).

4. Send the laboratory an empty sample container and ask for a check on any contamination that may be due to the sample containers. Since all physical and chemical sample treatments are potential sources of contamination (e.g., see Potts 1987), seek information about the quality-control procedures carried out by the laboratory.

ETHICS

A final brief word on general aspects of scientific analysis concerns the ethics of the workers. In many situations there are conflicts between the requirements of good science, funding available to perform the science, and social or political pressures. The scientist must give first priority to the best performance of his or her profession, regardless of these considerations. Otherwise, not only the individual but the profession as a whole will decay in the long term. Clear statements of all constraints and limitations imposed on the analyses and proposed solutions to problems should be made in every report, and the scientist should clearly differentiate his opinion from the data and analytical results. Insofar as the longest-term goal should be the advancement of knowledge, the scientist should also be prepared to change his opinions as new data become available and to avoid heated conflicts

based on opinions. Other aspects of ethical behavior that should be observed are courtesy to colleagues and clear acknowledgment of the contributions of others (although all science builds on the contributions of other scientists, acknowledgments should be made when contributions are specific to the project or database).

SELECTED BIBLIOGRAPHY

General

Abbey, S., 1983, *Studies in "Standard Samples" of Silicate Rocks and Minerals 1969–1982.* Geological Survey of Canada Paper 83–15.

Allman, M., and D. F. Lawrence, 1972, *Geological Laboratory Techniques.* Arco Publishing, New York, 335p.

American Geological Institute, 1992, *Planning for Field Safety.* AGI Publication Center, Annapolis, Md., 197p.

Armour, M. A., L. M. Browne, and G. L.Weir, 1984, *Hazardous Chemicals Information and Disposal Guide,* 2d ed. University of Alberta, Edmonton.

Bretherick, L. (ed.), 1986, Hazards in the Chemical Laboratory, 4th ed. Royal Society of Chemistry, London, 604p.

Carver, R. E., 1971, *Procedures in Sedimentary Petrology.* Wiley-Interscience, New York, 653p.

Compton, R. R., 1985, *Geology in the Field.* Wiley and Sons, New York.

Furr, A. K. (ed.), 1990, *CRC Handbook of Laboratory Safety,* 3d ed. CRC Press, Boca Raton, Fla., 704p.

Gardiner, V., and R. V. Dackombe, 1983, *Geomorphological Field Manual.* George Allen and Unwin, London, 254p.

Govindaraju, K., 1984, 1984 compilation of working values and sample description for 170 international reference samples of mainly silicate rocks and minerals. *Geostandards Newsletters* 8 (special issue).

Green, W. R., 1991, *Exploration with a Computer: Geoscience Data Analysis and Applications.* Pergamon Press, Oxford, U.K., 1225p.

Griffiths, J. C., 1967, *Scientific Method in Analysis of Sediments.* McGraw-Hill, New York, 508p.

Hutchison, C. S., 1974, *Laboratory Handbook of Petrographic Techniques.* Wiley-Interscience, London, 527p.

Jackson, M. L., 1958, *Soil Chemical Analysis.* Prentice-Hall, Englewood Cliffs, N.J., 498p.

Krumbein, W. C., and F. J. Pettijohn, 1938, *Manual of Sedimentary Petrography.* Appleton-Century-Crofts, New York, 549p.

Lefevre, M. J., and S. Conibear, 1989, *First-Aid Manual for Chemical Accidents,* 2d ed. Van Nostrand Reinhold, N.Y., 256p.

Lenga, R. E., 1988, *The Sigma-Aldrich Library of Chemical Safety Data,* 2d ed., 2 vol. Sigma-Aldrich Corporation, New York.

Mahn, W., 1990, *Academic Laboratory Chemical Hazards Handbook.* Van Nostrand Reinhold, New York, 600p.

Manufacturing Chemists Association, 1972, *Guide for Safety in the Chemical Laboratory,* 2d ed. Van Nostrand Reinhold, New York, 505pp

Martin, A., and S. A. Harbison, 1986, *An Introduction to Radiation Protection,* 3d ed. Chapman and Hall, New York.

Maxwell, J. A., 1968, *Rock and Mineral Analysis.* Wiley-Interscience, New York, 584p.

Mueller, G., 1967, *Methods in Sedimentary Petrology,* H.-U. Schmincke (trans.). Hafner, New York.

Potts, P. J., 1987, *A Handbook of Silicate Rock Analysis.* Chapman and Hall, New York, 622p.

Sax, N. I., and R. J. Lewis, 1986, *Rapid Guide to Hazardous Chemicals in the Workplace.* Van Nostrand Reinhold, New York.

Strakhov, N. M., 1957, Methode d'etude des roches sedimentaires. *Service Information Geologique, Annales* 1, 2 vols.

Tucker, M. (ed.), 1988, *Techniques in Sedimentology.* Blackwell Scientific Publications, Oxford, 394p.

Twenhofel, W. H., and S. A. Tyler, 1941, *Methods of Study of Sediments.* McGraw-Hill, New York, 183p.

van der Linden, W. J. M., 1968, *Textural, Chemical and Mineralogical Analyses of Marine Sediments.* New Zealand Oceanographic Institute Miscellaneous Publication 39, 37p.

Computing in Sediment Studies

Franseen, E. K., W. L. Watney, C. G. St. C. Kendall, and W. C. Ross (eds.), 1991, *Sedimentary Modeling: Computer Simulations and Methods for Improved Parameter Definition.* Kansas Geological Survey, University of Kansas, Lawrence.

Haines-Young, R., D. R. Green, and S. Cousins (eds.), 1992, *Landscape Ecology and GIS.* Taylor and Francis Group, Washington D.C., 300p.

Hanley, J. T., and D. F. Merriam (eds.), 1990, *Microcomputer Applications in Geology II.* Pergamon Press, New York.

Isdale, P., 1991, Three-dimensional representation of coral reefs: Generation of submarine terrain images on personal computers. *Marine Geology* 96:145–150.

Jones, T. A., 1989, The three faces of geological computer contouring. *Journal of Mathematical Geology* 21:271–283.

Jones, T. A., D. E. Hamilton, and C. R. Johnson, 1986, *Contouring Geologic Surfaces with the Computer,* Van Nostrand Reinhold, New York, 314p.

Koch, G. S., Jr., 1990, *Geological Problem Solving with Lotus 1-2-3.* Pergamon Press, New York, 234p.

Lindholm, R. C., 1979, Utilization of programmable calculators in sedimentology. *Journal of Sedimentary Petrology* 49:615–620.

Merriam, D. F., 1988, Bibliography of computer applications in the earth sciences, 1948–1970. *Computers and Geosciences* 14(6): 719–964.

Middleton, G. V., 1990, Fitting cumulative curves using splines. *Journal of Sedimentary Petrology* 60:615–616.

Moretti, I., and M. Larrere, 1989, Computer-aided construction of balanced geological cross-sections. *Geobyte* 4(5):16–24.

Raper, J., 1990, *Three Dimensional Applications in GIS.* Taylor & Francis Group, Washington, D.C., 189p.

Rhind, D., J. Raper, and H. Mounsey, 1992, *Understanding Geographic Information Systems.* Taylor and Francis Group, Washington, D.C., 250p.

Robinson, J. E., 1982, *Computer Applications in Petroleum Geology.* Van Nostrand Reinhold, New York, 288p.

Simaan, M., and F. Aminzadeh (eds.), 1989, *Artificial Intelligence and Expert Systems in Petroleum Exploration.* Advances in Geophysical Data Processing, vol. 3. JAI Press, Greenwich, Conn.

Slatt, R. M., and D. E. Press, 1976, Computer program for presentation of grain size data by the graphic method. *Sedimentology* 23:121–131.

Tetzlaff, D. M., 1990, SEDO: A simple clastic sedimentation program for use in training and education. In T.A. Cross (ed.), *Quantitative Dynamic Stratigraphy.* Prentice-Hall, Englewood Cliffs, N.J., pp. 401–415.

Tetzlaff, D. M., and J. W. Harbaugh, 1989, *Simulating Clastic Sedimentation.* Van Nostrand Reinhold, New York, 202p.

Wadge, G., 1993, Computers and geological information. The depth and width of it. *Geoscientist* 3:10–13.

2
Aerial Photographs and Maps

This chapter briefly reviews usage and basic principles of aerial photographs and some of the varieties of maps most widely used by sedimentologists. Aerial photographs have always been an essential aid to fieldwork, as a device for locating features and routes to get to the features of interest; in addition, many maps have been prepared by transferring information from transparent overlays onto base topographic maps. The variety of maps used by the sedimentologist is enormous, both for fieldwork and as final compilations to express ideas on the distribution or interpretation of only partially exposed sedimentary units.

AERIAL PHOTOGRAPHS AND SATELLITE IMAGERY

The use of aerial photographs and satellite imagery has evolved rapidly over the past 50 years from simple low-resolution black-and-white photographs, to stereographic pairs of high-resolution black-and-white photographs or color photographs, to digitally recorded satellite images that can be manipulated by computers to enhance resolution of selected features. Stereo pairs of aerial photographs are now one of the basic tools in mapping: cartographers use them with photogrammetric plotters to generate topographic contours; land use and military planners use them in regional and strategic planning; botanists use them in species distribution mapping; geologists use them as a base for recording field data, to detect large-scale features difficult to recognize in the field, and to trace structural features and lithologic units across inaccessible areas or areas with poor exposure.

Aerial photographs can be either vertical or oblique; the former are used most widely. Normally the aircraft flies a run at a constant speed in as straight a line as possible with the camera set such that successive photographs have an overlap

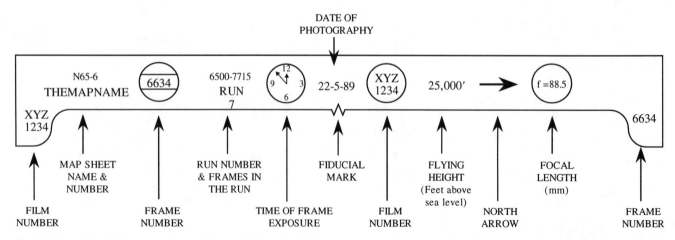

Figure 2-1. Information commonly recorded on aerial photographs. Note that when calculating scales flight height data are still recorded in feet whereas focal lengths are recorded in mm.

of at least 60%; adjacent runs are planned to give a side lap of about 30%. In addition to the photographic image, each photograph in a run contains important numerical information (Fig. 2-1), some of which is necessary to determine the *average* scale:

$$1: \frac{\text{flight height above the terrain (= flying height – terrain height)}}{\text{camera focal length}}$$

In Fig. 2-1, if the *average* terrain height was 2000 ft the *average* scale would be 1:79,214 (note that conversion of one or the other number may be necessary if both imperial and metric units are provided, as in the example!). This is an *average* scale, because if the land surface in the photograph represented a 5000-ft ridge that sloped down to sea level, the scale on the ridge would be 1:68,881, but at sea level it would be 1:86,102. There is always a scale distortion on aerial photographs due to differential land relief, unless the area is completely flat. Even where the ground surface is completely flat, there is a radial distortion because only the principal point (the point at the intersection of lines connecting opposite fiducial marks on the photograph) is vertically below the camera; as the distance away from the principal point increases, the effective distance between the camera lens and the ground surface changes, resulting in a scale change.

It is possible to correct for both scale distortions when preparing maps from aerial photographs (e.g., Moseley 1981), but for much photogeological work the scale distortions are not serious and most workers ignore them. However, note that distance measurements between points using aerial photographs may not agree exactly with that determined using a standard map or measured on the ground. In addition, when viewing pairs of aerial photographs stereoscopically, hills and dips appear steeper than they are. This vertical exaggeration (between two and five times for most stereoscopes) arises because the distance between the viewer's eyes is effectively increased to the distance between the principal points on adjacent photographs. Although the resulting increase in depth perception clarifies subtle features, corrections are necessary when estimating dips and heights. If the amount of vertical exaggeration (*VE*) is known, apparent dips (*Da*) can be converted to true dips (*D_t*) using the relationship:

$$D_t = \tan^{-1} [(\tan Da) / VE]$$

As well as providing base maps for recording field data, the use of air photos in geologic mapping is enhanced by interpretation of patterns, shapes, tone (or color), and texture (e.g., Moseley 1981; Boulter 1989; and see Table 2-1). How-

Table 2-1.   Selected Photogeological Features and Their Possible Interpretation

Photo Feature	Possible Interpretation
strong distinct lineament	commonly a fault or dike
group of subparallel lineaments	set of major joints, dikes, or faults
two sets of subparallel lineaments intersecting at a high angle	conjugate set of major joints or faults
radially oriented set of lineaments	radial dike swarm or fractures around a dome
irregular discontinuous concentric lineaments	fractures due to radial tension around an igneous or sedimentary dome
boundary between two uniform but distinct textural zones	contact between two massive rock types or between a massive rock and surface soils or sediments
boundary between a uniform textural zone and a zone with a parallel banded texture	unconformity between a massive rock or surficial sediment unit and a layered sedimentary sequence
boundary between two zones with parallel banded textures but different orientations	angular unconformity between two layered sedimentary sequences
circular or oval features	calderas, batholithic domes, impact structures, basin and dome interference folds, explosion structures
homogeneous tongue-like areas within texturally distinct zones	lava or debris flow over older strata
parallel banded zones where bands are distinctly curved or folded	folded layered sedimentary sequence where beds are not lying horizontally
dendritic drainage pattern	drainage developed on texturally uniform strata
trellis drainage pattern	drainage developed in an area of dipping layered sedimentary strata with minimal folding
offset river channels	progressive fault-controlled displacement of the river
anastomosing or braided drainage	alluvial and fluvial deposits where sediment supply exceeds the carrying capacity of the river
highly sinuous or meandering rivers	fluvial systems where land relief and sediment supply are fairly low and discharge is irregular

Geomorphological features such as volcanic vents, deltas, cheniers, raised beaches, eskers, moraines, kames, sand dunes, karst topography, alluvial fans, crevasse splays, chutes, levees, point bars, channel bars, ox bows, braided channels, spits, tombolos, and many others are usually clearly recognizable

Table 2-2.   Wave Bands for Thematic Mapping from Digitally Enhanced Satellite Photographs and Some Potential Applications

Band Number and Name	Band Width (μm)	Application
(1) Blue/green	0.45–0.52	Good water penetration suitable for coastal water mapping. Strong vegetation absorbance for differentiating soil and rock from vegetation.
(2) Green	0.52–0.60	Good for studying vegetation and water turbidity. Computer-enhanced band 1, 2, 3, and 4 data and ground truthing, enable discrimination between areas dominated by different plant species.
(3) Red	0.63–0.69	Absorbs reflections from vegetation and detects ferric oxides and oxyhydroxides; useful in assessing the underlying geology in lateritic terrains.
(4) Near infrared (IR)	0.76–0.90	Provides high contrast between land and water and very strong reflectance from vegetation.
(5) Near–middle IR	1.55–1.75	Very moisture sensitive; distinguishes between clouds and snow.
(6) Thermal IR	10.4–12.5	Useful in stress analysis, soil moisture, and surface temperature studies; particularly useful in high heat flow areas such as geothermally active sites.
(7) Middle IR	2.08–2.35	Used to discriminate between rock types; reveals hydrothermally altered zones around some mineral deposits.

Source: After Boulter (1989).

ever, despite the fact that some features may be relatively easy to interpret (e.g., traces of large faults and dikes), training to develop interpretive skills and field verification are vital. In particular, colors on photographs can be deceiving.

Since satellite imagery became available in the late 1960s, its value has grown from investigating large-scale features on the Earth's surface to various investigations of small-scale features, as the resolution of the imagery has improved with each new generation of satellite. The current French SPOT system has a ground resolution of about 10 m and further improvement is likely in the near future. Satellite imagery can be interpreted using patterns, shapes, tones, and texture in a similar way to conventional aerial photographs, but satellite data have major advantages over normal aerial photographs, both because of minimal scale distortion and because the data are recorded digitally for separate color bands. The colors are integrated using a computer to produce the final images; thus, they can be manipulated by computer to produce enhanced false color images tailored to particular investigations. The original Landsat multispectral scanner (MSS) recorded data for four wavelength bands, but the more recent thematic mapper (TM) system uses seven bands (Table 2-2). Bands recording in the infrared region (particularly band 7, in the middle infrared) are particularly informative as a result of the different heat-absorbing and heat-emitting characteristics of different rock types. Thus, for example, a computer can be used to determine the color combination that most clearly distinguishes salt diapirs or potentially diamondiferous kimberlite pipes from surrounding strata; this combination then can be applied over large areas to identify possible targets for field investigation. False color images can be used

to distinguish even between carbonate and detrital sediments or ferruginous and non-ferruginous sediments (e.g., McConchie and Lawrance, 1991). There is little doubt that applications of satellite imagery will increase as personal computers become more powerful.

Use of digitally recorded satellite images has led to the development of geographic information systems (GIS). GIS technology (e.g., Burrough 1986; Maguire 1989; Peuquet and Marble 1990) has already had a substantial impact on land use planning and resource management (e.g., Burrough 1986; Belward and Valenzuela 1991) and its application in geology is expanding rapidly. The value of GIS technology lies in its capacity to integrate and manipulate any form of mappable data that can be digitized and added to the database and in its capacity to store and make available a vast amount of any kind of data. Combinations of data types are almost unlimited; e.g., satellite imagery can be integrated with data from a simplified geological map, a geochemical soil survey, and a magnetic anomaly map as an aid in exploration programs.

The development of satellite and microelectronic technology provides another benefit for field geologists in the form of GPS (global positioning system) navigational aids, which are capable of fixing the user's position anywhere on earth to within about 50 m; when signals are available from four satellites, they can also be used to obtain an elevation fix to within about 3 m. Modern portable (will fit a large pocket) GPS instruments are reliable and available for under U.S. $2,000. Where an earth-based station at a precisely known location is available, many GPS instruments can be used in differential mode and will then provide a fix of the user's position and elevation to within l m. A revolution in mapping techniques is under way at the time of writing.

MAPS

Maps are a convenient and widely used means of presenting observations, inferences, and conclusions concerning often complex spatial relationships involving three- or four-dimensional data that may be either real (e.g., the location of a bridge or a contact observed in outcrop) or abstract (e.g., a paleogeographic map or time lines through a prograding delta sequence). Because their visual impact is far greater on most people than a thousand words, they are very effective ways of communicating. Maps can be used to convey information about any feature or attribute that can be fixed in time and space; hence everyone has some familiarity with at least one type of map. Road maps, general geographic maps, and weather maps are familiar to most; detailed topographic maps are important to climbers, civil engineers, land use planners, and many others. Topographic and bathymetric maps also form the base for (superimposed) scientific maps such as species distribution and diversity maps widely used by biologists. Geologists use an extensive variety of maps to convey or obtain such information as (1) the relative positions of surface and subsurface geological features; (2) the distribution of paleoenvironments; (3) the thickness of lithologic units, regional and stratigraphic variation in grain size, trace elements, or mineralogy; and (4) the nature and distribution of tectonic features.

Guidelines in Map Preparation

Whether preparing or studying maps, certain fundamentals should always be provided by the creator or viewed first by the user:

A scale. This vital piece of information must be shown graphically as a scale bar so that it remains meaningful should the map be reduced or enlarged. Small-scale maps should also show at least two lines of latitude and two lines of longitude. Ensure that the scale chosen is appropriate for the data presented.

A north arrow or other form of orientation information. The north arrow should indicate true north; if magnetic north has been used in the demarkation of map features, then its relationship to true north should be indicated (and the map dated because the magnetic declination varies with time).

A title. The title should indicate the name of the area mapped and the type of map (e.g., Hydrogeological Map of Australia).

An explanatory legend or key. As implied by Boulter (1989), calling the explanatory key to map interpretation a legend is a reminder that, although maps contain a wealth of factual information, most also rely heavily on interpretations that may change as knowledge and ideas evolve. The map legend must explain the meaning of all colors, shading

patterns, and symbols shown on the map. Legends that show rock types or stratigraphic units on geological maps must be constructed so that they are listed with the oldest at the bottom and the youngest at the top.

Grid lines. On small-scale maps, or where the precise location of multiple features is required, latitude and longitude lines may be shown (in degrees, minutes, and/or seconds); by convention, when giving a grid reference latitude is given before longitude. Many maps (e.g., most topographic and geologic) have grid lines indicating the distance east and north of an arbitrary or false origin (the false origin is placed outside the map so that all coordinates are positive); in these cases, eastings (the vertical lines) are given before the northings (the horizontal lines). When reporting false origin grid references it is also necessary to give the map sheet number. In the United States, township and range grids and grid-locating systems are commonly used (see the description on each map).

Many maps also provide:

Index maps—small maps inserted in a corner of the main map to show the relationship of the mapped area to a larger, more recognizable, area.

A reliability diagram—to indicate the relative reliability of data presented on different parts of the map. For instance, a reliability diagram on a geologic map may show the difference between data derived from a foot traverse (well-exposed section) and that derived from the use of air photos (poor or nil exposure).

Map projection—where large areas are depicted, it may be desirable to indicate the type of map projection used, because the curvature of the earth may have a significant influence on the form of the map (see Dent 1990).

Maps should be highly informative yet easy to read. Achieving both of these goals requires considerable attention to their design. For example, symmetry and balance are important in designing the layout of a map: the legend, north arrow, scale, and other features should be arranged to maximize the visual impact of the map itself. Guidance on general techniques and preparation can be found in Robinson and Sale (1969), Mailing (1989), Keates (1989), and Dent (1990).

Contour Maps

A contour map is a plan-view representation of the configuration of a real or imaginary surface. A contour line passes through all points of equal numerical value. Put another way, they are lines drawn at the intersection between the surface (defined by the data) and an imaginary plane parallel to—and spaced a stated numerical distance or value from—a reference datum. For example, a −150-m bathymetric contour is the line formed by the intersection of the seafloor and an imaginary plane 150 m below and parallel to mean sea level (reference datum = 0 elevation); or a map may show depths

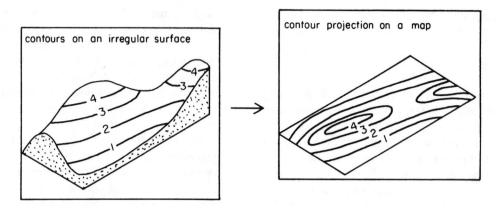

Figure 2-2.   Example of the projection of topographical contours.

or elevations from an unconformity to the earth's surface (or a stratigraphic boundary); or it may show the percentage of a particular mineral or rock type (the reference datum being in this case imaginary, i.e., "nothing"—the absence of this material). Any data are mappable that can be quantified and related to any consistent reference datum.

Contour maps can be used in a number of ways:

1. To determine the relationships between isolated data points
2. To determine the relationships between facies, environments, or other groups of like or unlike data
3. To extend inference from areas in which relationships are known into those where data are sparse or absent (i.e., to forecast trends)
4. To depict the three-dimensional configuration of units
5. To provide a means of estimating volumes of irregular masses (such as economic deposits)

The distance between contour lines in plan view represents the slope of the surface and the course of the contour as an expression of the irregularity of the surface being mapped (e.g., Fig. 2-2). The average slope angle between any two points on selected contour lines can be obtained by measuring the length of a line drawn on the desired bearing between the selected contours A and B (where A is greater than B). The length of the line is converted to distance (*d*) using the map scale and the height difference (Δv) equals (A – B); ensure the same units of length are used for (Δv) and (*d*). Then:

average slope angle (on the selected bearing) = $\tan^{-1}(\Delta v / d)$

When estimating the mean rate of change in the abundance of a mineral or element on a selected bearing, a slope angle is not normally calculated and the measurement units for (Δv) and (*d*) need not be the same; instead, the rate of change is recorded as a ratio and given by:

rate of change (on the selected bearing) = $\Delta v / d$

Gradients (or rates of change) for mineral or element abundance contours will have values expressed in units such as %/100 m or mg/kg/m, depending on the concentrations and distances involved. The slope angles or concentration gradients determined between two selected contours must of course be *average* values, because even if the height or concentration values at the contours are known exactly, no data are available on the nature of the change between the contours (e.g., Fig. 2-3).

All contour maps are interpretive. The positions of all points of a given value are never known with certainty, with perhaps exceptions in the case of three-dimensional photogrammetric displays of topographic, seismic reflection, or sonar data (but even these may have distortions or involve interpretive assumptions). Hence, several different patterns are possible with the same data, particularly where data points are widely spaced and the data are inaccurate. Figure 2-4 illustrates different patterns that result from several methods (discussed below) applied to two different sets of selected points from the same set of data (blank maps are provided for you to produce yet other versions). Because sudden changes in pattern and closed contour lines (marking high or low values within the map) generally indicate areas of interest or anomalies, the technique selected can be made to show either conservative or liberal patterns, depending on the purpose of the worker. For aesthetic reasons, all patterns should be smooth, parallel, and spaced so as to fill in the whole of the map. When interpreting a contour map, always bear in mind the nature of the reference datum, the kind of data you are mapping, and the data character (e.g., a bathymetric contour map can be easily misinterpreted if one is thinking "topographically"—high numbers on the former indicate depressions, whereas in the latter they signify hills).

Contour patterns can be drawn in three effective ways (see Fig. 2-4 for examples of each):

1. *By mechanical spacing.* Where data are accurate and data points relatively close, the location of contour lines between two points is mechanically interpolated (contours are drawn

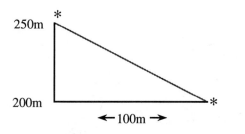 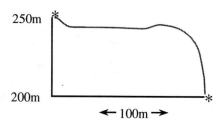

Any surface form could exist between the (*) contour points provided neither contour is crossed

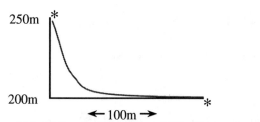 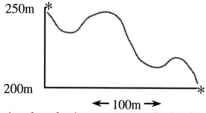

Figure 2-3. Examples of the possible differences in configuration of a surface between two control points for contour lines.

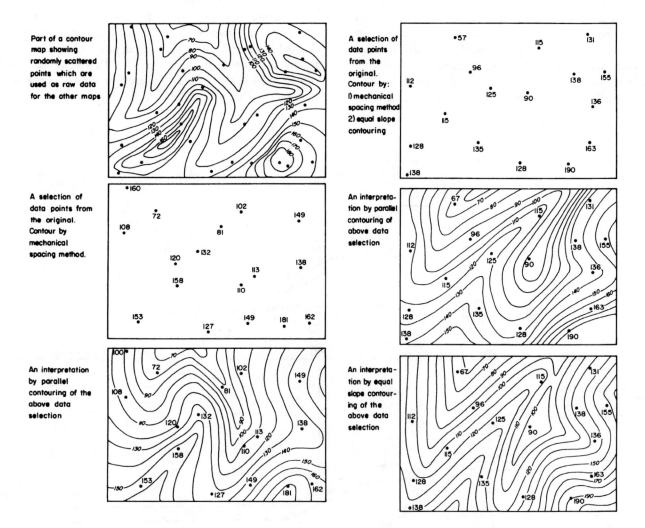

Figure 2-4. Examples of different contour patterns that are possible from a single real surface, depending on the selection of data points and the method of contouring.

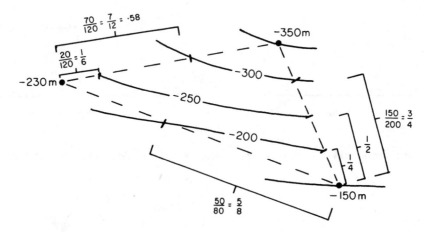

Figure 2-5.   Example of the method of mechanical contouring (50-m intervals).

between points at distances that are directly proportional to the difference in data values between the points; e.g., Fig. 2-5).

2. *By equal slope spacing.* Where data are more inaccurate or data points are widely spaced, a common technique is to assume that the surface being mapped has a similar slope throughout the area (i.e., contours are drawn with equal spacing throughout the area). The initial spacing should be determined by interpolation between two data points that are relatively close and show the greatest difference in values

(where the surface slope appears to be maximum). This procedure tends to maximize the possible anomalies and can be misleading (Fig. 2-6).

3. *By parallel spacing.* Where data are very inaccurate and data points are widely spaced, smooth patterns are constructed by keeping the contours parallel and varying the spacing between them where necessary. This method can be very useful when experience in nearby or similar areas suggests that a particular pattern is likely.

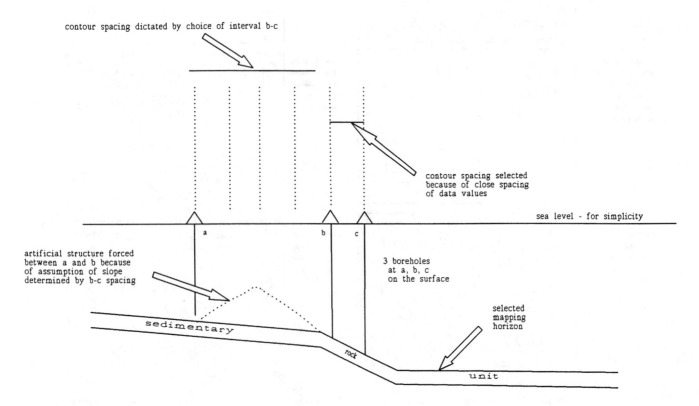

Figure 2-6.   Example of the construction of an equal-spacing contour map showing the reason why anomalies tend to be maximized.

In any contouring, all these methods are used in varying proportions. Computers can be programmed to construct contour maps, but the results are no more valid than the data fed in, and the pattern reflects the program used (commonly a modified version of the above); such maps are not inherently better than those constructed by a trained person!

When constructing a contour map:

1. Do not use ink until the pattern is finalized; draw initial contours lightly and keep an eraser handy.
2. Select the contour interval according to the map scale, the amount of data variation, and the detail required for the mapping purpose. The same contour interval should be maintained throughout the map (rare exceptions exist, such as where a log scale is used).
3. A contour must never end within a map—it should be closed or extended to the map edge.
4. Where the surface being mapped reverses slope, two contour lines of the same value must appear; that is, a single contour cannot mark the axis of reversal.
5. Contours should form rounded, not sharp, corners.
6. Contours of the same kind should not cross each other; in the exceptions (such as where there are overhanging cliffs), dash or dot the lower contours in the region of overlap.)
7. Begin contouring between data points that show a large difference in value; start a number of contours and take them as a group around the map (rather than trying to trace individual lines).
8. Break every second, third, or fourth contour line occasionally to insert its value.

To be conservative, avoid closed contours unless closure is forced by the data or you have other evidence to indicate that closures are likely (e.g., if mapping geochemical anomalies related to a point source of contamination, the probability of closed contours increases as the map area increases).

Any kind of map that includes contours should depict contour interval markers. How many contour lines should be labeled on a map depends on their density. As a general rule, a numerical value should be provided for at least every fifth contour line on maps, at regular, even intervals. The labeled contour is normally thicker than the others to aid readers to follow it around the map.

Varieties of Contour Maps

Topographic contour patterns show the elevation of the land surface above (rarely, below) sea level. Studies of topographic maps can provide much information on the geological history of an area (e.g., from a study of drainage patterns), as well as on the present morphology and relief. These maps are produced at various scales by federal, state/province, and local county or urban organizations. When planning work in a particular area, it is important to find and obtain topographic maps with scales that suit the type of work planned. Topographic contours are also necessary on most specialist

maps both for location and the use of the maps for other purposes (e.g., geologic maps themselves, or preparation of structure contours from geologic maps).

Bathymetric contour patterns show the submarine relief in negative values relative to sea level. Far fewer maps at far fewer scales exist than in the case of topographic maps; in addition, nearshore bathymetry is more likely to differ from the map because most seafloors comprise sediment that shifts in response to seasonal or storm influences on currents and waves. Major oceanic features, such as submarine canyons or submerged volcanoes, are well depicted, but beware of the date of large-scale maps of sediment-covered seafloor.

Structural contours depict the configuration in space of a surface on a rock unit, a fault plane, or an unconformity (topographic contours are equivalent to structure contours on the unconformity that constitutes the present earth surface). As with bathymetric maps, when sea level is the reference datum large negative data values imply depressions, not highs. Although structure contours are often highly interpretive, they are a powerful tool in subsurface studies because they reveal the effects and extent of postdepositional deformation. By extrapolating them above the present land surface, they can be used to determine pre-erosional positions of sedimentary units (e.g., Fig. 2-7). A full discussion of the procedures for constructing and applying structure contours is beyond the scope of this book, but an excellent introduction is provided by Boulter (1989).

Lithofacies contours can reflect any of a variety of lithological data: percentages or ratios of minerals or rock types in a stratigraphic interval (such as detritals to nondetritals); textural variations over an area (e.g., sandstones to mudstones, coarsest or modal grain size variations). *Isoliths* are contours on equal thicknesses of a particular lithotype from within a stratigraphic unit that contains diverse interlayered lithotypes. Ingenuity can produce other varieties, and combinations are possible using different colors and ornamentation. Choice of upper and lower boundaries varies with the purpose, and many different scales of unit may be mapped to provide complementary expressions of data distributions.

Isopach contours represent lines of equal thickness of a sedimentary unit. Consistent sharp upper and lower boundaries to the unit are necessary; where there are gradational or interfingering contacts, they are difficult to construct. It is generally assumed (until proven to the contrary) that the upper boundary was originally horizontal (approximately parallel to sea level); the map then depicts the configuration of the depositional basin. The zero contour represents the depositional limit to the unit, or the edge of the unit as influenced by later erosion. Abrupt changes in pattern commonly indicate that erosion has subsequently removed part of the unit (or faulting may have displaced it). These maps are rarely used alone, but together with structure contour and lithofacies maps, they are widely used in subsurface ore exploration and exploitation. Isopach maps of a number of stratigraphic units (vertically superimposed or even overlapping) will show basinal evolution. (Data points for isopachs may be obtained by

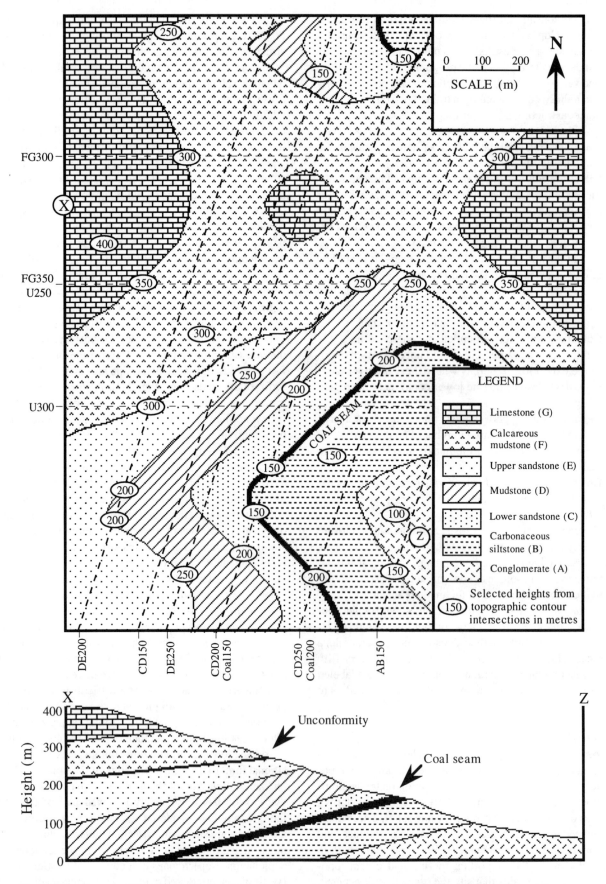

Figure 2-7. Example of the use of structural contours. Structure contours are shown as dashed lines and labeled with two letters to signify the beds in contact, or U to signify the unconformity, and a contact height AB125.

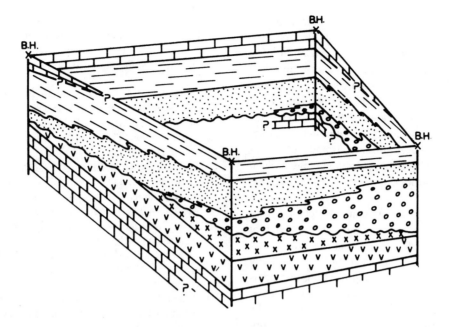

Figure 2-8.   Example of a *panel* (or *fence*) diagram.

simple calculations at intersection points of structural contours constructed for the upper and lower boundaries of any kind of unit; but the isopach map is then a third-order interpretation!)

Geophysical contours include gravity or magnetic anomalies, seismic reflection or refraction data, electrical self-potential or resistivity data, heat flow, and radioactivity. In particular, the field of seismic stratigraphy is of major importance in basin analysis as well as for the definition of most major features in the earth's crust. High-resolution seismic records are used to distinguish genetically related strata and to produce structure contour maps.

Geochemical contours depict the concentration of selected elements, the ratios of combinations of elements, or the difference between the concentration of an element and a standard reference concentration. They are essentially a very special type of lithofacies map, and they provide the foundation for most geochemical mineral exploration and for the assessment of environmental pollution. Although almost any type of chemical data from any type of sampled material (e.g., sediment, rock, soil, water, biota) can be contoured, it is essential that both the sampling and analytical procedures employed are standardized as closely as possible. Even small differences in sampling procedure or in the type of sample obtained can produce false anomalies; for example, it is not valid to contour data for soil samples where some samples are from the B horizon and some are not, or where some are for whole soils and some are for the sub-sand size fraction. Ensuring that sampling procedures are rigorously standardized and that samples are representative (sample size and type) with respect to the target elements are vital but frequently underemphasized steps in preparing these maps (e.g., see Joyce 1984; Hoffman 1986; McConchie et al. 1988;

and numerous articles, letters, and editorials in the *Journal of Geochemical Exploration* and *Explore*).

Other Geological Maps

The best-known *geological maps* (sensu stricto) are compilations on various scales of the distribution of rocks and sediments as they occur on the earth's surface. Data are obtained from fieldwork, analysis of aerial photographs, drilling, and indirect geophysical exploration techniques (such as seismic reflection and refraction). Rock units (such as formations) or time-rock units (such as series and stages) may be depicted. Most of these maps show the distribution of the units as *inferred* from scattered exposures or boreholes—it is very rare to find complete exposure of any unit. *Control* may be very good or very poor (extensive cover); the extent of control should be shown by outcrop overlays or some other means, but rarely is. *Geological cross-sections* are special types of geological maps that depict the mapper's concept of the vertical distribution of rock units in the subsurface, below a line of section shown on the plan-view map either between two points or along a drawn irregular line. Figure 2-7 illustrates a simple geological map and cross-section.

A *paleogeologic*, or *subcrop*, map shows the present distribution of rocks below a specific reference surface. The reference surface should be a time plane, whereupon the depiction is of a geological map at that earlier time. Isochronous surfaces can rarely be traced, however, and the practical reference surfaces that are selected (such as unconformities) impart constraints on the meaning of the subcrop map. Nonetheless, such maps are useful for locating truncations and pinch-outs of stratigraphic units.

Rock distributions above a reference surface (ideally isochronous) are depicted in *worm's eye*, or *lap-out*, maps, as viewed from below. Overlap relationships of younger units will be apparent (e.g., accompanying progradation of a coastal facies sequence over a surface of unconformity) and are useful in prospecting for stratigraphic settings in which placers may be concentrated or petroleum may be trapped.

Palinspastic maps attempt to show the distribution of rock units before structural deformation (e.g., removing the strike-slip movement along the San Andreas or Alpine faults; flattening out mountains such as the Appalachian ranges). Such maps are difficult to construct and are subject to much personal bias.

Paleogeographic maps most commonly depict only gross relationships between land and sea or plains and highlands at specified time intervals in the past. Most are not strictly maps, in that the position of the mapped features is unlikely to be accurate: they are conceptual reconstructions or models of what the available data suggest the area *probably* looked like at the selected time in the past. Hence they tend to be highly interpretive and do not depict features that can be objectively measured. Nonetheless, paleogeographic reconstructions are a major objective of many sedimentological studies, and where they are based on a large volume of data, well-prepared reconstructions can be very close to maps. On the smaller scale, there are rarely sufficient data to construct detailed paleotopographic maps, but exceptions occur (e.g., Asquith 1970; Dolly and Busch 1972). Problems are particularly great in areas where rocks of a particular age are absent—their absence may reflect nondeposition during the time interval (which may be suggestive of terrestrial conditions) or subsequent erosion of rocks once present.

Geological cross-sections are maps that depict the distribution of rocks and sediments in a vertical section through some part of the crust. The line of section is shown on a plan-view map; it may be a single straight line or a connected set of straight lines oriented variously. Many cross-sections are based on geological inference of the subsurface distribution of the rocks (usually using structure contours, e.g., Fig. 2-7); hence they are second- or third-order interpretations, since in transferring those inferences to a plan-view map, further inference is necessary. Others are constructed by interpreting the distribution of rocks laterally away from vertical control lines (borehole) in which the extent of the various units has been measured. A common example of this type of cross-section is the *fence* or *panel* diagram; see Fig. 2-8.

SELECTED BIBLIOGRAPHY

Asquith, D. O., 1970, Depositional topography and major marine environments, late Cretaceous, Wyoming. *American Association of Petroleum Geologists Bulletin* 54:1184–1224.

Belward, A. S., and C. R. Valenzuela (eds.), 1991, *Remote Sensing and Geographical Information Systems for Resource Management in Developing Countries.* Eurocourses, Remote Sensing 1, Kluwer Academic Publishing, Dordrecht, Holland.

Bishop, M. S., 1960, *Subsurface Mapping.* Wiley, New York, 188p.

Boulter, C. A. 1989, *Four Dimensional Analysis of Geological Maps.* John Wiley & Sons, Chichester, England.

Burrough, P. A. 1986, *Principles of Geographical Information Systems for Land Resources Assessment.* Oxford University Press. Oxford.

Busch, D. A., 1974, *Stratigraphic Maps in Sandstones—Exploration Techniques.* American Association of Petroleum Geologists Memoir 21, Tulsa, Okla., 174p.

Dent, B. D., 1990, *Cartography: Thematic Map Design,* 2d ed. Brown, Dubuque, Ia.

Dolly, E. D., and D. A. Busch, 1972, Stratigraphic, structural and geomorphologic factors controlling oil accumulation in Upper Cambrian strata of central Ohio. *American Association of Petroleum Geologists Bulletin* 56:2335–2368.

Hoffman, S. J. 1986, Soil sampling. In J. M. Robertson (ser. ed.), Exploration geochemistry: Design and interpretation of soil surveys. *Reviews in Economic Geology* 3:39–77.

Irvine, J. A., S. H. Whitaker, and P. L. Broughton, 1978, *Coal Resources of Southern Saskatchewan: A Model for Exploration Methodology.* Geological Survey of Canada Economic Geology Report 30, Geological Survey of Canada, Ottawa, 151p. and 56 plates.

Joyce, A. S., 1984, *Geochemical Exploration.* Australian Mineral Foundation, Adelaide, 183p.

Kay, M., 1945, Paleogeographic and palinspastic maps. *American Association of Petroleum Geologists Bulletin* 29:426–450.

Keates, J. S., 1989, *Cartographic Design and Production,* 2d ed.. Longman Scientific & Technical, New York, 261p.

Krumbein, W. C., 1952, Principles of facies map interpretation. *Journal of Sedimentary Petrology* 22:200–211.

Krumbein, W. C., 1955, Composite end members in facies mapping. *Journal of Sedimentary Petrology* 25:115–122.

Krumbein, W. C., 1959, Trend surface analysis of contour type maps with irregular control-point spacing. *Journal of Geophysical Research* 64:823–834.

LeRoy, L. W. (ed.), 1950, *Subsurface Geologic Methods—A Symposium,* 2d ed. Colorado School of Mines, Golden, 1166p.

Levorsen, A. I., 1960, *Paleogeologic Maps.* Freeman, San Francisco, 174p.

Low, J. W., 1958, Subsurface maps and illustrations. In J. D. Haun and L. W. LeRoy (eds.), *Subsurface Geology in Petroleum Exploration.* Colorado School of Mines, Golden, pp. 453–530.

McConchie, D. M., and L. M. Lawrance, 1991. The origin of high cadmium loads in some bivalve molluscs from Shark Bay, Western Australia: A new mechanism for cadmium uptake by filter feeding organisms. *Archives of Environmental Contamination and Toxicology* 21:303–310.

McConchie, D. M., A. W. Mann, M. J. Lintern, D. Longman, V. Talbot, A. J. Gabelish, and M. J. Gabelish, 1988, Heavy metals in marine biota, sediments, and waters from the Shark Bay area, Western Australia. *Journal of Coastal Resources* 4(1):37–58.

McKee, E. D. (ed.), 1956, *Paleotectonic Maps of the Jurassic System.* U.S. Geological Survey Miscellaneous Geological Inventory Maps 1–175.

McKee, E. D. (ed.), 1959, *Paleotectonic Maps of the Triassic System.* U.S. Geological Survey Miscellaneous Geological Inventory Maps 1–300.

Maguire, D. J., 1989, *Computers in Geography.* Longman Scientific & Technical, Essex, England.

Mailing, D. H., 1989, *Measurement from Maps: Principles and Methods of Cartometry.* Pergamon Press, Oxford.

Moody, G.B. (ed.), 1961, *Petroleum Exploration Handbook.* McGraw-Hill, New York.

Moseley, F., 1981, *Methods in Field Geology.* W. H. Freeman and Company, San Francisco, 211p.

Pelto, C. R., 1954, Mapping of multicomponent systems. *Journal of Geology* 62:501–511.

Peuquet, J., and D. F. Marble (eds.), 1990, *Introductory Readings in Information Systems.* Taylor & Francis, London.

Rees, E. B., 1972, Methods of mapping and illustrating stratigraphic traps. In R. E. King (ed.), *Stratigraphic Oil and Gas Fields.* American Association of Petroleum Geologists Memoir 16, pp. 168–221.

Robinson, A. H., and R. D. Sale, 1969, *Elements of Cartography,* 3d ed. Wiley International.

Schuchert, C., 1955, *Atlas of Paleogeographic Maps of North America.* Wiley, New York, 177p.

Sebring, L., Jr., 1958, Chief tool of the petroleum geologist: The subsurface map. *American Association of Petroleum Geologists Bulletin* 42:581–587.

3
Fieldwork

Successful and cost-effective fieldwork relies heavily on a clear identification of the objectives for the fieldwork and efficient planning before departing for the field. Plan operations in the time framework of the project as a whole and allow ample time with pessimistic assumptions with respect to difficulties and days lost—time left over from your estimate will be fully utilized in tidying up at the end. Continual review of progress is also necessary after initiating work in the field area. In the field it is vital to establish a systematic pattern of work for field observations, note taking, sample collecting, and storage (see Chapter 4)—and for evenings and rainy days. In this chapter we offer some thoughts on fieldwork strategies we have found effective; books and papers cited at the end of the chapter provide more comprehensive and detailed treatment of techniques and procedures.

FIELD EQUIPMENT

Ensure the appropriate general field equipment is on hand (e.g., checklist provided in Table 3-1): it is easy to overlook basic essentials when concentration is on readying major equipment for a project.

Various specialist accessories (e.g., geochemical testing equipment or stain preparations for identifying carbonate minerals; Chapter 8) may be desirable depending on the project. For example, in country with substantial relief, the ability to measure elevation and gradients may be necessary (e.g., with an altimeter, a level, or a sighting compass). A miner's (gold) pan may be desirable for concentrating heavy minerals. A mechanical counter will be useful if pace-and-compass traverses or pebble counts are to be made. Checklists or report forms designed for specific purposes may be useful (e.g., **PS** Fig. 1-8). Many geologists are now finding that a small voice-activated tape recorder is a useful aid for recording field observations, particularly when support staff are available who can transcribe the notes later. This method

of recording field observations can be very efficient, but unlike the simple notebook, it may be useless if spare batteries or tapes are forgotten, or if the tape recorder breaks down. The recorder can be sealed in a plastic bag for use in wet weather. A GPS (global positioning system) instrument (Chapter 2) may be desirable for some work—it will not be long before it is part of general field equipment.

FIELD MAPPING

Precise location of your position in the field is of first priority when mapping; ensure you have the best possible maps and air photos (see Chapter 2). Initially, do not waste time in areas where accurate location is difficult; e.g., leave bush-covered areas to be examined after the general distribution of units is known. Where bush cover is extensive, it may be necessary to take bearings on easily identified reference points and either locate by triangulation, pacing the distance to the locality, or utilizing GPS equipment (also desirable in extensive flat areas that lack distinctive morphological features, such as much of central Australia). An altimeter can help in location on heavily forested hillsides and in steeply incised streams (e.g., to distinguish which bend or tributary of a stream you are crossing); be sure to verify your elevation with Trig stations (fixed survey points with accurately known elevation, latitude, and longitude) twice a day when variable weather is likely to influence the barometric pressure. If a distinctive feature is at your own elevation, the location can be found easily on a topographic map by sighting along a leveled compass (see later in chapter). When taking bearings on hilltops, be sure that the true top can be seen.

Adopt a systematic numbering system for each locality and mark sample sites on the field map as a backup for the records in your field notebook (or data record cards) and as a quick visual indication of your sampling frequency and sample distribution. Make sure the sample numbers on

Table 3-1.   Checklist for Field Equipment

Topographic maps and aerial photographs (or air-photo mosaics) should be in hand. An essential requirement for any fieldwork is the ability to locate oneself precisely. Portable GPS equipment (see Chapter 2) should be included if available.

A map board or case, plus plastic tracing film and masking tape for air-photo overlays.

A notebook and/or data record cards on a clipboard—preferably waterproof paper; when wet, the book or cards may be enclosed in a sufficiently large sample bag in which you can also insert your hand and a pencil.

Pencils (of different hardness), a good eraser, and a sharpening device.

A hand lens (normally 10×—lower or higher power ones can prove useful depending on the sediments involved).

Tracing film and masking tape for playing about on maps (e.g., sketching projected structural contours).

A pocket knife preferably with a surgical steel blade that keeps its edge and resists rusting. Many geologists prefer a knife with a built-in corkscrew and bottle opener for incidental activities, but the extra fittings commonly get in the way and complete cleaning is more difficult.

A grain size comparator (e.g., a *properly scaled* set of photographic images or sieved grains mounted on cards or in containers such as Fig. 3-1, or see Lewis, van der Lingen, and Jones 1970).

A cm/mm scale and/or (vernier) calipers are needed for precise measurement of structural features and pebble dimensions.

An assortment of sample bags appropriate to the sediments and purpose of their subsequent analysis (e.g., samples for geochemical analysis require higher-quality bags than those for most other purposes).

Several indelible markers for labeling (Murphy's Law states that if you have only one, it will run dry). Note that many markers based on organic solvents become unusable if they get wet or are used to write on a wet surface.

A good camera is essential to produce photographic images of exposure detail and for illustrating the final report; a zoom lens with macro capability is desirable. Ensure that appropriate scales are present in every photograph; e.g., a set of aluminum arrows (which can indicate the way up or specific features) with length specified on each and each with one side stained black (to permit either side to be photographed as

the background requires), or a rectangular piece of aluminum with a scale along the edges and a black surface on which to chalk locality number or details. Do not use coins or similar items as scales because not everyone (e.g., overseas scientists reading the paper) viewing the picture may be familiar with the size of the object used as a scale. Excellent protection against rain and air-mobile debris can be provided by half a car chamois used as a wrap when the camera is out of its environment-proof bag. A fine photographic blower brush is desirable to clean the lens of dust (a cloth will grind dust into the lens!). Use a waterproof camera case if there is any chance of falling into streams or when working on a coastline. Black and white film provides the best resolution for prints to be published, but color slides (or prints) are best for studying and presenting at talks. Many workers carry two cameras so that both 35-mm slides and black and white prints can be prepared. Also consider taking one of the new breed of cameras that prepares digitized data on a minidisk that can be fed into a computer for image enhancement. Clean the camera regularly, particularly when even small amounts of salt spray have been encountered.

A small whisk broom can assist in cleaning exposures of loose cover (e.g., preparatory to photography).

A hammer is necessary if working on rocks—preferably with a flat end (for prying layers apart) rather than a pick end (a bricklayer's hammer generally has too short a handle; Lewis has found the long-handled Estwing "wrecker's" hammer to be ideal). Round off sharp edges on any hammer to minimize the chance of chipping, and wear safety glasses if hammering well-indurated rocks.

A cold chisel is also desirable for sample extraction when working with indurated rocks.

A flat cement-layer's trowel and/or long-bladed flexible knife (e.g., Fig. 3-2) or a sharpened spatula is necessary when working with loose sediment for scraping surfaces and sampling (stainless steel blades are desirable to avoid rust, particularly if dealing with seawater, but even stainless steel will rust if enclosed in a leather case for any period of time).

Dilute hydrochloric acid (10%) is desirable if carbonate sediments are likely to be encountered, but beware that most containers leak (especially when warmed up by the sun or when sat upon), to the detriment of clothing and other equipment!

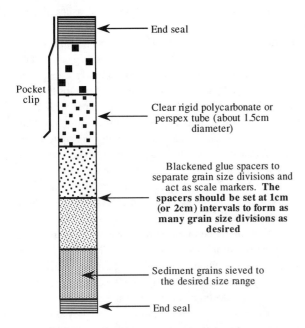

The side of each grain-size compartment can be engraved to indicate the size range of the grains in the compartment

Figure 3-1.   Examples of simple grain size comparators and scales.

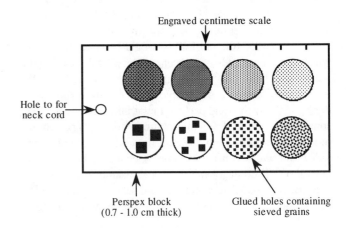

Table 3-1. *(Continued)*

A good-quality compass/clinometer should always be carried for taking bearings on landmarks and also for making measurements of the rock layers and orientational or directional structures in the sediment (see below).

A waterproof tape is needed (about 50 m minimum) if any sections are to be measured.

A standard (Munsell) color chart should be carried for unbiased descriptions of fresh, weathered, wet, and dry sediment color (e.g., Goddard et al. 1951 and see **PS** Chapter 1). If one is not available, restrict yourself to primary color names—terms like *purple* mean very different colors to different people!

A small spray bottle full of water is useful to enhance subtle differences in sediment color (e.g., when studying trace fossils).

A waterproof parka (especially if going into the field with Lewis)—and don't forget your lunch!

Additional Equipment for the Base Camp

Protractor

Scale ruler/straightedge

Hard pencils and an array of colored pencils

Drafting pens and ink

Drawing board and paper of adequate size for final field map

Graph and tracing paper

Spare base maps, sample bags, notebooks, etc.

White paint for labeling specimens

Containers for packing samples for transport

Stereonets if dealing with paleocurrent data and tilted strata (see Chapter 6)

Spare parts and basic tools for servicing any field analytical equipment

Laptop computer (and printer) for writing and storing data

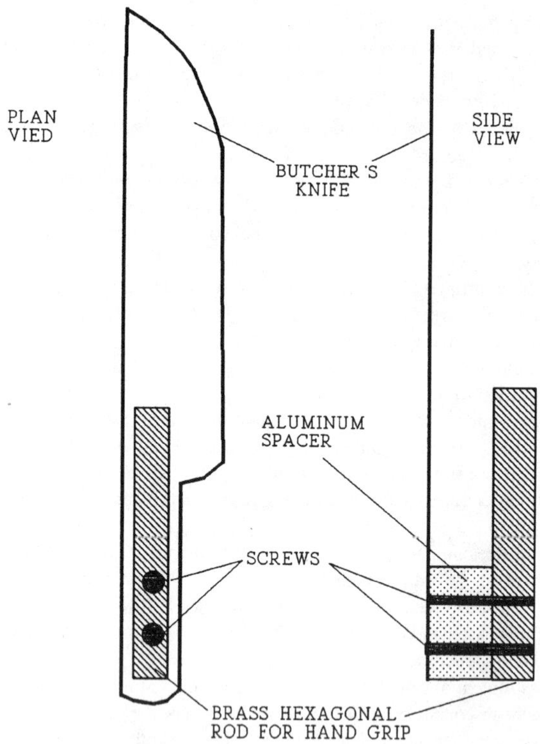

Figure 3-2. Sketch of a sediment knife constructed by attaching a handleless butcher's blade (stainless steel) to a side handle (aluminum will do, but it leaves the hand black). The flexible blade is drawn across the sediment without scraping fingers (as with an ordinary knife or spatula) to prepare a flat surface for peels (e.g., Chapter 4), photography, or mere viewing of a fresh face.

sample bags, on the map (or photo overlay), and in the notebook (or data card) are *exactly* the same; it is very irritating to have to match up different sample numbers back at base. When establishing a name/number code for samples, measured sections or sites where specific observations were made, the inclusion of a locality identification system will assist later data sorting and preparation (particularly when a computer database with a sorting function is used). For example, a series of samples collected on a traverse up Long Creek could be labeled LC100, LC110, LC120, etc.; each number would refer to a locality shown on the map and numbers between LC100 and LC110 would be reserved for use when several samples have been collected at the locality or for use with laboratory subsamples. Another example is LM-91-8-2, meaning second sample collected from locality 8 during 1991 by the field party of Lewis and McConchie. Personal initials are useful in the notation if there is a possibility of mixing samples from different field parties; use of a date index can avoid confusing samples collected by the same person on different jobs. However the samples or sites are labeled, use a consistent and standard set of symbols and abbreviations; list these on the first or last page of your notebook both for yourself and for other potential users of your map or notebook.

When mapping, put as much information as possible on the map or photo overlay in the field with a hard, sharp pencil. Use tracing film (tracing paper is neither sufficiently transparent nor weatherproof); do not write directly on the photos to avoid obscuring features you may wish to examine later with a stereoscope (key localities may be "projected" through the print and labeled on the back). Also, register your overlay to reference points on the photo and label with the photo number—the overlay will not always be attached to the photo! Outline (or mark with an "X") all exposures. *Do not rely on your memory to insert these data later!* Sketch in contacts (solid lines where unequivocal, dashed lines where hidden, dotted where uncertain; remember that depending on the scale of the map, a line width may represent a band 10 m wide!). Plot other features such as sample localities, faults, strike and dip wherever measured. Where exposures are absent, note the character and extent of cover both so that you have a record that you have traversed the area and because that information may assist any later air-photo interpretations and attempts to interpolate contacts from adjacent areas of reasonable exposure.

Contact landowners in advance of fieldwork; give a full explanation of your purpose and assure them of your intention to leave gates as you find them and to avoid leaving any rubbish (such as lunch wrappings). Ask if they have particular concerns and express your gratitude for their permission to work on their land. A written or printed note giving your occupation and contact address can allay suspicion and antagonism. Take every precaution to avoid damaging the prospects of other workers gaining access to the area—our ability to carry out science depends on access to the data! The landowners may provide very useful information relevant to your project (if only for information on whether farm tracks that stand out well on air photos have been changed since the photo was flown), and they may keep an eye out in case you are injured.

Traverses

First traverses should be along the most continuous, accessible, and locatable exposures—coastlines, cliffs, streams, and/or roads. When dealing with rocks, ridge exposures can be ideal for initial traverses because landmarks are visible and localities are easily found on a contour map. Fence lines and power lines (which may deflect compass readings) are also easily located on modern air photos.

The initial traverses will identify the general distribution of sediment types in the area. On the basis of this information, a series of straight cross-trend traverses at a regular spacing, such as 1000 m, can be designed to delimit the lateral extent of units. In most cases, contacts can be interpolated between traverses using air photos and topographic maps. Sampling of sediment, soil, or rock at regular intervals on straight-line traverses, or in grid patterns based on them (see Chapter 4), forms the basis for many environmental and exploration geochemical studies.

Because contact relationships between units provide important information, it is desirable to walk them out wherever they are even partially exposed. Changes in vegetation, soil character, or mass wasting effects may mark different underlying units where exposure is absent. In the case of rock units and intermittent exposure, walk along the contact zone, using the distribution of eroded fragments (float) as a guide where there is no exposure. Debris transported by streams is usually lithologically diverse, but progressive upstream disappearance of particular rock types may indicate where to look for contacts. The construction of structure contours (Chapter 2) on contacts while building a field map can be very useful in projecting the location of hidden contacts and predicting where they should outcrop. (If they are not where expected, the cause may be obscure folds or faults.)

Selection of Units

In an area of arenites and limestones, it is clear that the arenites should be distinguished from the limestones for mapping purposes, but it may be much less clear how many rock units of each type should be distinguished. Each sample or exposure of each rock type may have different characteristics. Differences apparent in the field may be greater within one limestone unit than between two limestones separated stratigraphically by an intervening arenite. General suggestions for selecting and recognizing mappable units are provided below.

The larger the scale of the map and the more time available, the more subdivisions should be attempted (and vice versa). It is better to attempt subdivisions in the early stages because these can always be combined later, whereas if subdivision is attempted late, a return to many exposures may become necessary. Initially, keep subdivisions informal in the notebook and by light penciling on the map, until the lateral extent of the units can be clearly demonstrated. Local but very distinctive lithologic variants within a formation can be mapped as "members" or given distinctive symbols over the color used for the formation as a whole. Be sure to note how complete the exposure is for each unit—you will ultimately have to specify a type or reference section for any unit formally defined.

Distinguish mapping units on the basis of characteristics that are likely to persist within that unit. Color is rarely a useful criterion: differences may reflect localized diagenetic or weathering effects; fresh and variously weathered rocks may look quite different. Induration may reflect localized surface cementation (case hardening) by modern weathering processes. Changes in biota may occur because of selective transport or subtle differences in environmental conditions. Sedimentary structures and textures may vary greatly laterally or vertically because of local or short-term fluctuations in energy conditions at the time of deposition. In contrast, composition of detrital sediments reflects source rock characteristics and is unlikely to change rapidly. Hence composition is a better criterion for first-order distinction between units composed of detrital sediment. However, composition is not necessarily the best criterion for distinguishing between limestone units deposited at the same time in different adjacent environments. No firm rules can be given for the criteria to be used, but consider carefully the limitations of those selected!

Contacts are critical to the recognition, mapping, and interpretation of units. The best boundaries to units are distinctive differences across a sharp contact. However, because environmental conditions may change gradually, or because there are lateral shifts of environmental conditions back and forth during deposition (e.g., as a result of meandering channels or of repeated coastal progradation and retrogradation), contacts may be gradational or interfingering on any scale. In these situations, personal bias usually dictates the choice of boundary (e.g., "the first" as opposed to "the last" occurrence of a particular characteristic). In all cases, consider carefully the implications of the selection; e.g., a sharp contact may reflect the presence of an unconformity, other criteria for which should be sought (e.g., **PS** Table 4-4). Because the characteristics of a contact may change within the area under investigation, particular care is required in description at every locality, and full discussion must be provided in the final report.

Where there is irregular (or apparently regular) alternation of lithologies (such as interbedded sands and muds), try to record the range of bed thickness and texture in each exposure. Upon final review of data compiled from multiple localities, the areal or vertical pattern of changes (e.g., the ratios of sandstone to mudstone) may be sufficient to warrant subdivision of units, even if on the basis of approximate or arbitrary boundaries.

Do not be concerned with the formal naming of the rock units in the early stages of mapping. Common exceptions to this guideline occur when distinctive units can be recognized from the same or adjacent areas that have already been mapped. Although use of already defined formal unit names simplifies mapping and description, there is a danger that the criteria used by the original worker may not be as well suited or as meaningful as other criteria that you may choose. Hence, it may generally be desirable to use informal names of your own choice and decide on final names only at the end of the project.

Description of Units

Upon arrival at any exposure, the first step is to take an overview—stand back and look at the large-scale features, observe relationships between units, and any three-dimensional geometric or structural (e.g., large channel shape) configurations. Only after you have the exposure in its full perspective should you begin to take notes.

Follow an orderly sequence of observation and notation—e.g., structure then texture then composition; establishing a consistent pattern reduces time spent per exposure, establishes a routine for comprehensive description (and later, reporting), and should ensure sufficient description to avoid a return to the locality for details missed the first time. Table 3-2 provides an example of a field observation checklist (field data record cards such as **PS** Fig. 1-8 perform a similar role). Paleocurrent directions and orientations (see discussion in **PS** Chapter 4 and **AS** Chapter 6) should be measured wherever possible, and all at the same time for the sake of efficiency (i.e., while the compass is in hand; see also Andrews 1982 and Tucker 1982 for means of dealing with the resultant data).

Age of Units

Relative ages can always be determined by the principle of superposition (younger rocks lie on older). Assigning an absolute geological age to a stratigraphic unit in the field is not particularly important in mapping operations within a local area, and is generally not feasible unless you find and recognize time-diagnostic fossils. It is better to await laboratory-based age determinations than to guess from similarities with other areas. Be aware of the likelihood that deposition of any sedimentary unit spans a range of time and that age determinations from any one sample provide only a single indication of a time *interval* (resolution of any fossil assemblage is rarely more restricted than ±500,000 years)—it can be very misleading to assume the same age for the entire stratigraphic unit based on determinations for a single locality (or several, for that matter!). It is also worth remembering that with the exception of some chemically precipitated units and volcanic ash layers, time horizons do not coincide with lithologic contacts, although they may appear to do so in terms of temporal resolution capability. Because sedimentary facies shift position during transgression or regression, contacts between these units will be time transgressive, with the angle between time lines and lithologic contacts reflecting the rate of transgression. Consequently, a single contact may change in age by several million years over a distance of a few tens of kilometers.

Tectonic Features

Small faults and folds may be very useful in determining sedimentary or tectonic history, but they can be confusing if their origin is uncertain. Evidence for soft-sediment deformation (e.g., diffuse contacts, smearing along faults, flowage within folds) may indicate penecontemporaneous deformation as opposed to postlithification tectonism. Even if faults are too small to map individually, note the strike, dip, and direction of movement in your notebook; the data may be useful during general synthesis. With minor folds and crumples, map the trends of the fold axes and their average plunge (the attitude of the line formed by the intersection of a bed and the axial plane), the orientation of the axial plane, and the vergence (direction of closure) or sense of facing with asymmetric folds. Cleavage-bedding relationships can be used as stratigraphic way-up indicators and to relate localities to their relative position on major folds (cleavage is generally close to parallel to the axial plane of a fold, but fans outward from that plane). Large faults and folds usually become clear only as a result of detailed mapping of contacts and the synthesis of numerous dips and strikes on a complete map. Do *not* expect to see them in the field.

Minor folds and faults of tectonic origin generally are produced by the same regional stresses that produce major structures; hence they are usually consistent within the same exposure and show a similar pattern to those found nearby. Penecontemporaneous, soft-sediment deformation structures tend to be more irregular and diverse in pattern, and most are confined to a few beds in a sequence. Hence the origin of the structures may become apparent after compilation of all data from an area. However, there are overlapping regimes, and fracturing and/or distortion by modern mass-wasting processes also may result in small-scale folds or faults (observation will generally reveal a direct relationship to the topography).

Table 3-2.   Checklist for Field Observations

Preparation (remember you may be reading these notes years later, or your boss may be reading them tomorrow!):

Number pages in upper-right-hand corners; leave left-hand (or top) pages for sketches.

General title and location of project.

Field party (note each day if party varies).

Map and/or aerial photograph number(s).

Date; weather conditions.

Each Locality (look over the entire exposure first!):

Locality number (consecutive within project).

Plot locality number on field sheet; note exact location in book (grid intersection from map) *and* give brief notation of physiographic setting (e.g., stream cut in heavy bush).

Exposure character (size, shape, and quality of exposure).

Relationships of Units

Rock unit name (not necessarily a *formal* name, until you have finalized the stratigraphic terminology).

Approximate stratigraphic position within major unit (if possible).

Approximate stratigraphic thickness of each unit exposed.

Trends—vertically or laterally in bed thickness or character (e.g., thinning and fining upward); are there any repetitive vertical rhythms ("cycles") at any scale?

Description of boundaries of each unit recognized and spatial relationships with other units exposed.

Miscellanea—whatever seems useful for later report, such as physiography ("forms a waterfall across the stream"); stratigraphy ("similar units appear to be present 50 m above and below" or "unlike the same unit to the west in having. . ."); vegetation (do certain flora grow on, or avoid, the unit in preference to other lithologies?); soil (is there any apparent relation to bedrock?).

Description of Lithology (begin at the base of any unit and work upward)

Index number for any graphic log.

Color (wet vs. dry)—use primary colors (not buff, chocolate); preferably use color chart.

Weathering—degree and depth.

Permeability/porosity—qualitative observation of obvious holes or rapidity of water absorption.

Induration—is it due to any obvious cement?

Major stratification—scale and character of irregularities, sharp or gradational contacts, thickness of strata.

Internal structures—e.g., graded bedding, cross-bedding, channels, ripple marks, flute marks, concretions, mud cracks. Note characteristics of each—scale, extent of development, attitude, etc. Measure orientation or direction of paleocurrent where relevant.

In deformed rock sequences, record facing direction (which way was up?). What evidence is there? How convincing is it?

Texture—size (max., min., mode, proportions), sorting, shape and roundness, fabric, quantity of matrix, matrix- or clast-supported Textural classification.

Composition—identify components as possible; estimate percent of each. With conglomerates, note abundance of pebbles of each (specified) lithology. Note any relationship between composition and texture. Compositional classification.

Fossils—identify to level possible, collect unknown types or samples for later study; note relative abundance of each type, apparent associations, morphology of odd types. Sketch. Are they (generally or as individual species) uniformly distributed or concentrated in layers, lenses, concretions? Do they show a preferred orientation (e.g., individual valves concave upward)? Proportion of articulated forms? Degree of abrasion/fragmentation? Further data should be collected when paleoecological study is warranted (special checklists necessary).

Trace fossils—describe occurrence with respect to bedding, noting size, morphology, and relationship to sedimentary structures and body fossils. Define three-dimensional geometry of each trace wherever possible; give formal genus/species name if known. Infer behavior of progenitor organism if possible. Note any difference in sediment texture/composition within the trace relative to that surrounding it. Check for evidence of boring. Note concentrations of traces relative to lithology and sedimentary structures.

Structural Data

Attitude of principal bedding planes, variability in strike and/or dip. Use a standard procedure such as dip and dip direction, or dip and strike (using clockwise convention) when recording the attitude of bedding planes (see "Use of the Compass" section in this chapter).

Folds—sketch and annotate. Symmetry and direction of vergence, fold form, trend and plunge of fold axis, and strike and dip of the axial plane (line up a field notebook), or if not possible, trend and plunge of the axial plane trace, and trace of the crest or trough. Relationships to other structural features.

Faults—sketch and annotate. Attitude, extent and character of fault zone, direction of movement (apparent vs. real?), trend and plunge of any striations, and if stepped, sense of relative motion.

Cleavages—measure strike and dip, especially when widely developed and parallel to the axial plane of folds.

Joints—strike and dip if well and pervasively developed; the nature of the investigation dictates the level of attention directed to these features.

Flow banding in igneous rocks; presence or absence of baking at contacts with sediments; chilled margins; cross-cutting relationships.

Interpretations

Jotting down ideas on origin and possible interrelationships are well worthwhile, but they will generally change later, so keep them confined to a separate section in your notes rather than juxtaposed with observations.

Description of Coals

Although the megascopic (and microscopic) characteristics of coals vary considerably with age and location, resulting in diverse descriptive classifications, relationships between coal seams–associated sediments can be defined on the basis of features that have general rather than local relevance. Measurement of overall seam thickness and the thickness and frequency of mineral-rich horizons within the seam are essential. Seam roof (top) and floor (bottom) characteristics should also be recorded. A gradual transition from coal to underlying sediments, via a highly carbonaceous interval, is typical of local derivation of plant material growing in place *(autochthonous)* or nearby *(hypautochthonous)*. Coalified roots within sediments below the seam provide direct evidence of in situ plant growth; they are distinguished from detrital vegetation by a generally high angle to bedding, connection with carbonaceous sediments at the floor of the seam, or downward branching. Some woody root systems penetrate to 1 m below the bed of origin. If roots are absent, and the transition from sediment to coal is abrupt, a transported *(allochthonous)* origin for the vegetation should be considered, particularly if the sub-seam sediments are sandy. The upper contact between coal and overlying sediments can be either gradational or abrupt, and the sequence of lithologies at the contact should be described in detail, at centimeter scale. A gradual transition via carbonaceous mudstone suggests that peat has not been eroded before burial, whereas a sharp contact with silt or sand, carbonaceous or otherwise, indicates the possibility of erosion. Peat clasts in the overlying bed provide clear evidence of nearby erosion, and the seam roof may be visibly scoured. Coals deposited in a coastal setting sometimes have burrows at the roof that are commonly filled by clastic sediment and/or pyritized. Interpretations of depositional setting based on the sedimentary relationships of coal seams must often be tentative, and the character of the coal itself can sometimes provide important additional information (see Chapter 8).

Measurement of Geological Sections

Locations of representative stratigraphic sections for detailed measurement are best selected after areal mapping is almost completed; consider accessibility as well as the quality and extent of exposure.

Measurement of section by tape traverse along lines determined by compass bearings is the most widespread and useful technique wherever sediment variation is rapid and detailed description is necessary. If distances are great, lithological character does not vary greatly over short intervals, units dip more than 10°, and there is little relief along the line of section, *pace-and-compass* measurement is feasible for near approximations to the true values (discover your regular pace—distance from right toe to right toe after one full step—then maintain that stride along a compass bearing and record the number of paces). Where beds dip less than 10° and/or there is marked relief along the line of traverse, plane-table surveying and graphical solution for slopes and dips are commonly necessary (unless a very accurate, large-scale topographic map is available).

Because the results of section measurement will generally be presented in the final report as a columnar section, it is wise to attempt direct recording of data on *graphic logs* (e.g., Andrews 1982; Tucker 1982); Fig. 3-3 shows an example of the layout of one such log. If working as a field party, standardization of log symbols is vital (Fig. 3-4 shows a few suggested symbols). The completed sheets will provide a comprehensive "picture" of the sequences measured and show precise positions of samples as well as all salient features of the sediments. Sections measured may be composite—i.e., segments from various localities—so long as the localities are not excessively separated (e.g., by over 1 km)

and there is at least one horizon that can be traced between any two segments. Always begin plotting at or near the bottom of a page and work upward in the order of deposition. Use of standard scales is useful for easy comparison between sections—e.g., 1:100, 1:1000, or 1:5000, depending on the purpose and nature of the sections (all three can in fact be used on a single sheet); remember to specify the scale(s) used on each sheet!

Useful computer packages exist (e.g., MacSection™ or Logger™ from Rockware, see Chapter 1) to assist with preparing stratigraphic sections and to produce fence diagrams from multiple measured sections. If you intend using such a package, data transfer to the computer can be simplified by using the same patterns on the field logs as will be used later on the computer (with a portable computer, it may be possible to construct stratigraphic logs at field base camp while everything is fresh in your mind).

Use of the Compass

There are many compasses with a wide variety of features. Consideration must be given to the kind of bearings one wishes (or is required) to record. Compass scales can be expressed as degrees of azimuth (0–360°; generally three digits are given even if the bearing is under 90°, e.g., 050°), as degrees in a quadrant (4 × 90° quadrants; the bearing is given as number of degrees east or west of north or south—e.g., S45°E), as mils (640; mainly military use), or as grads (400; mainly European use). The former is most desirable in our opinion. Two essential features for compasses to be used in fieldwork are:

1. The ability to set the *angle of declination* (the difference in the bearing of magnetic north relative to true north). This angle varies with both longitude and latitude (and it varies annually by few seconds of arc, but this variation is generally ignored by geologists). Topographic maps generally show the angle of declination, and when initiating fieldwork that angle should be set on the compass. (If the compass cannot be adjusted and you are likely to forget to make the corrections at some later stage, get an adjustable type!) If using a compass that cannot be corrected (some of these have the advantage of quicker stabilization or greater accuracy when taking bearings), remember to correct at the plotting stage (e.g., by ruling the declination on a protractor).

2. The adjustment for *angle of inclination* of the magnetic lines of force, which go down to the north in the northern hemisphere and down to the south in the southern hemisphere. Unless the compass needle has a counterbalancing weight, it may not swing freely on its bearings when the compass is held level. Many compasses have the counterbalance set for the northern hemisphere; when imported into the southern hemisphere, there can be significant problems even at midlatitudes.

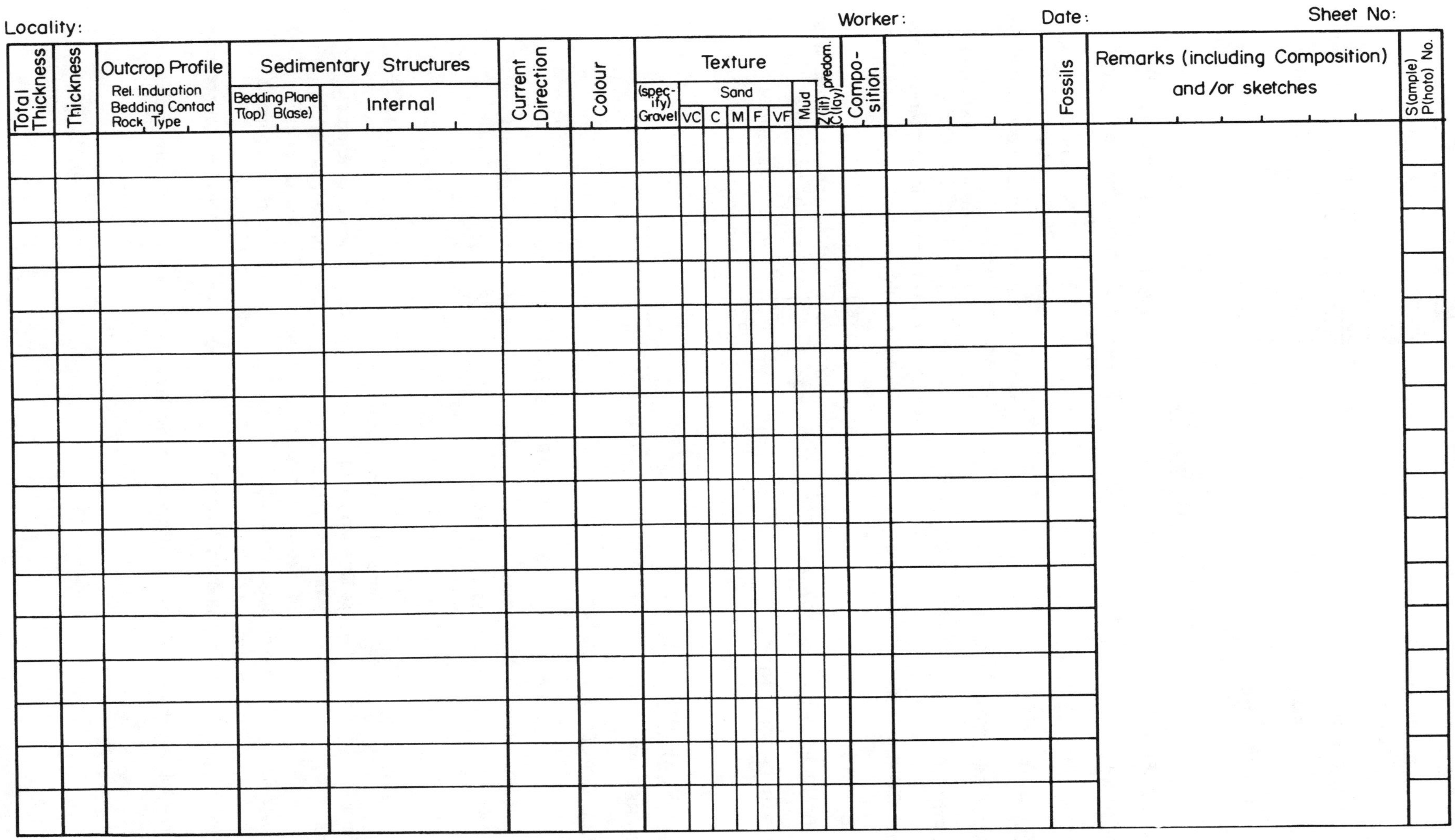

Figure 3-3. Example of a graphic log setup for description of individual lithologic units (for which thickness needs specification); the blank columns are for the worker to name as appropriate for the sediments involved. Alternative forms may be designed according to the preference of the worker and for particular situations (e.g., a useful form might have no horizontal lines and a scale marked on both sides). Whatever form is selected for recording data in the field, transferral of data to a properly scaled final log will be necessary.

Locality:

Worker:

Date:

Sheet No:

Total Thickness

Thickness

Outcrop Profile — Rock Type, Bedding Contact, Rel. Induration

Sedimentary Structures — Bedding Plane (Top) Base), Internal

Current Direction

Colour

Texture — Gravel (specify), Sand (VC C M F VF), Mud, C (clay) N (silt) predom.

Compo-sition

Fossils

Remarks (including Composition) and/or sketches

S(ample) P(hoto) No.

31

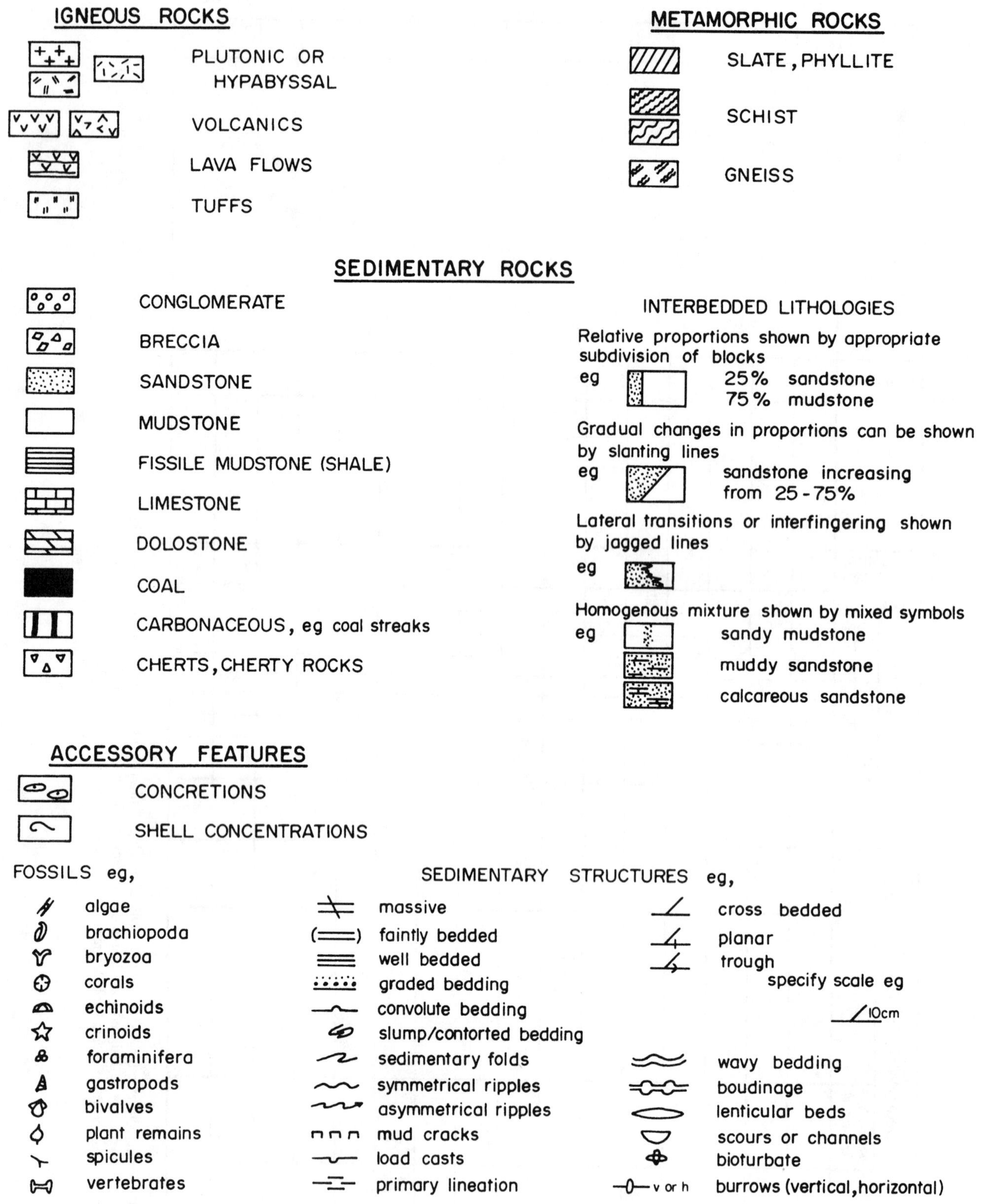

IGNEOUS ROCKS

PLUTONIC OR HYPABYSSAL

VOLCANICS

LAVA FLOWS

TUFFS

METAMORPHIC ROCKS

SLATE, PHYLLITE

SCHIST

GNEISS

SEDIMENTARY ROCKS

CONGLOMERATE

BRECCIA

SANDSTONE

MUDSTONE

FISSILE MUDSTONE (SHALE)

LIMESTONE

DOLOSTONE

COAL

CARBONACEOUS, eg coal streaks

CHERTS, CHERTY ROCKS

INTERBEDDED LITHOLOGIES

Relative proportions shown by appropriate subdivision of blocks

eg   25% sandstone
      75% mudstone

Gradual changes in proportions can be shown by slanting lines

eg   sandstone increasing from 25-75%

Lateral transitions or interfingering shown by jagged lines

eg

Homogenous mixture shown by mixed symbols

eg   sandy mudstone

      muddy sandstone

      calcareous sandstone

ACCESSORY FEATURES

CONCRETIONS

SHELL CONCENTRATIONS

FOSSILS eg,

- algae
- brachiopoda
- bryozoa
- corals
- echinoids
- crinoids
- foraminifera
- gastropods
- bivalves
- plant remains
- spicules
- vertebrates

SEDIMENTARY STRUCTURES eg,

- massive
- faintly bedded
- well bedded
- graded bedding
- convolute bedding
- slump/contorted bedding
- sedimentary folds
- symmetrical ripples
- asymmetrical ripples
- mud cracks
- load casts
- primary lineation

- cross bedded
- planar
- trough
  specify scale eg
  10cm
- wavy bedding
- boudinage
- lenticular beds
- scours or channels
- bioturbate
- burrows (vertical, horizontal)

Figure 3-4.   Example of symbols and patterns for representation of lithologic characteristics on graphic logs and/or geological cross-sections (see Chapter 2).

Highly desirable features include:

1. A centering bubble to ensure leveling (see applications below).
2. A sighting system to permit taking precise bearings on features at a distance (e.g., topographic landmarks, bedding planes).
3. A damping system to settle the needle to a fixed bearing in a short period of time. Three different methods are used: (1) *needle lock,* where a locking lever is used to stop movement temporarily; repeated manual locking by the operator is probably the fastest method of damping, and there is the additional advantage of being able to lock the final bearing for easier reading; (2) *liquid immersion,* where the compass is filled with liquid that slows the needle's perturbation; problems arise with bubbles forming in the liquid (e.g., with a change in altitude or temperature) that may interfere with the needle movement; (3) *induction damping,* where the magnetized needle swings within a coil that exerts an electrical induction braking effect. Many other compasses merely stabilize by the swinging card method, allowing gravity to damp movement.
4. A clinometer (means to indicate angle of inclination) to measure the attitude of sedimentary units and features

(beds, cross-beds, current lineations, long axes of imbricated pebbles, and so on).

Figure 3-5 depicts a compass with clinometer and sighting device.

Measurements with the Compass

Obviously the main use of the compass is to take bearings for location, on objects, in the direction of measured traverses, and so on. The sighting function permits the compass to serve as a leveling device for measuring elevations when the clinometer bubble is centered. For example, knowing your own height to eye level, you can measure multiples of that height merely by sighting on any distinctive small-scale feature such as a clump of grass, then moving onto it and resighting; if the clinometer is set to the dip of the beds and its bubble is centered while sighting, this procedure measures the thickness of the beds (Fig. 3-6). An attached prism (prismatic compass) or mirror on the compass lid (folded back more than 90°) is needed to permit viewing of the needle and azimuth scale when using the compass in these ways (e.g., Fig. 3-7). Devices such as the *Abney level* (which permits sightings on targets at measured angles relative to the eye of the user) and *Jacob's staff* (a rod with a scale painted on it) are designed to be used for these purposes with ordinary

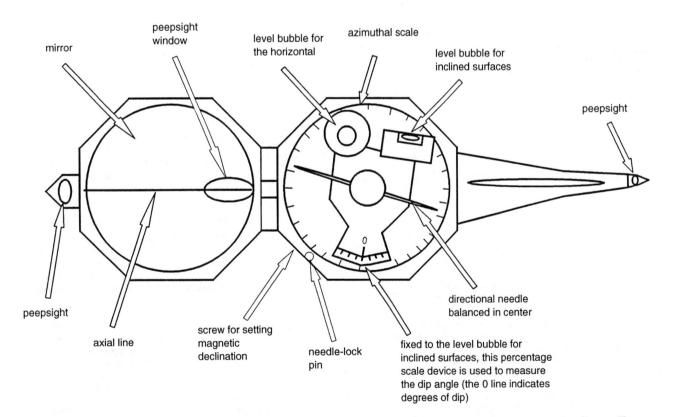

Figure 3-5.   Sketch of a combination sighting compass and clinometer (the Brunton Compass™); see Figs. 3-6 and 3-7 for examples of its use. The circular level bubble should always be centered when taking a horizontal reading.

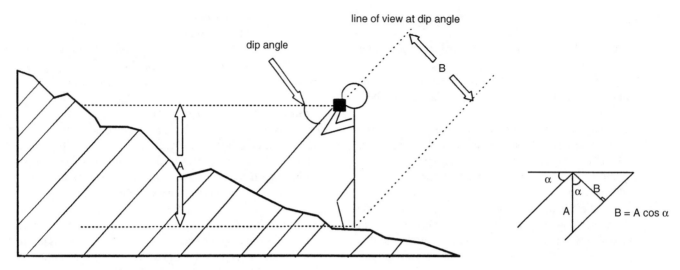

A — elevation difference = height (foot-to-eye) of worker
B — true stratigraphic thickness = (height of worker) × cosine (dip in strata), with compass clinometer or level set so that sighting is along the dip of the strata. Alternatively, a rod with scaled markings (e.g., Jacob's Staff) could be oriented perpendicular to the line of view and the intersection of the sighting on the rod = true stratigraphic thickness.

Figure 3-6.   Example of the use of a sighting compass in the field. A: to measure elevation differences; B: to measure thickness of dipping strata directly (multiple beds are depicted; cf. Fig. 3-8).

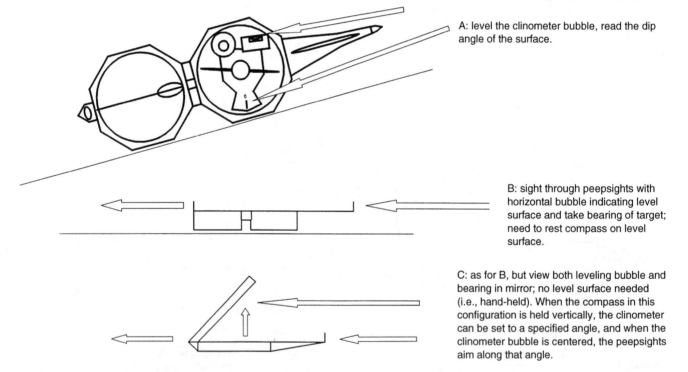

A: level the clinometer bubble, read the dip angle of the surface.

B: sight through peepsights with horizontal bubble indicating level surface and take bearing of target; need to rest compass on level surface.

C: as for B, but view both leveling bubble and bearing in mirror; no level surface needed (i.e., hand-held). When the compass in this configuration is held vertically, the clinometer can be set to a specified angle, and when the clinometer bubble is centered, the peepsights aim along that angle.

Figure 3-7.   Example of the use of a compass combined with clinometer. **A:** Measuring the dip of a surface by placing the straight side of the compass along a surface (e.g., a flat bedding plane or notebook in same attitude as the sediment surface) and centering the clinometer bubble (the clinometer in the Brunton will not move after adjustment; hence the compass can be rotated to the horizontal for easy reading of the dip angle). The dip must be measured down the steepest part of the surface, perpendicular to the strike of the surface, which can be determined by discovering the line formed by lying the compass edge along the surface while the horizontal leveling bubble is centered. For low dips (<20°), the strike and dip can be located more accurately by first using the clinometer set to 0° to mark a horizontal line on the face; **B:** Use of the sighting compass for taking bearings on a map or plane-table surface. The horizontal leveling bubble must be centered; **C:** Use of the sighting compass for taking bearings while the compass is held in the hand. The horizontal leveling bubble must be centered. The folded-back top of the compass has a mirror that permits the operator to both view the bubble and the compass face (although if the device is available, the needle may be locked when the bearing is taken, then held until the bearing is recorded). By holding the compass in this configuration vertically and setting the clinometer to the desired angle, when the clinometer bubble is centered (seen in the mirror) the peepsights show a target at the set angle from the eye (the principle utilized for measuring stratigraphic thicknesses in Fig. 3-6).

Determination of Thickness of Tabular Bodies

a) Influence of topography and dip on map pattern widths

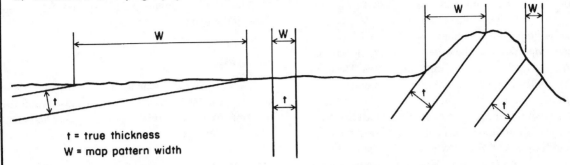

t = true thickness
W = map pattern width

b) Ground surface horizontal

c) Ground surface sloping in dip direction

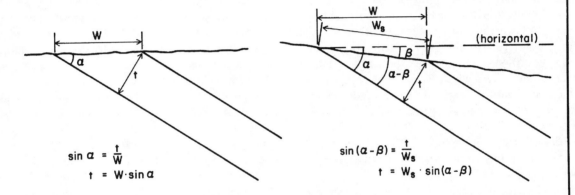

$$\sin \alpha = \frac{t}{W}$$
$$t = W \cdot \sin \alpha$$

$$\sin(\alpha - \beta) = \frac{t}{W_s}$$
$$t = W_s \cdot \sin(\alpha - \beta)$$

d) Ground surface sloping opposite to dip direction

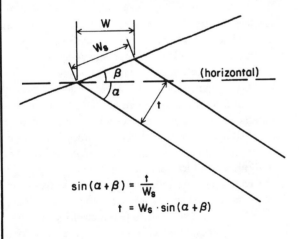

$$\sin(\alpha + \beta) = \frac{t}{W_s}$$
$$t = W_s \cdot \sin(\alpha + \beta)$$

e) Length of drill hole intersection

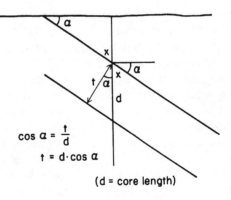

$$\cos \alpha = \frac{t}{d}$$
$$t = d \cdot \cos \alpha$$

(d = core length)

t = true thickness
W = breadth of map outcrop (projection on horizontal)
W_s = slope width of outcrop

α = true dip
β = angle of slope

Figure 3-8.   Examples of the relationship of the true and apparent thickness of stratigraphic units (a), showing trigonometric solutions for obtaining the true thickness from outcrop width (b, c, d) and apparent thickness as intersected in a borehole (e).

compasses (i.e., those without a built-in clinometer). A few sighting compasses use a stereographic system where one eye sights on the target, the other on the azimuthal scale (with practice the images become superimposed). The trigonometric relationships linking bed thickness, surface topography, dip, and outcrop width are shown in Fig. 3-8.

For the geologist, measurement of strike and dip is the second most important use of the compass—after location, if you are equipped with good base maps/air photos. The *dip* is the steepest angle of inclination from the horizontal of a surface; the *plunge* is the angle of inclination of a linear feature (lineation). To obtain the dip direction, rotate the clinometer until it reaches its maximum inclination (or pour some water onto the surface—it will flow down-dip). The vertical angle of inclination can be measured by sighting down the surface (or lineation) itself (with one's eye effectively *on* the surface), by placing the instrument along a straightedge on the surface (or lineation) and reading the instrument from the side, or by sighting at a distance by aligning a straightedge (either upper or lower) against a profile view of (i.e., at right angles to) the surface. The angle of inclination is generally read as degrees (0–90°), but may be expressed as gradient (horizontal distance over which the slope changes by one unit of the same measure—e.g., 1 in 20), or as gradient percent (the gradient expressed as a percent—e.g., 1 in 20 = 5 units in 100 = 5%).

The *strike* of a surface is the bearing of a horizontal line on the surface. (When measuring strike, a straightedge of the leveled instrument should be applied to a flat part of the surface to be measured—or an extension of it, like a notebook in the same spatial orientation.) *The strike bearing can be taken facing either direction, but at least the quadrant of the direction of dip must be stated:* e.g., Unit A strikes 140° and dips 32°NE—which may be recorded as 140°/32° NE. *If only the dip is recorded, then its azimuth must be given:* e.g., Unit A dips 32° at 050°.

When the dip and strike of a surface is recorded it may be expressed as 25N 085, meaning that the bed strikes 085° (or 265°) and dips 25° in a northerly direction. Because dip and strike data are so often processed mathematically by computers, there is a problem with the nonnumerical N needed to distinguish the general northward dip from a southward alternative. This problem can be resolved by use of the clockwise convention called the right-hand rule (i.e., the dip direction lies to the right—clockwise—of the azimuth of strike; or, the strike azimuth reported is that alternative which occurs clockwise from the dip direction; in this example, the bed attitude would be recorded as 25.085). A better solution to the problem, because it is easier to remember, is to record the dip and dip direction (in the above example the bed attitude would be recorded as 25.355) rather than the dip and strike. Either way both the ambiguity and the nonnumerical character are eliminated, but it will be necessary to indicate which method has been used and to be consistent in its use. Lineations have no strike, but their *trend*—bearing of the line when viewed from vertically above—is recorded along with their

plunge. (The azimuth of the trend is given in the direction of the plunge.)

Further details on compasses, clinometers, and applications to which they can be put are provided in Lahee (1961), Moseley (1981), Compton (1985), and other publications listed at the end of the chapter.

Surveying

A fundamental requirement of all geological fieldwork is knowledge of the horizontal position and elevation of sample sites, outcrops, etc. in relation to each other and in relation to other features in the field area. For some studies, it may be adequate to estimate site positions by taking compass sightings on distinct topographic features and plotting those bearings on maps or air photos, but for other studies sites may need to be located more accurately and some surveying is necessary. Surveying involves the use of angular and distance measurements to fix positions relative to a known reference point and includes a wide variety of techniques (e.g., Bannister and Raymond 1984; Ritchie et al. 1988).

The simplest, and least accurate, surveying technique is the pace and compass survey (e.g., Lahee 1961), which may be refined by replacing the pace by a tape (or surveyor's chain). A good steel tape should enable distances to be measured to ±0.005% over 30 m (after correction for temperature); a fabric tape will provide measurements accurate to about ±0.5% over 30 m. GPS (global positioning system) satellite navigation equipment (see Chapter 2) will indicate a horizontal position and elevation, but these are usually only reliable to about ±50 m and 2 m respectively; GPS equipment is very useful where high positional accuracy is unnecessary, or reference points are either not available or not visible from the site (e.g., on the sea, in heavily forested areas, or on featureless plains). For routine measurements of elevation data, a good barometer provides an alternative to GPS equipment as long as the weather does not change dramatically and you have a reference datum to which you can refer morning and night. For many studies, the level of accuracy provided by these procedures is sufficient and no real benefits would be obtained by devoting resources to improving accuracy, but in other studies greater positional accuracy is necessary.

When horizontal position and elevation need to be known with high accuracy, optical and/or electromagnetic surveying techniques are required. Standard instruments for this work include the surveyor's level, the theodolite, and a variety of electromagnetic distance-measuring devices (EDMs; e.g., the laser geodimeter or distomat; see Burnside 1982) all of which are used in conjunction with a staff or a reflector that is moved between the target sites. The instrument itself is accurately positioned at a starting reference point and used to measure distances and angles from that point to the staff (or reflector) positioned at the target site, or at a series of intermediate points between the starting point and the target site. Accurate positioning, leveling, and alignment of the instrument at the reference point are critical because any errors will

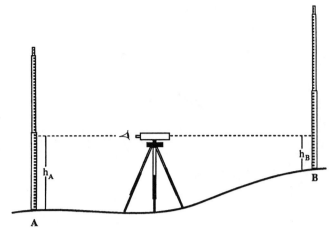

Figure 3-9. Determination of elevation using a level and scaled staffs (cf. Figure 3-6). Elevation at point B = elevation at point A + (h_A − h_B), where h_A and h_B are the heights at which the cross-hairs of the level meet the staff positioned at points A and B respectively.

View of staff (staff scale in metres)

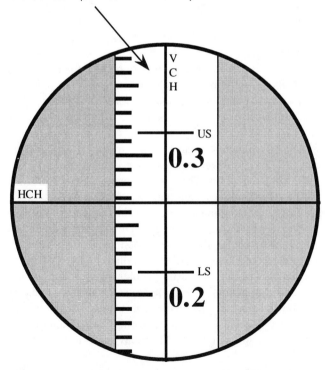

Figure 3-10. Use of stadia intercepts on a staff for determining horizontal distance, view through telescope. HCH is the horizontal cross-hair, VCH is the vertical cross-hair, US is the upper stadia hair, and LS is the lower stadia hair. The distance from the instrument to the staff is given by the length of staff between the upper and lower stadia multiplied by the stadia multiplying constant. In this example, the distance to the staff = (0.283 − 0.184) × 100 = 9.9 m.

be passed on to all subsequent measurements. Levels and most EDM instruments indicate distance and bearing (horizontal angle) to the target, and although they do not measure angles in the vertical plane, they can be used to measure small changes in elevation (e.g., Fig. 3-9). The theodolite, which is the most versatile and best known surveying instrument, can be used to determine the distance to the target and angles in both the horizontal and vertical planes; it can be used as a level if the vertical tilt is locked at 0°.

EDM instruments determine distance by measuring the time for a light beam to go from the instrument to the target and back, whereas the optical instruments determine distance by reading the upper and lower stadia intercepts on the staff and multiplying this length by a stadia multiplying constant (usually 100), which is a function of the optics of the instrument used (see Fig. 3-10). Angles in the horizontal (and vertical, if applicable) plane are measured by reading directly from the engraved circle corresponding to each rotation plane; accurate reading of the angle of rotation is aided by the use of vernier scales (older instruments) or optical micrometer scales (most modern instruments; e.g., Fig. 3-11). Average quality instruments should provide angular readings to ±5 sec of arc (roughly equivalent to 10 m over 1 km) or better. All other positional data required are obtained by applying trigonometric procedures to the basic angular and distance data obtained using the instruments.

PHOTOGRAPHY

Photographs are a particularly useful means of recording field (and laboratory microscope) data. In addition, the judicious use of photographs can save many words in a report. However, photographs need to be taken and processed carefully because later reproduction (e.g., in a journal) will almost inevitably result in loss of clarity and a sharpening of contrast. Some guidelines on photographic technique can be found in Heron (1986) and in miscellaneous "Kodak Professional Photoguides" published by Kodak at irregular intervals, but nothing beats experience. The best way to build on your experience is to note the camera settings, lighting, subject appearance, etc. when you take each photograph and to record this information when you record the notes on the subject location and frame number; checking this information against final photograph quality later will improve your knowledge of which settings give the best results for particular lighting conditions and subject types.

Although photographs are intended to communicate specific scientific information, their visual impact will be enhanced if some attention is given to artistic aspects of their composition. To a large extent the composition of the photograph will hinge on the eye of the photographer, but the following are some points to consider:

Try to compose the photograph so that the most important subject matter is near the center of the photograph.

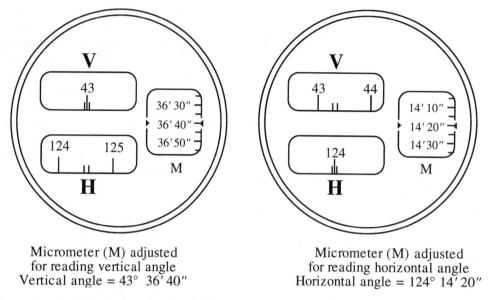

Micrometer (M) adjusted
for reading vertical angle
Vertical angle = 43° 36′ 40″

Micrometer (M) adjusted
for reading horizontal angle
Horizontal angle = 124° 14′ 20″

Figure 3-11.   Example of the view through the optical micrometer of a modern theodolite or of an electro-magnetic distance-measuring device (EDM). The angles shown are true to the point where the cross-hairs meet the staff if the instrument was correctly leveled and zeroed to north before making the readings.

Remember to include a scale that viewers will recognize (no foreign coins, etc.); a way-up indicator also may be desirable in some outcrop photographs.

Ensure that lighting is as uniform as possible across the photograph (i.e., avoid photographs which have one dark half and one light half, because the difference will be accentuated during reproduction). Contrast may need to be emphasized for some themes (generally achieved with smaller aperture size or a filter).

Try to keep as much of the photograph area in focus as possible—not just the principal subject area; i.e., make sure you understand the depth-of-field functions of your particular camera. This point is particularly important for slides, which may be shown on a projector with an autofocus facility.

Ensure that the subject of the photograph takes up as much of the field of view as possible within the limits dictated by artistic balance.

Try to exclude from the field of view other interesting features that might distract the viewer from the subject of the photograph.

Where the photograph incorporates a horizon, ensure that it is horizontal.

Low sun angles help to bring out subtle topographic features.

Give some attention to the overall impression to be created by the photograph, as well as to the clarity of the subject matter.

Know your printer well, and coordinate with him or her directly to provide feedback on contrast and cropping modifications that can improve the final product.

For work that will be presented in lectures and in print it may be useful to take photographs with one camera loaded with color slide film and another loaded with black and white print film; black and white prints can be made from color slides, but some detail is lost and contrast will be accentuated. Tweezer-like devices (e.g., Kodak ML135 Model M Film Extractor) that permit extraction of the film trailer from a completely rewound film cassette can be purchased from some camera shops; with such a device and with careful recording of the number of photos taken, it is feasible to exchange color and black and white film cassettes in one camera for duplicate sets.

Photomosaics are a special type of photograph that can be particularly useful in fieldwork, but some skill is required to prepare good photomosaics and they must be prepared before the final fieldwork is undertaken (see Wizevich 1992 for a discussion of this technique).

Digital technology is also having an impact on photography and two new developments deserve consideration by scientific photographers. The digital camera, which records the image on a floppy disk, has many benefits for those wanting to process photographic images in the computer. We have not yet seen reflex-action versions of digital cameras but they are probably not far off. It is also possible now to have slides or print photographs recorded digitally on CD-ROM disks. Each disk will store at least 150 color pictures, and the images can be manipulated (cropped, color adjusted, etc.) by computer before or after recording. CD-ROMs provide an increasingly popular means of storing many photographic images, but you will need a CD-ROM reader to view them.

BASE WORK—EVENINGS AND IMPOSSIBLE WEATHER

During evenings and on (impossibly) wet days, update and interpret the field map, compile detailed sample descriptions, and plan future field and laboratory operations.

Keep the field map and graphic logs up-to-date. Plot each day's operations on both your base map and air photos (see Chapter 2), complete details of any graphic logs, and index your notebook or field data record cards (e.g., **PS** Fig. 1-8) so that you and others can easily find data. Cross-index these records. Sort samples and photographs and confirm indelible, cross-indexed labeling. Go over all observations on the field map or air photo overlay with a fine drafting pen. Light coloring of distinctive lithologies may help show relationships.

Transfer data from your field sheets to the office map. The office map is the basis of the final map, and by the time you leave the field it should include virtually all data needed for the final map. Ink on raw data from the field sheets, and with a hard pencil, extend contacts and extrapolate data as you deem useful for interpreting relationships and planning future fieldwork. This step will assist in pinpointing problem areas and is particularly useful when you are working with a number of air photos.

Select lines for cross-sections and compile sketch sections before leaving the field area to ensure that you have the necessary information.

Prepare hard copy from tape recorders or computers and from field notebooks. Assemble descriptions on each lithological unit into coherent written records. Review notes to ensure that all necessary data are recorded and think about interpretations and the additional data needed to substantiate or disprove them. Prepare final graphic log sheets.

Compare all samples from the same units. Examine each sample in detail for properties missed in the field. Select those requiring laboratory analysis, check the labeling, and pack them for transport.

Plan the next day's fieldwork. Examine the map and air photos, mark probable exposures, plot intended traverses, and so on. Plan to spend maximum time in the problem areas pinpointed by the preceding review.

Field Report Preparation

Preparation of a field (or back-to-the-office) report may take as long as, or longer than, the fieldwork itself, particularly if supplementary laboratory and library work is necessary (see also **PS** Chapter 10). The following suggestions apply to most field reports; your employer may impose additional requirements:

Compose a superdetailed table of contents before writing; it will usually be revised several times during the course of writing, but initially it will help organize your thoughts as well as the report. Use plenty of headings and subheadings in the report to aid clarity and help readers find their way around (e.g., Table 3-3).

Be concise: make full use of diagrams, photos, and tables to conserve words. Organize carefully.

Be consistent: describe the same kinds of features in the same order for each rock unit; develop discussions in a parallel manner for clarity and ease of comparison. Beware of internal contradictions (both in the text and between the text and the map).

Write at least one rough draft; let it lie for some time, then reread, correct, shorten, and rewrite. Give it to a colleague to review; evaluate his or her comments and revise accordingly.

Proofread.

CHEMICAL FIELD ANALYSES

Analytical work on sample chemistry is normally carried out in a laboratory, but some must be conducted in the field because any attempt to remove the sample from its natural setting will almost inevitably result in changes to the required parameters. Eh (redox, see **PS** Chapter 3) measurements must be made in situ because Eh conditions in sediments can change substantially over a few millimeters, and the measured Eh for the sample is very likely to be affected if the sample is moved to any place where different Eh conditions prevail. Measurements of the dissolved oxygen (DO) content of water (often closely related to Eh conditions) must also be made in situ (or as soon as possible after the sample is collected) to avoid gas exchange with the atmosphere or modification by microbiota in the water. For the same reason, the initial measurement of DO for determinations of biochemical oxygen demand (BOD_5; standard method 5210; Clesceri, Greeberg, and Tressell 1989) or chemical oxygen demand (COD; standard method 5220; Clesceri, Greeberg, and Tressell 1989) must be made as soon as possible after the sample is collected. Similarly, determinations of pH should be made in situ because pH can be affected, over short time intervals, by gas exchange with the atmosphere, ion exchange with sediment in the water sample, or as a result of the activity of microbiota in the water.

Analyses also are carried out in the field to avoid the need to transport unnecessary extra sample material back to the laboratory (e.g., determinations of salinity in water samples) or because the on-site analyses are required as part of an interactive sampling program (see Chapter 4) or as a guide to sample site selection (e.g., sampling up the salinity gradient in an estuary). Although interactive sampling strategies offer some obvious benefits (see Chapter 4), their widespread application has been constrained by the lack of suitable field-portable analytical equipment. However, many instrument manufacturers (e.g., Chemtronics, Western Australia) are now beginning to provide a rapidly expanding range of suitable equipment capable of providing on-site analyses every bit as good as those that can be obtained in a laboratory. Some analysts have questioned the reliability of field-based analyses and stressed the risk of contamination, but studies that have compared field and laboratory analyses of the same samples (e.g., McConchie et al. 1988; McConchie and Harriott 1992) indicate that there need be no difference in accuracy, precision, or sensitivity if the field analyses are properly executed. Furthermore, field analyses are not vulnerable to the changes in concentration that can affect some

Table 3-3.   Example of Field Report Organization

Title: Combine brevity with clarity and precision.

Abstract: Single-spaced; not over 150 words. Give salient facts and conclusions only. Write this section after completing the rest of the report.

Table of contents: Headings and subheadings in the same form as in the text; page numbers where discussion begins in text.

List of figures: Includes drawings, photographs, and map(s). Titles in same format as in text.

Introduction
  Purpose
  Methods (field party, time spent, maps and aerial photos, methods of study)
  Physiographic setting (location of area, topography, climate, drainage, vegetation, access, nearest population center(s), size and limits of map area)
  Geological setting (very brief resume of broad relationships to major tectonic, time, and widespread or well-known stratigraphic units).
  Previous work (brief review of work specifically on the area concerned)
  Terminology (used in present report where it differs from that used by other reputable geologists, e.g., *graywacke* or term used in broader or more restrictive sense than originally defined)

Geomorphology: Description and explanation of all significant landforms (discussion of this section may be more appropriate elsewhere; in some cases it need not be included).

Stratigraphy. Will have numerous headings and orders of subheadings—be consistent in order and format. Treat intrusive igneous rocks and tectonic structures in sections separate from sedimentary rocks. Treat systematically from oldest to youngest units within each section.

  Stratigraphic unit (defined; extent/best exposures; thickness; list of recognized subunits)
    (Oldest subunit—as above)
    Primary structures
    Lithology—color, induration, texture, composition; variability within area
    Paleontology
    Inferences—geologic history (portions of whole; may be deferred)
    Contact with underlying rocks—nature, recognition criteria, rough time span represented

Igneous rocks (intrusives): An introductory paragraph delineating relationships may be useful.
  Lithology and name (definition, how and why defined; extent; characteristic exposures; broad lithological characteristics; subdivisions)
  (Subdivision) Petrology (include features evident in field—e.g., foliation, jointing, weathering, variation in properties)
  Relation to surrounding rocks (contacts, form of body; relative age)
  Petrogenesis—genetic inferences

Structure
  Folds; faults (including resultant features such as schistosity and lineations); major vs. minor structures; field evidence, areal extent; form and characteristics; age relationships; units involved and not involved; degree to which presence is inferred vs. observed
  Joints—sets, systems; consistency; relation to folds and faults
  Interpretation—casual factors; depth of formation; integration with structure outside of map area

Quaternary deposits (to extent warranted by purpose of project).

Economic geology: Describe mineral deposits, workings, claims, etc. Subdivide as suggested by economic history of region or type of deposits.

Geological history: Synthesis of proceeding sections oriented to sequential development of geological features of region. May take the place of *Summary.*

Summary: (may precede *Geological History* to take the place of *Conclusions*).

Conclusions: salient discoveries of project (may be replaced by *Geological History* or *Summary,* depending on project).

Final map: Clean, neat, inked; show scale, north line, declination, latitude and longitude, reference grid system, legend, title, date, name of geologist. Preferably include an exposure map (overlay) as well as the interpretive map (see Kupfer 1966). Color in units; use reference symbols where useful. Use solid lines where geological features are certain. Use different line widths for faults, fold axes, unconformities, etc.

Cross-sections (aside from title and scale, no extra legend should be necessary): One or more are generally necessary with the geological map to show the inferred geological relationships. Draw cross-sections through the regions most in need of clarification; the lines of section need not be straight throughout the area. The north or east end of the section should be to the right. Choose the *vertical exaggeration* with care (and do not forget to record it); avoid using any if at all possible, and if any is used, do not forget that dip values cannot be used as recorded in the field but must be corrected for the exaggeration factor. Do not extrapolate *doubtful contacts*—terminate them with a small question mark.

Columnar section: A profile columnar section (e.g., see Lahee 1961) should be drawn for the sedimentary rocks. Age and formation names appear on the left, lithology and fossils on the right. Breaks and interruptions in the record are indicated by jagged breaks in the columns. Intervals of known thickness but unknown lithology (e.g., covered in the field) may be left blank in the unbroken column. With careful preparation, the columnar section may serve as the lithologic legend for the map.

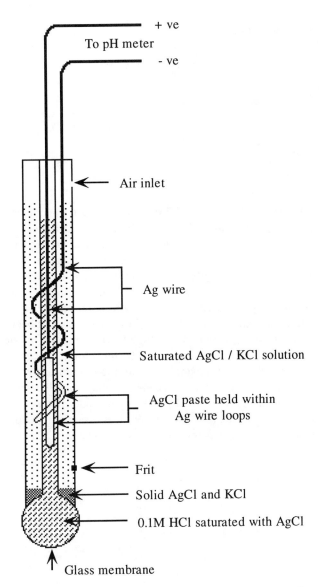

+ ve
To pH meter
- ve

← Air inlet

← Ag wire

← Saturated AgCl / KCl solution

← AgCl paste held within
   Ag wire loops

← Frit

← Solid AgCl and KCl

← 0.1M HCl saturated with AgCl

↑ Glass membrane

Figure 3-12.   Schematic diagram of a common glass combination pH electrode used with field pH meters.

analytes in samples that are stored and transported to a distant laboratory (e.g., Fourie and Peisach 1977; Thomson et al. 1980; McConchie and Harriott 1992). Also, in contrast to storage containers and laboratories that contain a new set of potential contaminants, the only likely contaminants in the field are those to which the samples are naturally exposed.

pH

pH is an easily measured environmental parameter, but it is easy to record incorrect values if the electrode is not correctly used and maintained (Bates 1964). A wide variety of small field-portable instruments are available; most instruments are fitted with a glass combination pH electrode (e.g., Fig. 3-12). Because their reliability, stability, and sensitivity differ, it is necessary to ensure that the instrument used suits your particular needs. Because pH readings are temperature sensitive, most pH meters will be fitted with a temperature sensor that is inserted with the electrode to correct for temperature effects automatically; others will have a knob that must be set to the appropriate temperature, which must be measured separately. For sedimentologists working with highly saline brines (e.g., in salt-lake systems), there is also the problem that standard pH electrodes are designed for use in low-ionic-strength solutions, and it may be necessary to employ nonstandard procedures (e.g., Marcus 1989; Knauss, Wolery, and Jackson 1990) to obtain accurate pH data. Additional points to note when using pH meters are listed in Table 3-4. Some buffer solutions that can be used for calibration of the field pH meter are listed in Table 3-5.

Redox (Eh)

Eh (see **PS** Chapter 3) indicates the redox potential in water or subaqueous sediment; in combination with pH (e.g., in Eh/pH diagrams), it is very useful in describing chemical conditions in sedimentary environments and for indicating

Table 3-4.   Considerations When Making pH Determinations in the Field

Do not allow electrodes to dry out! Keep the sensor end of the electrode clean and wet with pure water; if it dries out, it may need to be reconditioned by soaking in water for several hours.

Calibrate the meter regularly using the two-buffer system (ideally the buffers used should bracket the range of pH values being measured). If an unexpectedly high or low reading for any particular sample is obtained, check the calibration immediately. Commercially prepared pH buffers are readily available; if you wish to make your own, some examples are given in Table 3-5.

The best procedure to calibrate an instrument varies with the design of the instrument; check the manufacturer's instructions. However, most modern instruments can be calibrated using the two-buffer system as follows:

1. Clean the electrode with pure water.
2. Place the electrode in a buffer solution with a pH near 7, set the appropriate temperature if necessary, allow the electrode to equilibrate for at least 1 min, and adjust the "calibrate" knob to give the appropriate pH reading.
3. Rinse the electrode thoroughly with pure water and place it in a second buffer solution (e.g., with a pH near 4 or 9), allow it to equilibrate for at least 1 min, then adjust the "slope" knob to give the appropriate pH reading.
4. Repeat the above steps until the instrument reads the correct pH for both buffer solutions.

Table 3-5.   Buffer Solutions Used for Calibration of the Field pH Meter

Buffer Solution	pH at 20°C	pH at 25°C
Saturated KH-tartrate	—	3.557
0.05 molar KH_2-citrate	3.788	3.776
*0.05 molar KH-phthalate	4.002	4.008
*0.025 molar KH_2PO_4 and 0.025 molar Na_2HPO_4	6.881	6.865
*0.01 molar borax ($Na_2B_4O_7 \cdot 10H_2O$)	9.225	9.180
0.025 molar $NaHCO_3$ and 0.025 molar Na_2CO_3	10.062	10.012

*Most commonly used buffers.

the types of chemical species likely to be stable under those conditions. Eh is easy to measure in the field because it is simply a matter of recording the voltage (or mV) difference between a Pt test electrode and a reference electrode (usually Ag/AgCl), and if necessary, correcting the reading for temperature (see the Nernst equation in **PS** Chapter 3). However, as with pH, it is also easy to record incorrect values if the electrodes are not correctly used and maintained.

Manufacturers of analytical electrodes offer a range of Eh electrodes (many are combination Eh/pH electrodes), but because most of these comprise glass bodies that break easily when pushed into sediment and reference electrodes that cannot be reconditioned easily, we have found it better to make our own electrodes (e.g., Fig. 3-13). The electrode design in Fig. 3-13 is only one example and many variations on the theme can be developed with a little imagination; e.g., the electrode could be fitted to a telescopic pole for reaching into deeper water, or it could be made with multiple Pt sensors (connected in turn using a multiposition switch) arranged at regular intervals down a spear that can be forced into the sediment to record the Eh at a series of predetermined depths. Although there is nothing complex about recording Eh, some important points to consider are listed in Table 3-6.

Dissolved Oxygen

The dissolved oxygen (DO) content of water samples (reported in mg/L or % saturation) is closely related to Eh and has a major influence on chemical and biological processes at and near the sediment/water interface. DO concentrations are commonly measured electrochemically using a DO probe and meter, but good-quality equipment is expensive and many scientists find DO probes to be unreliable (they are excellent when working well, but you cannot be sure you will know when they are faulty!). The alternative approach, which we prefer, is to use a Winkler titration (e.g., Adams 1990), which can be carried out immediately after the sample is collected. This approach is cheap and reliable. Because it is inconvenient to carry normal burettes and ancillary equipment in the field, we have developed a microtitrator (Fig. 3-14), which is easily constructed and provides highly reproducible results accurate to ±0.01 mg/L. The microtitrator is designed to be small, robust, and to consist of components that are easily replaced. The reaction vessel and all

solutions are completely sealed against dust and water, thus a wide range of normally laboratory-based titrimetric analyses (e.g., Table 3-7) can be carried out in the field with a precision equivalent to that obtainable with a laboratory burette. The instrument is calibrated once to determine the solution volume dispensed by one full turn of the dispenser knob (determined by the syringe size and the pitch of the screw thread), and all readings are made by counting the number of turns to the end point and multiplying that number by the calibration factor to determine the titrant volume.

The procedure for using the microtitrator to measure DO is shown in Table 3-8. Because the maximum dissolved oxygen content is limited by the temperature and salinity of the water being tested (e.g., Table 3-9), it is essential to record both these parameters for every sample on which a DO deter-

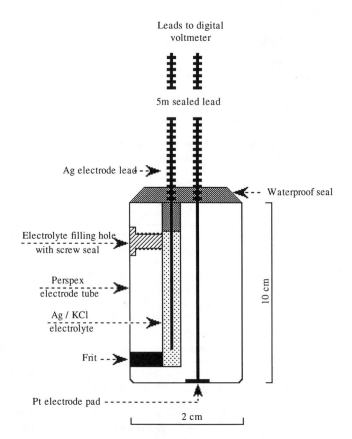

Figure 3-13.   Schematic diagram of Eh electrode that can be constructed for field measurement of Eh.

Table 3-6.   Considerations When Making Eh Determinations in the Field

1. Ensure that the voltmeter is not a type that draws a significant current from the electrodes, and that it will give a stable reading down to at least 0.01 V. Most modern digital voltmeters meet these requirements.

2. Try to obtain equipment that has a means of adjusting the voltage reading between the reference and Pt electrodes so that your sensor can be calibrated (the alternative is to correct all readings mathematically). Calibration is usually achieved using Zobell's solution (Zobell 1946) which consists of equal volumes of 1/300 molar potassium ferricyanide $\{K_3Fe(CN)_6\}$ and 1/300 molar potassium ferrocyanide $\{K_4Fe(CN)_6 \cdot 3H_2O\}$. *Caution: Cyanide compounds must be treated with care! Note: The potassium ferrocyanide solution is light sensitive and oxidizes readily, and the mixture of the two solutions will only last for a few hours, so make fresh solutions frequently.* Check the calibration regularly because in many settings (particularly sulfidic sediments) the performance of the electrodes can be affected by chemical species in the sediment or water being measured. At 25°C Zobell's solution will give a reading of +244 mV relative to the Ag/AgCl reference electrode, which converts to an Eh of +441 mV when +197 mV are added to convert to voltages relative to a hydrogen reference electrode (see below).

3. When recording final Eh values, ensure that the measured voltages are corrected to convert from the reference electrode used to the hydrogen electrode standard. For the Ag/AgCl reference electrode (the most widely used type), add 197 mV to all readings at 25°C (theoretically the conversion should involve adding +222 mV, but because the chloride ion activity in the reference electrode is not equal to unity, the value of +197 mV is more accurate). The Nernst equation can be used to make any necessary temperature corrections, but over the temperature range normally encountered by sedimentologists, these are small (a bit less than 1 mV/°C) and are usually neglected.

4. When measuring the Eh of sediments, ensure that the electrodes are kept clean at all times; even a small amount of one sediment type adhering to the electrode may influence subsequent readings. The readings can also be affected if the Pt electrode is abraded when pushing it into the sediment.

5. In some settings, chemical species in the sediment or water can "poison" your electrodes (particularly the reference electrode); incorrect readings result. When this problem arises, the reference and Pt electrodes and the frit must be cleaned (or replaced), the Ag electrode must be cleaned and recoated with AgCl, and the reference electrode electrolyte must be replaced; the difficulty in reconditioning many commercial electrodes is a major problem.

6. Unless you are very familiar with what readings to expect, check the condition and calibration of your electrodes regularly; many useless data have been recorded by operators who did not notice progressive deterioration of their electrodes.

7. When recording Eh data, it is useful to record the time of day the measurements were made, because in some sediments Eh may change in response to diurnal changes in the photosynthesis/respiration balance.

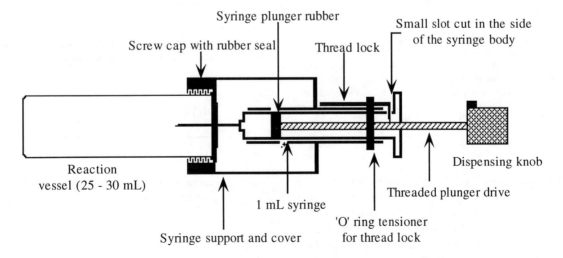

Figure 3-14.   Schematic diagram of a microtitrator for determination of dissolved oxygen content of water in the field.

Table 3-7.   Additional Chemical Analyses That Can Be Performed in the Field with the Microtitrator Shown in Fig. 3-14

Determination of Water Hardness

1. Measure 10 mL of the water sample into a clean reaction vessel.
2. Add about 0.4 mL of ammonia/ammonium chloride buffer solution (prepared using 17.5 g ammonium chloride + 142 mL concentrated ammonia made up to 250 mL).
3. Add 2 drops of Erichrome BT indicator solution, screw on the cap, and shake the reaction vessel gently. The solution in the reaction vessel will now have a purple color.
4. Load the syringe with a 0.1 molar solution of disodium EDTA and fit it to the microtitrator.
5. Titrate the EDTA solution into the reaction vessel, counting the number of turns of the dispensing knob, until the liquid in the reaction vessel has a *pure blue* color. The microtitrator should be shaken gently during the titration to ensure that all titrant drips are washed off the syringe needle and mixed with the solution in the reaction vessel.
6. The total water hardness (in mg/L $CaCO_3$ equivalents) can now be calculated by converting the number of turns (and decimal parts of a turn) to the equivalent volume for your titrator and using the relation $1.0~\mu L = 1.0$ mg/L $CaCO_3$ equivalents. For very hard water it may be necessary to dilute the water to 1/10 or 1/100 of its initial concentration before testing.

Determination of Total Chloride

1. Filter about 20 mL of the water sample (a syringe and disposable filter is ideal for this purpose) and test the pH of the filtered sample using a pH meter or indicator paper. The pH should be between 6.5 and 9. Adjust the pH as necessary using dilute nitric acid or sodium bicarbonate solution as appropriate.
2. Measure 10 mL of the filtered and pH adjusted water sample into a clean reaction vessel.
3. Add 0.1 mL of potassium chromate indicator solution (prepared by dissolving 4.2 g of "Analar" (analytical grade) potassium chromate and 0.7 g of "Analar" potassium dichromate in 100 mL of deionized water) to the reaction vessel and screw on the cap and syringe support.
4. Load the syringe with 0.1 molar silver nitrate solution, and fit it to the microtitrator.
5. Titrate silver nitrate solution into the reaction vessel, recording the number of turns of the dispensing knob, until a faint reddish-brown color persists in the liquid after shaking. The microtitrator should be shaken gently during the titration to ensure that all titrant drips are washed off the syringe needle and mixed with the solution in the reaction vessel. *Note:* If you are unfamiliar with this titration it may be wise to test a few blanks and standards because the end point of the titration can be difficult to spot; the liquid at the end point for the standards and blanks can be kept as a color reference for the sample tests.
6. Using a clean reaction vessel, test a pure water blank.
7. The chloride concentration (in mg/L Cl^-) is calculated by converting the number of turns (and decimal parts of a turn) to the equivalent volume for your titrator and using the relation $1.0~\mu L = 0.355$ mg/L Cl^-. For high-chloride water it may be necessary to dilute the sample to 1/10, 1/100, or 1/1000 of its initial concentration before testing. For low-chloride water, diluted silver nitrate solution can be used.

Determination of Residual Free Chlorine

1. Measure 10 mL of the water sample into a clean reaction vessel and add 0.1 mL of acetic acid (sufficient to reduce the pH to 3–4) and 0.02 potassium iodide.
2. Screw on the cap and dispenser unit and load the syringe with 0.0025 molar sodium thiosulfate solution.
3. Titrate sodium thiosulfate solution into the reaction vessel, counting the number of turns of the dispensing knob, until the liquid in the reaction vessel has a pale straw color. The microtitrator should be shaken gently during the titration to ensure that all titrant drips are washed off the syringe needle and mixed with the solution in the reaction vessel.
4. Unscrew the cap of the reaction vessel with the syringe unit still attached, add about 1 mL of starch indicator solution (prepared as for DO determinations in Table 3-8) to the reaction vessel, then screw the cap and syringe unit back on. The mixture in the reaction vessel should now have a bluish-purple color; if the solution is clear, you have already passed the end point.
5. Recommence adding the sodium thiosulfate solution starting the count of the number of turns from where you left off at step 3 and stopping when the solution becomes clear.
6. Repeat the above procedures for a blank prepared using high-purity water in place of the sample solution.
7. The residual free chlorine concentration (in mg/L Cl_2) is calculated by converting the number of turns (and decimal parts of a turn), minus the number of turns for the blank, to the equivalent volume for your titrator, and using the relation $1.0~\mu L = 0.00925$ mg/L residual-free Cl_2.

Determination of Total Alkalinity

1. Filter about 20 mL of the water sample, place 10 mL into a clean reaction vessel, and add 3–4 drops of methyl orange indicator solution.
2. Screw on the cap and dispenser unit and load the syringe with 0.25 molar sulfuric acid solution.
3. Titrate sulfuric acid solution into the reaction vessel, recording the number of turns of the dispensing knob, until the liquid in the reaction vessel changes color from orange to red. The microtitrator should be shaken gently during the titration to ensure that all titrant drips are washed off the syringe needle and mixed with the solution in the reaction vessel.
4. The total alkalinity (in mg/L $CaCO_3$ equivalents) is calculated by converting the number of turns (and decimal parts of a turn) to the equivalent volume for your titrator and using the relation $1.0~\mu L = 2.5$ mg/L $CaCO_3$ equivalent. Sample dilutions can be used if necessary.

Determination of Carbonate Alkalinity

Carbonate alkalinity is determined by the same procedure as total alkalinity, but phenolphthalein is used as the indicator solution.

---

mination is made. Similarly, because temperature and salinity have such a marked influence on the amount of oxygen that can be dissolved in water, DO concentrations are widely reported as a percentage of the saturation concentration for the prevailing conditions. DO concentrations below saturation indicate that biological or chemical processes are removing oxygen faster than it is being added, and DO concentra-

tions above saturation indicate that biological or physical processes are adding oxygen faster than it is being removed.

Salinity

Salinity (reported in mg/L or g/L) is normally measured by filtering a water sample through a 0.45-μm filter, evaporat-

Table 3-8.   Determination of Dissolved Oxygen Using the Microtitrator Shown in Fig. 3-14

Operation of the Microtitrator

Loading the Reaction Vessel

Unscrew the cap of the reaction vessel and add a measured amount of the sample to be tested (usually 10 mL) using a pipette, then replace the screw cap ensuring that it is screwed tightly onto the rubber seal. The reaction vessel and screw cap should be washed thoroughly between titrations to avoid cross-contamination.

Loading the syringe

Withdraw the syringe from the cap and support by sliding down the O-ring tensioner, holding back the thread lock, and pulling out the plastic syringe body. Rinse the syringe in deionized water and with the titrant solution to be used, or replace it if a new titrant solution is to be used; if the same titrant solution is to be used the syringe can be reloaded immediately. Load the syringe by placing the needle in the titrant solution and *slowly* pulling back the plunger. When the syringe is full it should be held needle upward and tapped a few times to ensure that all air bubbles rise to the top, then the plunger should be depressed until any air is expelled. The loaded syringe is now ready for use.

Dispensing titrant solution

Place the syringe support and cover over the reaction vessel cap, hold back the thread lock, and insert the syringe so that the needle pierces the rubber seal in the screw cap; take care to avoid pushing the plunger during insertion of the syringe. As the needle pierces the rubber any drips on the end will be wiped off and excluded from the reaction vessel. Allow the thread lock to click into position and roll up the O-ring tensioner to hold it there. The titrant can now be dispensed by rotating the dispensing knob.

Removing the syringe for reloading during a titration

If for some reason the syringe is emptied before the end point of the titration is reached, the syringe can be reloaded and titration can continue; reload as described above.

Determination of Dissolved Oxygen

1. Fill a 50-mL water sample bottle with the water to be examined and screw on the cap ensuring that there are *no* air bubbles trapped in the bottle (The bottle will normally need to be filled under water).
2. Open the sample bottle, add 1 mL of manganese sulfate solution (prepared by dissolving 446 g of $MnSO_4 \cdot 4H_2O$ in 1 litre of ultra-pure water) recap, and invert a few times to mix.
3. Add 1 mL of alkaline potassium iodide solution (prepared by dissolving 700 g of KOH and 150 g of KI in 1 litre of ultra-pure water) following the procedure in step 2.
4. Reopen the bottle and carefully add about 2 g of sulfamic acid powder. Recap the bottle and dissolve the sulfamic acid by gentle inversion.
5. The dissolved oxygen is now fixed. Reopen the bottle, measure out

10 mL of the fixed sample, and place it in the reaction vessel of the microtitrator.

6. Load the syringe with 0.025 molar sodium thiosulfate solution and fit it to the microtitrator.
7. Titrate sodium thiosulfate solution into the reaction vessel using the dispensing knob until the liquid in the reaction vessel has a pale straw color. Count the number of turns of the dispensing knob necessary. The microtitrator should be shaken gently during the titration to ensure that all titrant drips are washed off the syringe needle and mixed with the solution in the reaction vessel.
8. Unscrew the cap of the reaction vessel with the syringe unit still attached, add about 1 mL of starch indicator solution to the reaction vessel, then screw the cap and syringe unit back on. (To prepare the starch indicator, add a little cold water to 5 g starch [potato, arrowroot, or soluble] and grind in a mortar to a thin paste. Pour the paste into 1 L of boiling distilled water and allow to settle overnight. Separate off the supernatant liquid and add either 1.25 g salicylic acid or 4 g zinc chloride as a preservative.) The mixture in the reaction vessel should now have a bluish-purple color; if the solution is clear, you have already passed the end point.
9. Recommence adding the sodium thiosulfate solution starting the count of the number of turns from where you left off at step 7. Stop when the solution becomes clear and record the total number of turns required.
10. The DO concentration in mg/L O_2 can now be calculated by converting the number of turns (and decimal parts of a turn) to the equivalent volume for your titrator and using the relation 1.0 μL = 0.02 mg/L O_2. If high sensitivity is required at low DO concentrations, 0.0025 molar sodium thiosulfate can be used in the titrations such that 10 μL = 0.02 mg/L O_2.

Determination of 5-Day Biochemical Oxygen Demand (BOD_5)

1. Determine the DO content of the water sample, then fill a 50-mL water sample bottle with the water to be examined and screw on the cap, ensuring that there are *no* air bubbles trapped in the bottle.
2. Store the bottle in a dark cupboard for 5 days at about 20°C.
3. Proceed as for dissolved oxygen determinations starting at step 2.
4. The biochemical oxygen demand can be calculated by subtracting the DO concentration measured after storage for 5 days from the DO concentration recorded before storage for 5 days.
5. If the BOD_5 is likely to be high, a series of diluted subsamples (where high dilutions are used it will be necessary to dilute with nutrient-supplemented water; see Adams 1990) can be prepared prior to testing, but you will need to allow for the dilution in later calculations of BOD_5 and you will need to determine the starting DO on the diluted mixture as you would normally do for the original water sample.

ing a measured volume to dryness, and determining the dry residue weight. This method is necessary where precise data are required, but it is impractical for routine field use. One instrument for salinity measurements that is easily used in the field is the *refractive index salinometer,* which operates on the basis of refractive index changes that occur with changes in salinity. However, because refractive index is also temperature dependent, we advise the use of an automatic temperature-compensated refractometer; this instrument compensates for temperature differences using a hollow glass prism filled with a stable liquid that changes

its optical characteristics in the same way as the test solutions in response to temperature changes.

An alternative technique for salinity measurement is to measure conductivity of the test solution. Conductivity is easily measured (in siemens or millisiemens), using readily available conductivity probes, and is both sensitive and reliable provided that temperature corrections are correctly applied (conductivity varies systematically as a function of temperature) and the ion ratios do not change between test solutions. The problem for sedimentologists is that the conductivity of a 1000-mg/L solution with ion ratios similar to

Table 3-9.   Solubility of Dissolved Oxygen at Different Temperatures and Salinities

Temperature (°C)	Chloride Concentration (mg/L)				
	0	5,000	10,000	15,000	20,000
0	14.60	13.72	12.90	12.13	11.41
5	12.75	12.02	11.32	10.67	10.05
10	11.27	10.65	10.05	9.47	8.96
15	10.07	9.53	9.01	8.53	8.07
20	9.07	8.60	8.16	7.73	7.33
25	8.24	7.83	7.44	7.06	6.71
30	7.54	7.17	6.83	6.49	6.18
35	6.93	6.61	6.30	6.01	5.72
40	6.41	6.12	5.84	5.58	5.33
45	5.95	5.69	5.44	5.20	4.98
50	5.54	5.31	5.08	4.87	4.66

Table 3-10.   Properties of Common Specific Ion Electrodes, and Common Ions That Interfere with the Measurements

Ion Detected	Working Range (moles/L)	Operational pH Range	Common Interfering Ions
F^-	10^{-6}–1	5–8	OH^-
Cl^-	10^{-4}–1	2–11	CN^-, S^{2-}, I^-, Br^-
Br^-	10^{-5}–1	2–12	CN^-, S^{2-}, I^-
I^-	10^{-6}–1	3–12	S^{2-}
CN^-	10^{-6}–10^{-2}	11–13	S^{2-}, I^-
NO_3^-	10^{-5}–1	3–8	CN^-, HS^-, I^-, Br^-

seawater is not the same as, for example, a 1000-mg/L river water sample where Ca^{2+} and HCO_3^- are likely to be the dominant cation and anion. Conductivity is pH dependent. Hence, if conductivity is to be used as a means of evaluating salinity, it is essential that the instrument be calibrated using solutions with a similar ion ratio and pH to the water being tested; calibration can be achieved using standards, or if the water chemistry is unknown, it can be achieved by carrying out laboratory salinity determinations for a selection of samples and applying an appropriate correction factor to the recorded conductivity data.

Analyses for Elements and Compounds

A variety of specific ion electrodes are commercially available for field analyses, but the value of these electrodes for environmental and geochemical work is constrained by their vulnerability to interference by other ions that may be present (see Garrels 1967; Troll, Farzaneh, and Cammann 1977; Morf 1981). The problems of interference can be largely overcome for some samples by calibration using standard spike additions to ensure that the chemical environment for both samples and standards is as similar as possible. However, even the use of spike addition as a calibration procedure does not overcome the interference problems for all electrode or sample types. Some interference effects with common specific ion electrodes are summarized in Table

3-10. Our general recommendation is that unless you are very familiar with these electrodes, you would probably be better off collecting samples for later laboratory analysis by other methods such as ion chromatography (see Chapter 9).

Numerous spot tests, which provide a qualitative indication of the presence or absence of particular elements (the most widely used tests are for metals or phosphorus), have been developed for field use. Most of these tests involve reaction between the element of interest and organic reagents to form a colored complex that is specific to the relevant element, but because qualitative data alone are usually of little value in modern studies or exploration programs, these tests are rarely used.

Over the past few years, there have been major advances in analytical techniques and instrumentation, meaning it is now possible to analyze many elements in the field with a high degree of accuracy, precision, and sensitivity. These developments in real-time analysis have already enhanced the range of interactive sampling strategies, and with the advantages inherent in these approaches (see Chapter 4), there is no doubt that many new field-portable analytical instruments will become commercially available in the near future. Already available is electrochemical analytical equipment (using anodic stripping voltammetry, cathodic stripping voltammetry, differential pulse polarography, and square-wave voltammetry; e.g., see Mann and Lintern 1984*a*, 1984*b* and description of the equipment and procedures in Chapter 9) capable of high-precision analyses of a wide range of metals to sub-ppb concentrations (<1 µg/L) and used in an increasing number of environmental and exploration programs. This equipment can be used to analyze water samples or digests of sediment, rock, or biological material (e.g., McConchie et al. 1988), and if the work is carried out carefully, it provides field analytical data as good as can be obtained in a specialist laboratory.

Several small field-portable X-ray fluorescence instruments (see Chapter 9 for discussion of X-ray fluorescence equipment and applications) also have been developed for use in the field. These instruments use a radioactive source (usually americium), instead of the primary X-ray tube used in laboratory instruments, to cause the emission of fluorescent X-rays from elements in the sample; use of energy dispersion, rather than wavelength dispersion, is employed to analyze the fluorescent X-rays. The field-portable X-ray fluorescence instruments currently available are much less sensitive than laboratory instruments (the lower limit of detection is usually about two orders of magnitude higher), but with a strong source, careful sample preparation, and appropriate standardization they can provide precise and accurate data. Both of these types of instruments can rapidly provide the type of quantitative data that will ensure analytical geochemistry is incorporated into an increasing number of field research programs in the future; we expect that the use of field geochemical analyses will be one of the major growth areas in sedimentology over the next few decades.

SELECTED BIBLIOGRAPHY

General

Andrews, P. B., 1982, Revised guide to recording field observations in sedimentary sequences. *New Zealand Geological Survey Report 102,* Lower Hutt, 74p.

Bannister, A., and S. Raymond, 1984, *Surveying,* 5th ed. Pitman Publishing, London, 510p.

Berkman, D. A., and W. R. Ryall, 1976, *Field Geologists Manual.* Australian Institute of Mining and Metallurgy, Parkville, Victoria, Australia, 291p.

Burnside, C. D., 1982, *Electromagnetic Distance Measurement* 2d ed. Granada Publications.

Compton, R. R., 1985, *Geology in the Field.* Wiley and Sons, New York.

Dackombe, R. V., and V. Gardiner, 1983, *Geomorphological Field Manual.* George Allen & Unwin, London, 254p.

Goddard, E. N., D. D. Trask, R. K. de Ford, O. N. Rove, J. T. Singlewald, and R. M. Overbeck, 1951, *Rock Color Chart.* Geological Society of America, New York.

Heron, D. (ed.), 1986, *Figuratively Speaking (Techniques for Preparing and Presenting a Slide Talk).* American Association of Petroleum Geologists, Tulsa, Okla., 110p.

Kupfer, D. H., 1966, Accuracy in geologic maps. *Geotimes* 10:11–14.

Lahee, F. H., 1961, *Field Geology.* McGraw-Hill, New York, 883p.

Lewis, D. W., G. J. van der Lingen, and D. J. Jones, 1970, The construction of a transparent grain-size comparator. *Journal of Sedimentary Petrology* 40:472–4.

Low, J. W., 1957, *Geologic Field Methods.* Harper, New York, 489p.

Moseley, F., 1981, *Methods in Field Mapping.* Freeman and Sons, San Francisco, 211p.

Ritchie, W., M. Wood, R. Wright, and D. Tait, 1988, *Surveying and Mapping for Field Scientists.* Longman Scientific and Technical, Harlow, 180p.

Selley, R. C., 1968, Facies profile and other new methods of graphic data presentation: Application in a quantitative study of Libyan Tertiary shoreline deposits. *Journal of Sedimentary Petrology* 38:363–72.

Selley, R. C., 1970, Studies of sequence in sediments using a simple mathematical device. *Quarterly Journal of the Geological Society of London* 125:557–581.

Stephan, W. J., 1977, A one-man profiling method for beach studies. *Journal of Sedimentary Petrology* 47:860–3.

Tucker, M. E., 1982, *The Field Description of Sedimentary Rocks.* Halsted Press, New York, 128p.

Visher, G. S., 1965, Use of the vertical profile in environmental reconstruction. *American Association of Petroleum Geologists Bulletin* 49:41–61.

Wizevich, M., 1992, Photomosaics of outcrops: useful photographic techniques. In A. D. Miall and N. Tyler (eds.), *The Three-Dimensional Architecture of Terrigenous Clastic Sediments.* Society for Sedimentary Geology Concepts in Sedimentology and Paleontology, vol. 3, Tulsa, Okla., pp. 22–4.

Chemical Analyses in the Field

Adams, V. D., 1990, *Water & Wastewater Examination Manual.* Lewis Publications, Chelsea, Mich., 247p.

Bates, R. G., 1964. *Determination of pH.* Wiley & Sons, New York.

Clesceri, L. S., A. E. Greeberg, and R. R. Tressell (eds.), 1989, *Standard Methods for the Analysis of Water and Wastewater,* 17th ed. American Public Health Association, American Water Works Association and Water Pollution Control Federation Publication, Washington, D.C., 1466p.

Fourie, H. O., and M. Peisach, 1977, Loss of trace elements during dehydration of marine zoological material. *Analyst* 102:173–200.

Garrels, R. M., 1967, Ion-selective electrodes and individual ion activity coefficients. In G. Eisenman (ed.), *Glass Electrodes for Hydrogen and Other Cations, Principles and Practice.* Marcel Dekker, New York, pp. 344–61.

Knauss, K. G., T. J. Wolery, and K. J. Jackson, 1990, A new approach to measuring pH in brines and other concentrated electrolytes. *Geochemica et Cosmochemica Acta* 54:1519–23.

McConchie, D. M., A. W. Mann, M. J. Lintern, D. Longman, V. Talbot, A. J. Gabelish, and M. J. Gabelish, 1988, Heavy metals in marine biota, sediments, and waters from the Shark Bay area, Western Australia. *Journal of Coastal Research* 4(1):37–58.

McConchie, D. M., and V. J. Harriott, 1992, The partitioning of metals between tissue and skeletal parts of corals: Application in pollution monitoring. *Proceedings of the International Coral Reef Symposium, Guam, 1992,* pp. 95–101.

Mann, A. W., and M. Lintern, 1984a, Field analysis of heavy metals by portable digital voltammeter. *Journal of Geochemical Exploration* 22:333–48.

Mann, A. W., and M. Lintern, 1984b, Portable digital voltammeter for field analysis of trace metals. *Australian Water Resources Council Technical Paper* 83, 80p.

Marcus, Y., 1989, Determination of pH in highly saline waters. *Pure and Applied Chemistry* 61:1133–38.

Morf, W. E., 1981, *The Principles of Ion-Selective Electrodes and of Membrane Transport.* Elsevier, Amsterdam.

Thomson, E. A., S. N. Luoma, D. J. Cain, and C. Johansson, 1980, The effect of sample storage on the extraction of Cu, Zn, Fe, Mn and organic material from oxidized estuarine sediments. *Water, Air, Soil Pollution* 14:215–33.

Troll, A., A. Farzaneh, and K. Cammann, 1977, Rapid determination of fluoride in mineral and rock samples using an ion selective electrode. *Chemical Geology* 20:295–305.

Zobell, C. E., 1946, Studies on redox potential of marine sediment. *American Association of Petroleum Geologists Bulletin* 30:477–513.

4
Sampling

Samples are collected (1) for comparison with other samples to help identify similar stratigraphic units and to distinguish between different ones, and to show the lateral and/or vertical variation within the units; (2) for detailed laboratory analysis (e.g., for pollen and microfossil identifications, for textural or compositional study, or for geochemical analysis). Sampling occupies a key position in any sedimentological investigation because it is both where the investigation truly begins and because all conclusions are fundamentally tied to the properties of the samples collected. Nonetheless, in our experience (and see most textbooks!), this aspect of a project receives the least attention. Most workers believe that they can intuitively select representative samples—and most workers have inherent biases, albeit unconscious, that guide them to select what "looks" most interesting or promising for the purposes of their investigation: a potential source of error in the study. Altogether too few workers clearly relate and confine the conclusions of their study to the sampled population of data. (How often do we read papers that make worldwide generalizations based on a handful of samples from a single region? Some of these generalizations even find themselves in textbooks as accepted dogma!) Readers of reports have both the need and the right to know precisely what data and samples were examined, and how they were treated. In this chapter we present a summary of salient considerations from our experiences and a review of some common sampling devices; detailed discussions of various aspects of sampling are provided in the Selected Bibliography (for the broadest discussions, see Griffiths 1967, 1971; Cochran 1977; Keith 1988).

The first step in any sampling program is to select the *target population* of the study and the appropriate sampling strategy carefully (e.g., Fig. 4-1 and see Krumbein 1960)—i.e., to decide what kind of data are relevant and available in the context of the purpose of the project.

SAMPLING STRATEGIES

As a general rule, sampling should be as extensive as possible within the time, financial, and personnel constraints that apply to the project. However, because it is seldom possible to collect and examine all the desired samples, strategies are needed to ensure that sampling is representative and will lead to resolution of the questions the study was designed to answer. All sampling programs are designed to answer questions such as: What is there? How much is there? How is it distributed? Do the answers to these questions change with time? The key to planning a cost-effective sampling program is to know what data are required to answer these questions. Too few data may lead to unacceptable inaccuracies, whereas too many data add to the cost of the project without adding to the problem-solving value of the data. The skill is in knowing exactly the optimum amount of data required and the right samples needed to answer the questions the program was designed to resolve.

Three basic sampling strategies are used by geologists:

1. *Purposeful (selective) or opportunistic sampling:* Samples are collected as and where "suitable" material is found and it is up to the collector to decide which material is worth collecting, i.e., the scientist relies on his or her expertise to decide the validity of both the sample data and the results of the study. A subvariety of this strategy, sometimes considered separately, is *search sampling,* where the worker is seeking some data that will help formalize the purpose or design of the program (e.g., seeking an interval in which microfossils are common, when studying paleoecology); search sampling is replaced by the other sampling programs once the search is complete.

2. *Statistically planned (probability) sampling:* Samples are collected at points predetermined by statistical pro-

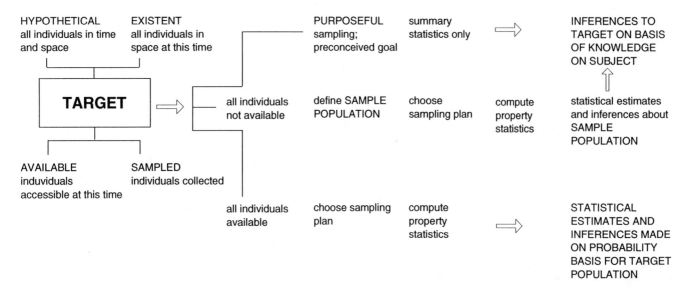

Figure 4-1. Schematic outline of potential sample targets and sampling procedures suitable for each target. The *ultimate population* refers to every individual of the objects or events under study (e.g., every sandsize particle in the world throughout time). A *hypothetical population* is that array of individuals which once existed but part of which are now eroded away or otherwise destroyed (e.g., an entire sedimentary unit, only part of which is now in existence). An *existent population* is the range of individuals in existence today (e.g., every sand grain in the world at this moment). The *available population* is the array of individuals accessible for sampling (sand grains obtainable by the sampling devices you possess be they spoons, shovels, or drilling rigs). The *target population* is the conceptual array of individual objects or events about which you wish to make statements—i.e., is dependent on the objectives of the study; while the target population may comprise any of the above populations, conclusions can be statistically evaluated only for an available and sometimes the existent populations.

cedures, and the collection is tested for representability and bias.

3. *Interactive sampling:* The sampling program is continually modified as data from previously collected samples become available; this program can involve either of the preceding sample strategies.

Only in the cases of strategies 2 and 3 can full-scale probability statistics be applied and quantified predictions made (Fig. 4-1); despite applications of summary statistical techniques to the samples from procedure 1, reliability of the results is uncertain because representability of the sample suite cannot be demonstrated.

Purposeful or opportunistic sampling is most familiar to, and most widely used by, geologists; all who carry out fieldwork have employed this approach on many occasions. For example, the collection of a particularly well-preserved fossil, an "interesting" or "representative" sample of a rock or sedimentary unit, and a sample containing uncommon minerals are examples of opportunistic or selective sampling. Because persons using this sampling strategy usually have some prior knowledge or purpose that will influence decisions on which samples to collect, this strategy is frequently very efficient and effective. Consequently, opportunistic or selective sampling provides the best and most practical approach to many studies. However, it is inappropriate for many environmental and exploration geochemical studies where quantitative statements are required.

Statistically planned sampling usually involves sampling at regular or random (e.g., using a table of random numbers) intervals on line transects such as along a traverse or down a drill core, at intersections on a grid, or at regular intervals up a river and its tributaries. In most cases, only *semi-probability* sampling programs can be carried out (rather than full probability sampling), because the available or accessible population is not the same as the target population (e.g., not all sands on a thick, wide beach may be accessible), and full statistical statements can only be applied to the available (or sampled) population. Sampling may be *simple random* (randomized coordinate system—low cost and easy to implement), *stratified* (used where different intervals exist, allowing each to be evaluated separately and to be contrasted with the entire population—best for determining variability and average values of measured properties), or *systematic* (samples are spread evenly over the population—best for interpretation of areal patterns; e.g., Fig. 4-2). The exact sample interval used will depend on the size, type, and likely homogeneity of target, on the type of statistical data required, and on the confidence limits imposed. Complications arise when, for example, the desired target is a three-dimensional unit that is only intermittently exposed for partial thicknesses at uncertain positions relative to the top/bottom of the unit (common situations in geology); without a drilling rig, the target will have to be redefined to something more realistic! Guidelines for setting up sampling programs suited to this type of strategy are widely published in the geological and environmental science literature (e.g., Krumbein and Graybill 1965; Kelley and McManus 1969; Cochran 1977; Dackombe and Gardiner 1983; Hoffman 1986; Keith 1988; Wolcott and Church 1991).

All probability sampling programs must be carefully planned and executed if the findings are to have the intended

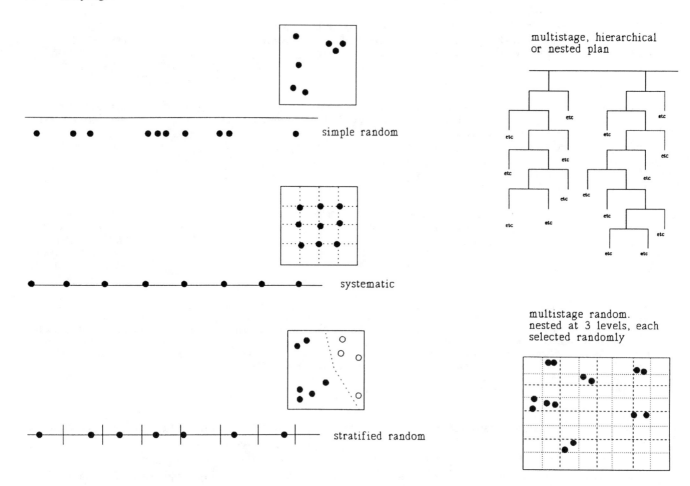

Figure 4-2.   Examples of systematic sampling plans. Stratified sampling refers to randomized sampling procedures applied to individual subunits of the sampled population. Systematic sampling refers to sampling at the intersections of grid patterns devised to cover the available population (shapes and spacings of the grid vary depending on the distribution of the population). Multistage, nested, or hierarchical sampling ensures randomization of selection at each level from layers of subunits of the available population.   (After Krumbein and Graybill 1965.)

statistical validity; tests for representability or bias must be applied (e.g., students' *T*-test, chi-square test, and others; e.g., Griffiths 1967 and others). For many studies (e.g., in ore reserve estimation, geochemical soil surveys, and many pollution investigations), a well-planned statistically based sampling program is essential. However, in many geochemical exploration and environmental analysis programs a statistically correct approach can result in collection and analysis of more samples than are necessary to answer the questions the program was designed to answer. Oversampling arises because even if "time to stop" criteria (see Chapter 1) were established during the initial planning stage, the necessary analytical data that determine when these criteria have been met must also be available. Undersampling can result when delays in the feedback of data to staff collecting the samples require return trips to areas that turn out to be of particular interest. Either undersampling or oversampling can add substantially to the cost of a project.

Interactive sampling is a strategy whereby analytical data are obtained in the field and used to modify the sampling program as the need arises. The classic example of an interactive

sampling strategy is the use of the gold pan in exploration. Despite the success of the gold pan, the wider application of interactive sampling strategies has been constrained by a lack of suitable analytical equipment. However, because modern analytical instrumentation is rapidly extending the range of components that can be accurately and easily analyzed in the field (e.g., Chapters 3 and 9), the use of interactive sampling strategies is likely to expand in the near future (e.g., McConchie 1989). The principal advantages inherent in an interactive sampling program are to:

1. Minimize the time involved in determining the appropriate sample type and size, minimize the number of samples required, and enable investigators to modify these requirements as the need arises.
2. Minimize the time involved in selecting the best method of compensating for dependent variables.
3. Provide rapid feedback on which key areas require further sampling (enabling leads to be followed up without the need for a return trip to the field).
4. Provide data to investigators in the field that may im-

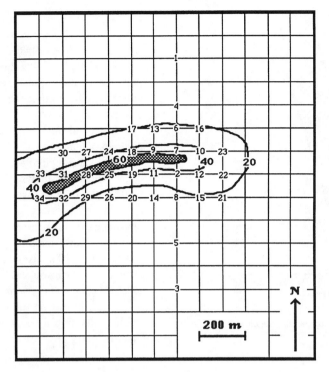

Item	Conventional sampling	Interactive sampling
Number of samples needed	168 (intersections on 100m grid)	c.a. 34 (as indicated)
Samples collected / day	60	20
Costs / day for 2 field assistants	$ 350.00	$ 500.00
Fieldwork labour costs	$ 1,050.00	$ 1,000.00
Analysis cost / sample	$ 20.00 (laboratory)	$ 12.00 (on site PDV2000)
Consumables cost	$ 100.00 (est. for containers etc.)	$ 200 (est.)
Total cost to define anomaly	$ 4,510.00	$ 1,608.00
Time needed to define anomaly	probably about 1 month (depending on lab. work load)	about 2 days

Figure 4-3.   A comparison of the relative costs and time requirements for the delineation of a hypothetical soil anomaly using conventional versus interactive sampling strategies. Costs (based on 1991 prices in Australia) and time requirements are for sample collection and analysis only, and do not include other field costs that are common to both exploration approaches. In the interactive alternative, the decision on where to obtain each sample is based on the analytical results for all the preceding samples. The numbers on the grid intersections represent the sequence in which the samples were collected for the interactive study.

prove their chance of recognizing links between parameters that might otherwise have been overlooked.

5.   Allow measurement of parameters that may change if samples are removed from their natural setting.

Fig. 4-3 shows a comparison of the relative costs and time requirements using conventional versus interactive sampling strategies.

SAMPLE COLLECTION

In the field notebook, record why each sample was collected, and note details of the relationship of the sample to the exposure (e.g., with a sketch); clearly note when you have selected particular samples that show unusual characteristics. Ensure that samples are adequately and indelibly labeled and located while you are in the field—*do not rely on your memory!*

When the goal of the investigation is interpretation of processes of deposition, sample a single sedimentation unit (a unit deposited under essentially constant physical conditions); this requirement may mean sampling individual laminations from some lithological units (e.g., see Erlich 1964; Emery 1978; Grace, Grothaus, and Erlich 1978; Macpherson and Lewis 1978). Wherever possible, ensure that samples are from fresh rather than from weathered exposure. For representative samples of rocks, an optimal sample size is approximately $10 \times 8 \times 5$ cm; for loose sand, about 200 grams is generally ample, whereas 50 grams may be adequate for muds. There are a variety of statistical approaches to determining minimum representative sample sizes (e.g., Krumbein and Graybill 1965; Ottley 1966; Griffiths 1971; Davis and Conley 1977; Dackombe and Gardiner 1983; see also Cochran, Mosteller, and Tukey 1954), but in practice most geologists collect more than necessary (just in case!) and sample size is primarily constrained by the ability to

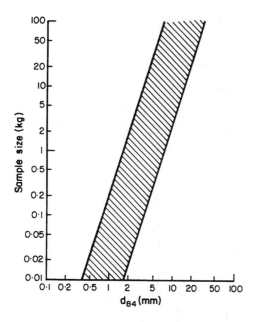

Figure 4-4. Chart for determining approximate sample size for grain size analysis of gravelly sediment. (After Jansen 1979.)

transport the collected material. Sampling of gravelly sediment is particularly difficult; few workers can carry representative samples in the necessary 44-gallon drums! Thus, analysis of gravelly sediments is best carried out in the field, although Fig. 4-4 provides a guide to the sample size needed for grain size distributions in coarse-grained sediments. Study by Wolcott and Church (1991) has shown that at least for river gravels the systematic or grid technique is superior to random sample procedures (see also Ibbeken 1974; Gomez 1983).

Sampling may be on the basis of *spot samples* (valid for the spots sampled only), *channel* or *compound samples* (cut through inhomogeneities of the population or made by homogenizing a variety of spot samples to obtain an "average" value; commonly used in economic mineral exploration/ evaluation), or *serial samples* (arbitrary but standard plan of spot sampling—determined from grid patterns and randomized selection of grid intersections). Serial samples are necessary for optimal statistical analysis, but are least used in geological studies.

Most geological sampling can be carried out using simple instruments such as a geological hammer, cold chisel, sampling knife, trowel, or spade, but many tasks require more complex techniques, some of which are described below. However, even the simple methods need to be used with caution if samples are to be used for geochemical studies of trace components (e.g., metals, pesticides, PCBs, etc.). Although hammers, spades, and other instruments are made of steel, they contain other metals at concentrations often far higher than exist in the sediment, and there is always the possibility that a fragment of the instrument may contaminate the sample. It is not possible to provide a set of rules for avoiding this type of sample contamination, but the risk must be evalu-

ated when deciding how to collect samples and precautionary steps should be taken to minimize it (e.g., if sediment samples are to be collected for metal analysis by divers, all the divers' lead weights should be nylon coated). When sampling for trace-component analysis, clean all equipment after collecting each sample to prevent cross-contamination of the next sample.

For general sample collections, consideration needs to be given to obtaining sufficient material to provide duplicate or reference subsamples that will not be destroyed during subsequent laboratory analyses. Plan for any required duplicates while in the field to ensure that enough material is collected. Once in the laboratory, initial samples can be subdivided into representative subsamples by a variety of simple to complex laboratory sample splitters of various sizes (e.g., Krumbein and Pettijohn 1938; Kennard and Smith 1961; Newton and Dutcher 1970; Brewer and Barrow 1972; Durden and Frohlich 1973; Ibbeken 1974; Cochran 1977). Before most such analyses, samples must be fully disaggregated and dispersed and soluble salts or grain coatings may have to be removed (see Chapter 5).

Coring

In many studies it is necessary to examine undisturbed samples of subsurface sediment. On land, some near-surface strata can be exposed and sampled by digging a trench either manually or using a machine, but at greater depths, underwater, or where strata are well indurated, some type of coring procedure is necessary. Coring associated with deep drilling (see Chapter 10) is not discussed in this book; procedures are complex, require specialized equipment, and are usually carried out by drilling engineers. However, cores of friable near-surface sedimentary sequences can be obtained using inexpensive equipment and simple procedures. The simplest procedure is listed in Table 4-1 (see Pearson 1978; Perillo et al. 1984; Gilbert and Glew 1985 for description of similar procedures). Cores collected in this way can be split along the length of the core tube, and logged in the field or returned to the laboratory in the tube for later splitting, examination, and analysis. A variety of techniques can be employed to split the core ranging from the use of a saw (which may contaminate the sediment) to the use of elaborate equipment designed specifically for core splitting (e.g., Hails, Gostin, and Wright 1979; Meisburger, Williams, and Prins 1980; Reddering 1981).

Relatively deep coring on land (up to about 30 m depending on sediment resistance) can be achieved by using a vibrator to drive in the core tube, but when this method is employed a winch will normally be necessary to extract the core tube, and if a plastic core tube is used, it will usually need an outer metal sleeve. Simple low-cost vibracoring equipment can be constructed in house; a range of designs for do-it-yourself vibracorers is described in the literature (e.g., Pierce and Howard 1969; Lanesky et al. 1979; Fuller and Meisburger 1982; Smith 1987). Many vibracorers can also

Table 4-1.   Example of a Simple Procedure for Collecting Core Samples of Unconsolidated Sediment in the Field

1. Measure a length (L) of core tube (rigid PVC, polycarbonate, or steel, usually 50 mm diameter) and drive it into the sediment to the required depth using a slip hammer or a vibrator (pencil vibrators used in concrete laying are available at a reasonable rental rate). A slip hammer consists of a length of heavy steel pipe with a cap welded onto one end and two handles welded onto the outside (the inside diameter must be larger than the outside diameter of the core tube); the length of the hammer pipe should be sufficient to keep it over the core tube during use. If the hammer is to be used underwater by a diver, it will be necessary to drill a large hole in the center of the cap (e.g., see Martin and Miller 1982).

2. When the core tube is hammered to the desired depth, or additional blows make no further progress, remove the hammer (or vibrator) and measure the distance from the sediment inside the tube to the top of the tube (D_I) and the distance from the sediment outside the tube to the top of the tube (D_O). The percent compaction of the core caused by the coring is then given by:

$$100 \times (D_I - D_O) / (L - D_O)$$

   Note: This calculation indicates average compaction only; some cores, particularly those obtained by gravity coring methods, may be nonuniformly compacted (e.g., Weaver and Schultheiss 1983).

3. Fill the upper (empty) part of the core tube with water and seal the top with an expanding bung (a plumber's dummy is ideal for this purpose) so that when the tube is pulled out the core will remain in the tube. (There may be some core loss when sampling well-sorted sands, but most sediment types can be recovered with negligible loss.)

4. Pull short cores from the sediment using lifting handles made by welding 10-cm sections of steel pipe (with an inside diameter about 5 mm greater than the outside diameter of the core tube) to lengths of steel pipe (about 2–3 cm in diameter and 0.5–1 m long). Longer cores, which cannot be lifted by hand, can be extracted using an endless-chain winch hung from a steel tripod fitted with base plates to prevent the legs from sinking into the sediment).

5. Seal the lower end of the core tube, either with a cap or by taping a plastic bag over the end.

6. Measure the length D_I down the tube from the top and cut off the excess tube about 1 cm from the top of the sediment with a hacksaw.

7. Seal the upper end of the core tube and mark it clearly with a core number and a way-up arrow.

be operated underwater by divers, or from a ship or small boat/barge fitted with a lifting winch. If operating from a small boat or barge, the weight of the drilling equipment can cause stability problems; these can be overcome by using a second vessel or pontoon as an outrigger and mounting the coring and lifting equipment on the platform between the two hulls (e.g., Hoyt and Demarest 1981; Irwin, Pickrill, and Main 1983).

Short cores of marine and lake sediments can also be obtained using ship-mounted piston or gravity coring methods (e.g., the Phleger, Alpine, or Benthos gravity corers; see also Sly 1969). These corers consist of a heavy steel tube (often with additional lead weights) with an inner sleeve and core catcher to receive the core, a sharpened tip to penetrate the sediment, a nonreturn valve at the top (to allow water to pass when the corer is dropped and to seal the tube when it is raised), and external fins to guide the corer vertically into the sediment and/or to cause it to spin during descent for improved penetration. Such equipment is heavy because coring relies on the mass of the corer to penetrate the sediment when it is dropped over the side of a ship; hence, gravity coring procedures require a ship fitted with a heavy-duty winch.

All coring systems have the advantage that they obtain undisturbed samples (piston corers can cause some surface disturbance) from precisely positioned sites and penetrate the sediment to a known depth, but the trade-off is that only a small area is sampled. An alternative coring technique that permits a greater area to be sampled uses Bouma and similar box-coring devices (e.g., Bouma 1969); these methods allow only small penetration depths.

Grab Sampling

On land, grab samples can be obtained using a spade or similar implement; grab sampling underwater sediments requires the use of more elaborate sampling equipment. The simplest means of sampling underwater sediments involves using a pipe dredge or similar "underway" dredge samplers. Dredge samplers consist of a pipe or box sample catcher with a lip in front, angled to cut into the sediment, and a towing cable (e.g., Fig. 4-5). These dredges have the advantages of small size, cheap production, and deployment while the ship is underway; disadvantages are that sample sites are not precisely located, depth of penetration is limited and not known exactly, and the surface area sampled is unknown. A more serious problem for a sedimentologist is that dredge samplers disturb the sediment being sampled and there is usually substantial fine fraction loss.

An alternative bottom sediment sampling approach is to use a Van Veen, Smith-MacIntyre, Shipek, or similar grab sampler (e.g., Förstner 1989). Small grab samplers can be used by hand from a small boat (or from a structure such as a

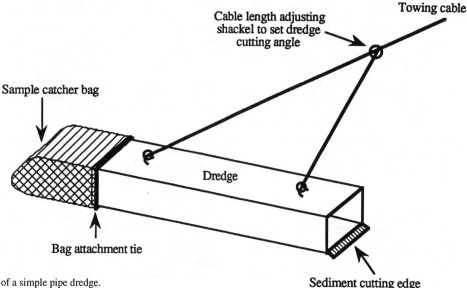

Figure 4-5.   Sketch of a simple pipe dredge.

bridge or wharf), but large samplers can be used only from a larger vessel fitted with a winch. This sampling equipment uses one or two rotatable jaws to scoop the sediment to an approximately known depth from a known area of the bottom. The sampler is lowered with the jaws open; a catch is released when the sampler hits the bottom, causing the jaws to close, either immediately or when the sampler is raised. Such grab samplers are easy to operate and have the advantages that the sample site can be located exactly, they sample a larger area than alternative methods, the approximate depth of penetration is known, and they sample a known area of sediment. However, they have the disadvantages that the vessel must be stopped, the amount of sediment sampled decreases with depth below the sediment surface, the sample is disturbed, and some of the fine fraction is likely to be lost.

Explosives

When used by an experienced shot-firer, explosives provide an efficient means of sampling indurated rocks and can be particularly useful when fresh sample material must be obtained from beneath a weathered crust. (In most countries and states a license is required for the purchase and use of explosives; extreme care must be exercised at all times in storage and transport, as well as at the time of sampling!)

When dealing with rocks, the most widely used sampling technique involving explosives is plaster shooting (e.g., Fig. 4-6) because no drilling is required, the procedure is rapid, and no major equipment is required. Plaster shooting is used to split boulders or cleave fragments from outcrops. Because of the inefficient manner in which explosives are used in plaster shooting (i.e., much of the explosive energy is wasted as concussion and noise), more explosives (up to 10 times) are required than if charges were placed in holes drilled in the rock; the charge required can be minimized by positioning the explosives over a crack or depression in the rock.

When dealing with soft sediment, explosives can be used to expose profiles of up to about 4 m below the water table (e.g., in a dry salt lake or on tidal flats where the water table is near the sediment surface); digging holes by hand in such settings is precluded by collapse of the edges of the hole and by the tendency for water to fill the hole as rapidly as it is dug. The blasting technique (developed for use on salt lakes by McConchie) is quick and easy to use and produces a hole that can remain dry for 15 min to 1 hr (depending on the sediment texture), even if the water table is only a few centimetres below the sediment surface. The hole remains dry for so long because the explosion forces the water into the sediment and compacts enough sediment near the hole to temporarily seal the pores and prevent the water from flowing back. When the hole has been formed, the sides can be scraped for examination and sampling. In salt lake sequences where there are often different water bodies separated by aquicludes, the individual water bodies can be sampled as water from each layer starts to seep back into the hole. The procedure that we have found effective for producing holes about 3 m in diameter and about 2.5–3 m deep in salt lake and tidal flat sediments is:

1. Make a small hole 2–3 m deep in the sediment using a metal or wooden spear.
2. Load medium-velocity (i.e., 2500–4000 m/s) explosives as shown in Fig. 4-7, fit a detonator to the PETN (pentaerythritol-tetranitrate) detonating cord, and fire when safe. If using electric firing methods, ensure that any electrical contacts are sealed against moisture and try to fire when a gust of wind is blowing to minimize the amount of material that will fall back into the hole.
3. When the shot firer gives the "all clear," place a short ladder down the hole and scrape the sides for examination, photography, and sampling.

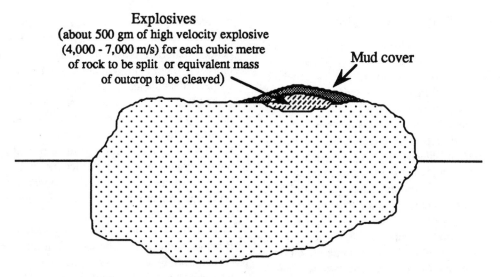

Figure 4-6.   Sketch of the simplest layout for plaster shooting with explosives.

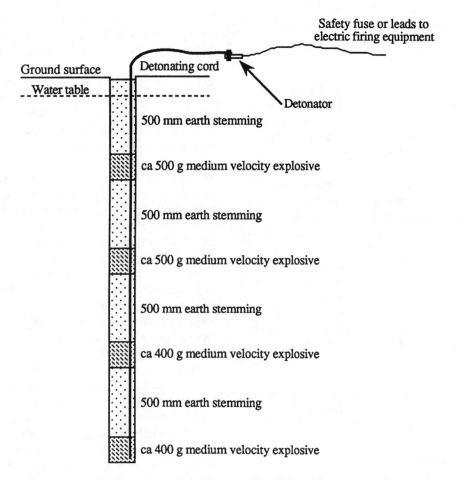

Figure 4-7.   Sketch of a suggested layout for deck-loading explosives to prepare a substantial pit in loose sediments.

Sediment Traps (and Rising Stage Samplers)

Determination of the suspension load in a water body can be achieved by collecting and filtering water samples, but these data will not provide a good indication of the rate at which suspended sediment is settling out at the sample point or of the average sedimentation rate over a period of days or weeks. Data on suspended sediment deposition rates are usually obtained using sediment traps that can be installed at multiple sites at different water depths or by rising stage samplers where different flood water depths are anticipated. In principle, a sediment trap is simply a container fixed at a predetermined position (usually at the bottom) in the water body, but the efficiency with which the traps collect and retain sediment depends on the shape and dimensions of the container and on the prevailing hydrodynamic conditions. Performance of alternative trap designs and various deployment strategies are reviewed by Gardner (1980) and Blomqvist and Hakanson (1981); the general conclusion is that the best guide to the actual rate of deposition of sediment from the water body is obtained from simple cylindrical traps with a 40–90-mm diameter and a height to diameter ratio of about 4. Funnel-mouth containers tend to undertrap, and narrow-mouth containers tend to overtrap, to unpredictable extents. Sediment traps should be deployed in sets to provide four or more replicates, and the position of each trap relative to the others and to any anchoring or other attachment structures should be planned to minimize interference with natural current-flow conditions.

Net sediment deposition rate at a site, representing accumulation of sediment transported by traction and saltation as well as suspension, cannot be monitored using traps; however, the average rate over a particular time interval can be recorded using simple procedures. One approach is to fix graduated poles into the sediment at the site and to record changes in the height of the sediment surface relative to the pole graduations. This approach is limited because fixed reference poles can alter water flow conditions, and hence depositional rates, in their vicinity; they are also vulnerable to damage during floods or to removal by visitors to the site. An alternative approach, which is effective at sites with moderate to high deposition rates, is to place a fine layer of red brick dust over an area of sediment (sieved fractions should be mixed to give the brick dust a similar particle size distribution to the sediment at the site) and to return at regular intervals to measure the thickness of sediment deposited over the red marker horizon. The main limitations to this strategy are dispersion of the marker horizon in areas where bioturbation is high, or loss of the marker horizon during a period of erosion.

Real-time sediment movement can be monitored using underwater time lapse or video cameras, or by using sedimeters such as that described by Erlingsson (1991).

Turbidity and Suspended Solids

Turbidity and suspended solids load (or nonfilterable residues) are both due to suspended matter in the water, but whereas suspended solids load represents the mass of suspended material, turbidity is a function of the volume of suspended material. Where the suspended matter is primarily inorganic there is a good correlation between turbidity and suspended solids load, but where biological material such as algae (with a high volume to mass ratio) is present, turbidity increases faster than suspended solids load.

Suspended solids load is reported in mg/L and is measured by filtering a known volume of water and measuring the dry weight gain on the filter (normally dried at 105°C); 0.45-μm filters are generally the standard size for this purpose, but even at this fine pore size, some ultra-fine (colloidal) particles will pass through the filter.

Filtration is usually achieved by vacuum filtration using a Buchner funnel in the laboratory. Filtration in the field is better achieved by using a syringe to force a known volume of water through a filter mounted in a sealed in-line holder (with a Lauer lock fitting); hand-powered vacuum pumps and filtration units are available for field use, but in our experience they are slow and the need for very regular pumping prevents moving on to other, more interesting, tasks. The principal limitation on any of these methods of recording total suspended solids load is the precision with which filter weights can be measured before and after use.

Turbidity can be estimated using a Secchi disc—a steel disc with quarters alternately painted black and white. The disc is lowered into the water by a graduated line attached to the middle of the disc until it is no longer possible for an observer positioned above the water to distinguish between the black and white segments (the depth to this point is recorded). More sophisticated measurements of turbidity are made using turbidity meters—either transmissometers (which record optical density in a straight line between a source and a detector) or nephelometers (which record light scattered at a set angle from a source); nephelometers operate more effectively over a wider range of turbidities than do transmissometers. Turbidity meters cost considerably more than Secchi discs, but they provide more precise data and are the only system that can record variations in turbidity at depth in a water column; depth probes capable of measuring turbidity are fitted either with a transmissometer or a nephelometer.

Acoustic transmission and back-scattering techniques (e.g., Vincent, Hanes, and Bowen 1991) can also be used to examine suspended sediment load. Acoustic methods are sensitive to both the suspended sediment mass and the density of the suspended particles, but the equipment is expensive. An alternative technique that we have found to be very serviceable is to record the amount of water that can be forced through a sealed 0.45-μm filter (26 mm diameter) before the filter is blocked to the extent that water passes through at a rate of only 1 drip every 5 seconds or less. This method is simple and economical; surprisingly, the operator's strength appears to have a negligible effect on the volume of water required to block the filter, but it is necessary to calibrate the procedure against alternative methods.

SAMPLING COAL

Sampling strategy for coals is selected according to objective. In all cases the coal seam is closely inspected for visible signs of variability, and logged. Megascopic features include mineral-rich ("high ash" or "dirty") horizons, which typically appear dull and sometimes hard in comparison to the "clean" coal. Clean coal may also be dull, and is then distinguished from dirty coal on the basis of specific gravity, which for clean coal is approximately half that of siliciclastics. Traditional megascopic classification defines fossil charcoal as *fusain*, which is easily identified by its softness, sheen, and well-preserved structure. Coal that is dull due to abundant waxy and/or oxidized material is called *durain*. Very bright homogeneous bands or lenticles are wood fragments, called *vitrain* because of their glassy character. Coal that is bright but inhomogeneous consists principally of woody fragments mixed with smaller amounts of oxidized and waxy constituents, and is called *clarain*. This classification was developed for Paleozoic coals, which are distinctively bright/dull banded. Cenozoic coals are comparatively featureless due to a lack of dull bands, and new terms have been informally adopted to describe the more subtle variability of these "all bright" coals. *Bright nonbanded* indicates that vitrain bands are too thin to be distinguished, resulting in a uniform "semi-bright" appearance. *Bright banded* indicates that megascopic vitrain bands are embedded in the semi-bright matrix. The size and frequency of the vitrain bands can be estimated, or quantified by an area count. Banding characteristics can provide important paleoenvironmental information, e.g., proximity to mire margins.

A detailed seam log based on the above characteristics will suggest natural positions for sample boundaries within the seam profile. Subseam intervals are called *plies,* and are usually positioned to minimize intraply variability. For example, a continuous interval of bright nonbanded coal might be sampled separately from an adjacent interval containing frequent vitrain bands, and a zone containing mineral-rich horizons—common near the base of seams—could constitute a third ply. In the case of a seam comprising uniformly clean, featureless semi-bright coal, arbitrary ply intervals can be defined. Important chemical and industrial variability may be defined by sampling sequentially from roof to floor, even though the plies are megascopically identical.

When the ply intervals have been decided, representative channel samples are taken. Supplementary samples of coherent blocks from each ply can also be obtained for specialized petrological analysis. All other analyses should be undertaken on representative portions of the plies, which should be ground to a suitable grain size before splitting.

Traditional chemical analysis of coal provides information on both depositional and burial history. Very careful interpretation is needed to accurately differentiate these effects on the bulk properties of a sample. An important advantage of serial sampling—taking multiple plies from a single seam profile—is that such samples have a common burial history.

Therefore, differences between the samples can be reliably attributed to depositional controls. Initial characterization should include a proximate analysis, and determination of sulfur content and specific energy. These parameters are all affected by outcrop weathering, and surface coal should therefore be cut back as far as possible to expose fresh material. Characterization of the inorganic fraction of coal samples ideally includes both major element (e.g., by XRF or AA) and mineralogical analysis (XRD of low-temperature ash; see Chapter 8). The resulting data have both paleoenvironmental and industrial significance.

SAMPLE STORAGE AND TRANSPORT

Many samples, for geochemical analysis in particular, have special sample storage and transport requirements that must be considered before sample collection (e.g., McManus 1985). For work involving trace components, such as heavy metals, it is important to test the type of sample storage containers to be used before initiating the sampling program; many plastics contain traces of heavy metals used as cross-linking agents and many glass containers are a potential source of lead contamination. Most sedimentary geochemists can relate stories of samples that were contaminated with trace metals leached from sample containers such as PVC (polyvinyl chloride) bags that interacted with the sediments placed in them. Do not rely on manufacturer's specifications: tolerances may differ from those you require. Contamination from sample containers is particularly important in relation to samples collected for environmental pollution studies, because these studies often lead to litigation: even when the containers have no contamination, you must be able to establish that fact to the satisfaction of a court. If water (surface or interstitial) samples are collected, they should be collected and stored in containers that have been both acid rinsed and sample rinsed.

Samples collected for nutrient (phosphorus and nitrogen species) studies should be frozen as soon as possible after collection and stored in frozen condition until ready for analysis because bacterial activity (e.g., by nitrogen-metabolizing bacteria) can rapidly modify the chemical form and concentration of the nutrient components in the sediments. Alternatively, if the necessary equipment is available, microbial alteration of chemical species in the samples can be prevented by freeze drying or oven drying in the field before storage in sealed bags or other containers. When metal speciation studies of the sediment are planned, it may also be necessary to freeze or dry (and store in sealed containers) samples to prevent bacterially mediated or inorganic oxidative transformation of one chemical species (e.g., sulfides) to another (e.g., sulfates or oxyhydroxides).

PEEL SAMPLES

Adhesive substances can be applied as surficial films to friable sediment or indurated rock surfaces of virtually any area; when peeled off, they retain a replica of the surface texture of that sediment, and in the case of friable sediments, they also retain a layer of grains in their original spatial relationships. Hence, in the field, peels can effectively sample grain populations and show structures, as well as the fabric of the sediment. In the laboratory, the most common application of peel studies is in the replication of etched surfaces on limestones (see Chapter 5).

Bouma (1969) and Klein (in Carver 1971) provide thorough descriptions in detail of a variety of peel techniques for use in both field and laboratory (see also Moiola, Clarke, and Phillips 1969 for a rapid field method; Hattingh, Rust, and Reddering 1990 for structures in dry, clast-supported gravels; Bull 1977 for peels of silt and clays; Brown 1986 for peels to be used with scanning electron microscopy). Davies (1968) discusses the laboratory preparation of peels from recent and old impregnated sediments of various compositions. We detail here only a generalized description of peel techniques.

For peels of any kind to be applied to field exposures of loose sediment, the surface must be flat and neither too dry nor too wet (but see Barr, Dinkelman, and Sandusky 1970 for a technique that will work for saturated sediment). A trowel, spatula, or other device such as a wire offset from a rigid frame (like an oversized cheese slicer) may adequately smooth friable sediment surfaces after the initial spadework; be careful to avoid smearing clayey sediment. A fine-spray applicator (e.g., gardening type) should serve to prevent excessive drying of the surface and consequent gravitational collapse. Excessive moisture at the cut face may be the main problem, particularly if pores are filled with salty water; it may be necessary to wait for a dry day and/or low tide, or to spray with acetone (water and acetone are fully soluble and the solution evaporates much more rapidly than pure water; do not have an open flame or lit cigarette nearby!); a range of other techniques may be applicable depending on the situation—ingenuity could be necessary! A variety of compounds that dry rapidly can be sprayed or painted onto the variously prepared smooth surface; they are absorbed differentially according to the permeability of the sediment, and after drying, a volume of undisturbed sediment is available for laboratory study that shows differential relief enhancing original structures and textures. Variations on the theme are many, and experimentation is generally necessary with whatever compound is locally available and affordable. Basic steps are:

1. Prepare smooth (vertical) surface. (In the laboratory, relief peels can be made from horizontal faces by applying compounds such as paraffin or plaster of paris to sediments that are dry and from which the salt has been removed.)
2. Pin a sheet of gauze to the smooth face over the area desired, ensuring that sediments do not crumble as the gauze is applied.
3. Spray (or paint) an adhesive compound, such as lacquer, onto the gauze. It is this step that requires the most experimentation—depending on the moisture, salt, and

organic content of the sediment, some of the compounds may not penetrate and bind adequately.
4. Wait for the compound to dry, spray or paint again several times; after the first coat, a different compound may be used (e.g., latex rubber on lacquer).
5. Label, roll up, and carry away the peel!

Yasso and Hartman (1972) report success using a GRS synthetic rubber base and aliphatic-type solvent (e.g., 3M Corporation's "Spray Adhesive Type 77") directly on a planar moist (freshwater or saltwater) surface. Spraying from a distance of about 0.3 m is continued until there is a slight excess on the surface; immediately after spraying, place a rigid backing board (e.g., plywood or Plexiglas) on the surface; gently remove the backing board with adhered sediment after a few seconds. Complete drying takes about 15 min; thereafter, spraying with a clear lacquer can improve the bonding and preserve the surface.

SELECTED BIBLIOGRAPHY

General

Bouma, A. H., 1969, *Methods for the Study of Sedimentary Structures.* John Wiley & Sons, New York, 458p.

Cochran, W. G., 1977, *Sampling Techniques,* 3d ed. John Wiley & Sons, New York.

Dackombe, R. V., and V. Gardiner, 1983, *Geomorphological Field Manual.* George Allen & Unwin, London, 254p.

Davis, J. C., and C. D. Conley, 1977, Relation between variability in grain size and area of samples on a carbonate sand beach. *Journal of Sedimentary Petrology* 47:251–6.

Emery, K. O., 1978, Grain size in laminae of beach sand. *Journal of Sedimentary Petrology* 48:1203–12.

Erlich, R., 1964, The role of the homogeneous unit in sampling plans for sediments. *Journal of Sedimentary Petrology* 34:437–9.

Förstner, U., 1989, *Contaminated Sediments.* Lecture Notes in Earth Sciences vol. 21. Springer-Verlag, Berlin, 157p.

Gomez, B., 1983, Representative sampling of sandy fluvial gravels. *Sedimentary Geology* 34:301–6.

Grace, J. T., B. T. Grothaus, and R. Erlich, 1978, Size frequency distributions taken from within sand laminae. *Journal of Sedimentary Petrology* 48:1193–1202.

Griffiths, J. C., 1967, *Scientific Method in Analysis of Sediments.* McGraw-Hill, New York, 508p (Chapter 2, pp.12–30).

Griffiths, J. C., 1971, Problems of sampling in geoscience. *Bulletin Institute of Mining and Metallurgy* 780:B343–56.

Hoffman, S. J., 1986, Soil sampling. In J. M. Robertson (ser. ed.), Exploration geochemistry: design and interpretation of soil surveys. *Reviews in Economic Geology* 3:39–77.

Jansen, P. L., 1979, *Principles of River Engineering.* Pitman, London.

Jopling, A. V., 1964, Interpreting the concept of the sedimentation unit. *Journal of Sedimentary Petrology* 34:165–72.

Keith, L. H. (ed), 1988, *Principles of Environmental Sampling.* American Chemical Society Professional Reference Book, 458p.

Kelley, J. C., and D. A. McManus, 1969, Optimizing sediment-sampling plans. *Marine Geology* 7:465–71.

Krumbein, W. C., 1960, The "geological population" as a framework for analyzing numerical data in geology. *Liverpool-Manchester Geological Journal* 2:341–68.

Krumbein, W. C., and F. A. Graybill, 1965, *An Introduction to Statistical Models in Geology.* McGraw-Hill Co., New York, 475p (Chapter 7 on Sampling).

McCammon, R. B., 1975, On the efficiency of systematic point-sampling in mapping facies. *Journal of Sedimentary Petrology* 45:217–29.

McConchie, D. M., 1989, Interactive sampling in coastal environments. In A. V. Arakel (ed.), *Proceedings of the Workshop on Coastal Zone Management, Lismore, 1988, Queensland University of Technology Publication,* pp. 134–53.

McManus, J., R. W. Duck, A. H. A. Alrasoul, and J. D. Thomas, 1985, Effects of storage before analyzing suspended sediment samples from lakes, rivers and estuaries. *Journal of Sedimentary Petrology* 55:613–5.

Macpherson, J. M., and D. W. Lewis, 1978, What are you sampling? *Journal of Sedimentary Petrology* 48:1341–3.

Ottley, D. J., 1966, Gy's sampling slide rule. *Mining and Mineralogical Engineering* 2:390–98.

Otto, G. H., 1938, The sedimentation unit and its use in field sampling. *Journal of Geology* 46:569–82.

Till, T. R., 1974, *Statistical Methods for the Earth Scientist.* Macmillan, London.

Tucker, M., 1988, *Techniques in Sedimentology.* Blackwell Scientific Publications, Oxford, 394p.

Wolcott, J., and M. Church, 1991, Strategies for sampling spatially heterogeneous phenomena: The example of river gravels. *Journal of Sedimentary Petrology* 61:534–43.

Sampling Devices

Benggiom, B. L., and T. A. Nelsen, 1984, "SLABBER": A useful device for subsampling box cores. *Sedimentology* 31:879–82.

Blomqvist, S., and L. Hakanson, 1981, A review on sediment traps in aquatic environments. *Archives of Hydrobiology* 91:101–32.

Brewer, R., and K. J. Barrow, 1972, A microsplitter for subsampling small particulate samples. *Journal of Sedimentary Petrology* 42:485–7.

Durden, T. W., and H. Frohlich, 1973, A splitter for unconsolidated cores taken in plastic liners. *Journal of Sedimentary Petrology* 43:521–4.

Erlingsson, U., 1991, A sensor for measuring erosion and deposition. *Journal of Sedimentary Petrology* 61:620–3.

Fuller, J. A., and E. P. Meisburger, 1982, A simple, ship-based vibratory corer. *Journal of Sedimentary Petrology* 52:642–4.

Gardner, W. D., 1980, Sediment trap dynamics and calibration: A laboratory evaluation. *Journal of Marine Research* 38:17–52.

Gilbert, R., and J. Glew, 1985, A portable percussion coring device for lacustrine and marine sediments. *Journal of Sedimentary Petrology* 55:607–8.

Hails, J. R., V. A. Gostin, and W. W. Wright, 1979, A cutter for plastic liner tubes. *Journal of Sedimentary Petrology* 49:646–8.

Howard, J. D., and V. J. Hendy, Jr., 1966, Sampling device for semiconsolidated and unconsolidated sediments. *Journal of Sedimentary Petrology* 36:818–20.

Hoyt, W. H., and J. M. Demarest, 1981, A versatile twin-hull barge for shallow-water vibracoring. *Journal of Sedimentary Petrology* 51:656–7.

Ibbeken, H., 1974, A simple sieving and splitting device for field analysis of coarse grained sediments. *Journal of Sedimentary Petrology* 44:939–46.

Irwin, J., R. A. Pickrill, and W. deL. Main, 1983, An outrigger for sampling from small boats. *Journal of Sedimentary Petrology* 53:675–6.

Kennard, M. C., and A. J. Smith, 1961, A simple micro-sample splitter. *Journal of Paleontology* 35:396–7.

Krumbein, W. C., and F. J. Pettijohn, 1938, *Manual of Sedimentary Petrography.* Appleton-Century-Crofts, New York, 549p.

Lanesky, D. E., B. W. Logan, R. G. Brown, and A. C. Hine, 1979, A new approach to portable vibracoring underwater and on land. *Journal of Sedimentary Petrology* 49:654–7.

Martin, R. J., and E. A. Miller, 1982, A simple, diver operated coring device for collecting undisturbed shallow cores. *Journal of Sedimentary Petrology* 52:641–2.

Meisburger, E. P., S. J. Williams, and D. A. Prins, 1980, An apparatus for cutting core liners. *Journal of Sedimentary Petrology* 50:641–2.

Newton, G. B., and R. R. Dutcher, 1970, An inexpensive student sample splitter. *Journal of Sedimentary Petrology* 40:1051–2.

Pearson, M. J., 1978, A lightweight sediment sampler for remote field use. *Journal of Sedimentary Petrology* 48:654–6.

Perillo, G. M. E., M. C. Albero, F. Angiolini, and J. O. Codignotto, 1984, An inexpensive, portable coring device for intertidal sediments. *Journal of Sedimentary Petrology* 54:654–5.

Pestrong, R., 1969, A multipurpose soft sediment sampler. *Journal of Sedimentary Petrology* 39:327–30.

Pierce, J. W., and J. D. Howard, 1969, An inexpensive portable vibracorer for sampling unconsolidated sands. *Journal of Sedimentary Petrology* 39:385–90.

Reddering, J. S. V., 1981, A pre-split soft sediment core tube. *Sedimentology* 28:483–5.

Savage, E. L., 1984, Sediment channel sampler for concurrent collection and storage of uniform samples. *Journal of Sedimentary Petrology* 54:651–2.

Sly, P. G., 1969, *Bottom sediment sampling.* International Association for Great Lakes Research, Proceedings 12th Conference on Great Lakes Research, pp. 230–9.

Smith, D. G., 1984, Vibracoring fluvial and deltaic sediments: tips on improving penetration and recovery. *Journal of Sedimentary Petrology* 54:660–3.

Smith, D. G., 1987, A mini-vibracoring system. *Journal of Sedimentary Petrology* 57:757–8.

Vincent, C. E., D. M. Hanes, and A. J. Bowen, 1991, Acoustic measurements of suspended sand on the shoreface and the control of concentration by bed roughness. *Marine Geology,* 96, 1–18.

Walker, J. D., and J.B. Southard, 1984, A sticky-surface trap for sampling eolian saltation load. *Journal of Sedimentary Petrology* 54:652–4.

Weaver, P. P. E. and P. J. Schultheiss, 1983, Detection of repenetration and sediment disturbance in open-barrel gravity cores. *Journal of Sedimentary Petrology* 53:649–54.

Peels

Barr, J. L., M. G. Dinkelman, and C. L. Sandusky, 1970, Large epoxy peels. *Journal of Sedimentary Petrology* 40:445–9.

Bjorlykke, K., 1966, The study of arenaceous sediments by means of acetate replicas. *Sedimentology* 6:343–5.

Buehler, E. J., 1948, The use of peels in carbonate petrology. *Journal of Sedimentary Petrology* 18:71–3.

Bull, P. A., 1977, A simple peel technique for silts and clays. *Journal of Sedimentary Petrology* 47:1361–2.

Burger, J. A., G. D. Klein, and J. E. Sanders, 1969, A field technique for making epoxy relief-peels in sandy sediments saturated with saltwater. *Journal of Sedimentary Petrology* 39:338–41.

Davies, P. J. and R. Till, 1968, Stained dry cellulose peels of ancient and recent impregnated sediments. *Journal of Sedimentary Petrology* 38:234–7.

Frank, R. M., 1965, An improved carbonate peel technique for high-powered studies. *Journal of Sedimentary Petrology* 35:499–500.

Hattingh, J., I. C. Rust, and J. S. V. Reddering, 1990, A technique for preserving structures in unconsolidated gravels in relief peels. *Journal of Sedimentary Petrology* 60:626–7.

Honjo, S., 1963, New serial micropeel technique. *Kansas State Geological Survey Bulletin 165,* part 6, pp. 1–16.

Katz, A., and G. M. Friedman, 1965, The preparation of stained acetate peels for the study of carbonate rocks. *Journal of Sedimentary Petrology* 35:248–9.

Lane, D. W., 1962, Improved acetate peel technique. *Journal of Sedimentary Petrology* 32:870.

McCrone, A. W., 1963, Quick preparation of peel prints for sedimentary petrology. *Journal of Sedimentary Petrology* 33:228–30.

Moiola, R. J., R. T. Clarke, and B. J. Phillips, 1969, A rapid field method for making peels. *Geological Society of America Bulletin* 80:1385–6.

Ostler, J., and I. P. Martini, 1973, Instant peels using polyester resin. *Journal of Sedimentary Petrology* 43:291–4.

Sternberg, R. M., and H. F. Beldong, 1942, Dry peel technique. *Journal of Paleontology* 16:135–6.

Yasso, W. E., and E. M. Hartman, Jr., 1972, Rapid field technique using spray adhesive to obtain peels of unconsolidated sediment. *Sedimentology* 19:295–8.

Young, H. R., and E. L. Syms, 1980, The use of acetate peels in lithic analysis. *Archaeometry* 22:205–8.

5
Sample Treatment in the Laboratory

Safety is the prime consideration when performing any work in the laboratory. Chapter 1 reviews salient considerations. Once samples have been collected and taken to the laboratory, they can be treated in a variety of ways depending on the type of sample and the information required from subsequent laboratory studies. Because the sample treatments required depend on the type of analytical work planned, it is important to consider what analytical work is necessary before deciding how to treat the samples. A flowchart showing some of the alternative analytical sequences that could be applied is shown in Fig. 5-1. The following sections of this chapter describe some of the most common sample treatments.

SAMPLE SPLITTING

Bulk sediment samples often need to be subdivided in the laboratory so that subsamples can be used in different tests and stored as reference material. It is important that each subsample be entirely representative of the bulk sample (see Chapter 4). For finely ground sample material, ensuring representative subsamples is not a problem because the powder will be fully mixed and homogeneous after grinding. For solid rock samples, subsampling usually requires the visual determination of representative sections to be cut. However, in loose-grain sediment samples, larger and higher specific gravity grains will segregate from smaller and lower specific gravity grains in a container, such that it is seldom possible simply to pour or scoop out a representative subsample. Sample splitters are commonly used to overcome this problem (see references in Chapter 4), but if this sort of equipment is not available *coning and quartering* can be used:

1. Pour the sample onto a flat sheet (or overlapping sheets) of glazed paper, then use a knife or piece of card to split the cone from the apex down into quarters.

2. Carefully move the quarters apart;
3. Combine opposite quarters to produce two equivalent subsamples from the original sample.
4. Repeat as necessary to produce the appropriate volume of sediment.

The method of coning and quartering works on the principle that as a result of the way in which grains will be distributed when poured to form a cone, each quarter has an equal probability of containing grains of any particular size, shape, or specific gravity; the combining of opposite quarters provides further confidence that the two subsamples are as similar as can be achieved.

WATER CONTENT AND BULK DENSITY

Original water content and bulk density measurements are important for many engineering (e.g., for sediment strength) and soil studies; also it is commonly desirable to determine these properties for samples to be used for other types of analysis. The initial samples should be in sealed plastic bags to ensure retention of original water content (see Chapter 4), and once unsealed, the remainder of the analysis should be performed rapidly to avoid significant evaporation. Table 5-1 lists a procedure for these determinations. Solid density of a sediment sample can be calculated by the method outlined in Table 7-6.

SEDIMENT DISAGGREGATION AND DISPERSION

Both rocks subjected to crushing and unconsolidated sediments commonly require treatment to disaggregate grains that are stuck to one another by various cements, and to dis-

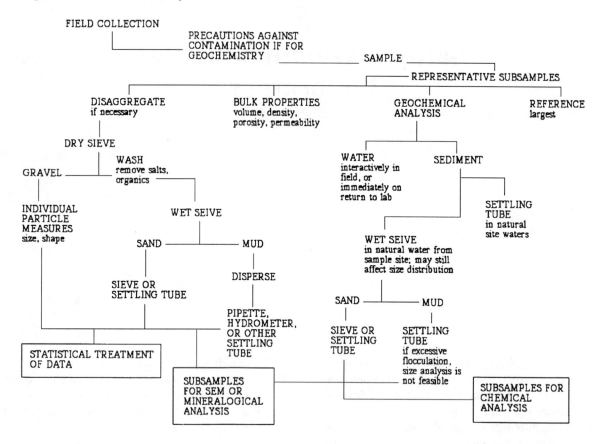

Figure 5-1.   Flowchart showing some of the alternative analytical sequences that could be applied to samples in the laboratory. Abbreviations are defined where appropriate in the text.

perse fine particles that are aggregated by interparticle charge attractions. Depending on sample characteristics, any one or a combination of the following techniques may be necessary to create a completely dispersed assemblage of grains, which may then be subjected to various methods of size, shape, and mineralogical analysis. Initial, then intermediate-stage, inspection with a binocular microscope will be necessary to evaluate results with the sandy sediments. Often, disaggregation can be accomplished only with a concomitant loss of the ability to analyze certain properties (e.g., even gentle disaggregation techniques may modify the size and shape of soft or brittle grains, such as shell fragments and natural aggregates). In general, use the gentlest method necessary and take precautions not to lose (or add) any material (e.g., wash fingers, containers, pestles, etc.). Sieve or decant at various stages in the procedure to separate fractions requiring further or different treatment. All samples should be subjected to the same treatment(s) in any suite of analyses.

Although some degree of disaggregation and dispersion of sample material is a necessary first step in grain size analysis, it is important to remember when interpreting grain size data that disaggregation will break up both natural flocs and natural aggregates. Many fine sediments are transported and deposited as flocs or aggregates and not as their constituent mineral grains. Thus, disaggregation of natural aggregates

commonly results in grain size data that record the presence of "artificially" fine particles.

Physical Disaggregation

Table 5-2 lists physical methods applicable to the breakdown of (essentially detrital) sediments into their constituent grains. *Note that the more rigorous techniques listed are likely to break grains along internal lines of weakness.*

Dispersion of Clays

Clay size particles are the most abundant sediment, and even though they do not normally settle in natural environments as individual particles, complete dispersion is necessary for textural analysis in the laboratory to avoid inconsistent results and to yield comparable data. (Even then, the "true size" of the clays is not measured because the measurement techniques do not differentiate between variations in size, shape, and density, which depend on mineralogy and the extent of cation adsorption). Because clays are commonly present with other size fractions, a dispersion treatment generally accompanies any sediment preparation.

For size analysis of clays, strict standardization in pretreatment and measurement technique is necessary for all

Table 5-1. Procedure for Determination of Water Content and Density of Sediment Samples

1. Use a small coring device (e.g., a large cork borer) to extract a sediment sample with a known volume (V_w cm^3); rotate the corer during insertion to minimize compaction. Alternatively, fill a vial with a known volume of sediment and pack with a spatula to approximate the degree of compaction characteristic of the sediment in its natural depositional setting.

2. Empty the corer or vial into a preweighed beaker and determine the weight of sediment (M_w g) as: weight of beaker plus contents minus weight of the beaker.

3. Place the beaker and contents in a drying oven at 105°C for 24 hr, then cool and reweigh. Determine the dry weight of sediment (M_d g) as: weight of beaker plus contents minus weight of the beaker.

4. Calculate the in situ water content of the sediment (%) as: $100 \times (M_w - M_d) / M_w$.

5. Calculate in situ bulk density as M_w / V_w.

For sediment samples that were originally water saturated (i.e., the volume of trapped air was minimal) the approximate dry density of the solids can be determined by continuing as follows:

6. Add 200 mL of ultrapure water to the dried sample in the beaker (if necessary transfer to a larger beaker, taking care to rinse any salts present in the original beaker into the larger beaker), then stir or shake for 15 min and allow to settle. Determine the salinity of the liquid (S_x g/L) by one of the methods described in the section on salinity in Chapter 3. Calculate the salinity (S_p g/L) of the pore water in the original sample (assuming a specific gravity of 1 g/cm^3 for pure water) as:

$$S_p = 200 \times S_x / (M_w - M_d)$$

7. Determine the dry density of solids in the sample (assuming dissolution of any salts that are present will have a negligible effect on the volume of water in the sample) as:

$$\text{Dry density of solids} = [M_d - \{S_p(M_w - M_d) / 1000\}] / [V_w - (M_w - M_d)]$$

For samples that are dry initially, or contain a significant amount of air in pore spaces, determination of dry density requires a different approach as follows:

8. Pack dry sediment moderately firmly into a vial with a known weight (M_c g) and volume (V_c cm^3) until full; weigh the container and contents (M_s g).

9. Slowly add pure water to the sediment sample in the container, until the surface is just covered by a thin film of water and no more water is being adsorbed into pores in the sample; record the volume added (V_p cm^3).

10. Calculate the average solid grain density as:

$$\text{Solids density} = (M_s - M_c) / (V_c - V_p)$$

samples to be compared—be wary of comparing results of different analysts or laboratories and clearly state your own procedure in any report.

The most effective dispersion agents (peptizers) for clays are strong bases, weak acids, and large acidic anions. Any dispersant will cause flocculation (i.e., act as a coagulant) if its concentration is sufficiently high or if it is allowed to act on the clays long enough. In fact, dispersants cause rapid dispersion at first and slow, continuous dispersion (or coagulation) later; total dispersion may never be fully achieved during an analysis. The presence of abundant natural electrolytes in the rock sample may invalidate the use of some dispersants, and a preliminary washing and filtration procedure may be desirable (see below). In addition, the various minerals (or size ranges) present may dictate the type and concentration of peptizer used, as well as the time necessary for dispersion—trial and error experimentation for "best" results is commonly necessary. Table 5-3 lists some common peptizers with their concentrations in distilled water (see also Anderson 1963; Tchillingarian 1952). To test for flocculation, let the suspension stand for up to 12 hr. Tilt the beaker of suspension approximately 20°: if settled sediments or strati-

fied layers flow to maintain a horizontal level, flocculation has occurred. Or examine a drop of the suspension under a microscope; if not flocculated, particles will be separate and the finest show Brownian movement. For maximum dispersion, it is probably best to subject the suspension to 15 min in an ultrasonic device as a final step, although particularly if a high-power setting is used, delicate grains and microfossils may be destroyed (e.g., Edwards and Bremner 1965; Walker and Hutka 1973). *Note:* Keep a record of the quantity of peptizer used. For some kinds of analyses it may be necessary to allow for the weight added.

In general, avoid oven drying of clays if grain size analysis is to be attempted (brick does not analyze well). Dry at no more than 60–65°C in a photographic drying cabinet or with an infrared lamp. Also try to avoid boiling as an aid to dispersion (it may disrupt original clay particles), but doing so may be necessary in recalcitrant cases.

For Recent sediments, try to disperse the sample in the moist natural state in which it was collected and in water collected from the sampling locality (see Prokopovich and Nishi 1967), but washing to remove soluble salts may still be required. In addition to disaggregation methods discussed

Table 5-2.   Procedures for Physical Disaggregation of (Detrital) Sediments, in Order of Increasing Severity of Treatment

1. Crush with (rubber-gloved) fingers, either dry on glazed paper or with distilled water (and dispersant) in a basin.

2. Crush gently with a rubber bung under the same conditions.

3. Stir gently in (warm) distilled water with dispersant.

4. Shake in a bottle with (warm) distilled water and dispersant.

5. Rub with a stiff brush in distilled water with dispersant.

6. Boil in water with dispersant. "Quaternary 0" (a strong detergent) is a 20% liquid solution that may be particularly effective when added to the boiling water (Zingula 1968). Boil it for up to 1 hr, stirring occasionally.

7. Place the sample plus water and dispersant in an ultrasonic (over 20,000 cps) shaker (see Moston and Johnson 1964; Kravitz 1966). *This procedure is probably the "best" and could well be adopted as a standard final step in routine sediment dispersion, but particularly at high power it can destroy delicate grains such as microfossils.* It is generally wise to check for potential destruction of delicate components such as microfossils, and purchase of a device with variable cps control is desirable, so that lower cavitation rates can be set. An ultrasonic probe device may be applied directly to resistant fragments (*be sure to use a plastic beaker!* Savage 1969).

8. Pound with a rubber pestle in a porcelain mortar (in water plus dispersant).

9. Use a porcelain mortar and pestle to pound the sample (remove fines periodically to avoid excess crushing). Avoid rotary grinding, which crushes grains.

10. Dry small aggregates of the sample thoroughly in an oven. Saturate the sample in kerosene (or gasoline). Decant the fluid. Cover the wet sample with water or place saturated sample in just-boiled water. The force generated by water molecules displacing the hydrocarbons (or the liquid-gas transformation in the case of the hot treatment) should be sufficient to disintegrate mudrocks. A small amount of detergent added to the water appears to hasten disintegration. After saturation, immersion into a boiling 50% solution of sodium carbonate (washing soda) and continued boiling has been suggested for some porous limestones as well as well-indurated mudrocks.

Caution: Fumes are dangerous, particularly in the presence of an open flame!

11. Heat the sample, then plunge it into cold water. Some grains may fragment; others may oxidize; size analysis, and probably mineralogical analysis, of clays cannot be performed subsequently.

12. Alternately freeze and thaw samples immersed in water. While effective, this technique usually requires a long time and is useless with impervious rocks such as well-indurated mudstones. It may damage some crystal structures.

13. Cover the crushed fragments with an equal amount of sodium acetate, add a few drops of water, and heat. The acetate will melt and saturate the sample; adding a crystal of acetate during cooling causes general crystallization and should disintegrate porous rocks.

14. Dry the sample and saturate it with either a concentrated solution of sodium hyposulfite or 14% sodium sulfate decahydrate (Glauber's salts). Crystallize the compound by drying.

15. Dry the sample and immerse it in a solution (made by adding the compound to warm water) or a melt of sodium thiosulfate ("hypo"). Boil gently. Remove the sample and allow it to cool thoroughly and evaporate (hypo crystallizes and disaggregates the sample). Recycle as necessary—grains will accumulate in the container. (If you use a beaker, do not allow hypo to crystallize in it —reheating will break the glass.) Sodium thiosulfate may cement the disaggregated particles but is very soluble in water and repeated washing will remove it.

16. Saturate the sample with calcium bicarbonate solution, or boil it for a few moments in a sodium carbonate solution. Subsequent immersion in dilute hydrochloric acid causes effervescence, which may disintegrate the rock.

17. Using an iron mortar and pestle, pound the sample. (Remove fines periodically, and remember that iron fragments must be removed from the final sample either by using a magnet, which will also remove magnetite and ilmenite, or by dissolving them in hot nitric acid, which may affect other minerals.)

18. Crush the sample in a mechanical disc pulverizer. (This is a last resort; the sample will then be suitable only for such investigations as chemical analysis, trace element analysis within certain minerals, and limited mineralogical analysis. Fragments of the discs must be eliminated. Evaluate the results by comparing with portions of the original sample and thin sections).

Table 5-3.   Common Peptizers Used for Dispersion of Clays

ammonium (or sodium) hydroxide (a few drops of concentrated NH_4OH per liter). This peptizer will not work if calcium or magnesium carbonates are abundant, and it may alter the chemistry of the sample (e.g., adsorbed Fe, Al, and Mn may be removed and converted to insoluble hydroxides).
sodium carbonate (0.53–2.12 g/L, try 1.06 g/L first)
sodium oxalate (0.67 g/L, or 0.34 g/L).
sodium hexametaphosphate (0.25–0.6 g/L), known by the trade name Calgon. This compound is probably the dispersant most widely used by sedimentologists, but it can dissolve delicate carbonate microfossils and hence should not be used with calcareous samples. It may also cause a change in the clay mineral chemistry—as will the sodium and ammonium dispersants—because of cation exchange.

above, the techniques listed in Table 5-4 have been applied to clay-rich samples, which are particularly difficult to disaggregate.

REMOVAL OF SALTS

Sediments from marine and brackish environments, and samples that have been chemically treated by processes such as those listed above, commonly have a large amount of salts trapped with water in pore spaces or remaining as residues after the water has evaporated. These salts need to be removed prior to most chemical analyses and may inhibit the dispersion of particles required for grain size analysis. The removal of salts is achieved by washing until the conductivity of the water is ≤0.5 µS (or until deflocculation occurs) as described below (p. 67) for the cleaning of chemically pretreated samples. There are a few additional points to consider when planning to remove salts from a sample, including:

1. Soluble salt crystals may be an important natural mineral constituent of the sediment (e.g., in saline soils or playa lake deposits). When this situation is likely, examine samples before and after washing, and/or analyze sample solutions produced from the washing procedure.
2. Some cryptocrystalline sulfides (e.g., the hydrated iron monosulfides) may oxidize and decompose. When these minerals are likely to be present, examine samples before and after washing; for some samples it may be useful to retain and analyze the filtered washing water.
3. Some exchangeable ions may be stripped from minerals in the sample.
4. Some natural particle aggregates may be disaggregated during the washing process; this possibility can be assessed by examining the sample material with a microscope before and after washing.

REMOVAL OF SEDIMENT COMPONENTS BY CHEMICAL MEANS

Caution: Extreme care is needed when using some of these chemicals—always work under a fume hood, wear protec-

Table 5-4.   Additional Methods of Disaggregation (cf. Table 5-2) Applicable to Clay-Rich Sediment Samples

1. Boil the sample in a caustic soda (NaOH) solution (20%) for several hours (this procedure is likely to decompose some clay and a few other minerals).

2. Place caustic soda pellets on the sample, and allow hydration from atmospheric moisture to take place.

Caution: Hot and/or concentrated caustic soda solutions can cause burns to skin and eyes; caustic soda pellets generate considerable heat when water is added. Also, even small amounts of dilute caustic soda solution can make fingers very slippery, and it can then be hazardous to handle containers of other reagents!

3. Add boiling water to dry washing soda (Na_2CO_3), covering the sample.

4. Boil the sample in a 0.02-N ammonia solution or a 0.01-N sodium-pyrophosphate solution.

5. Soak the sample in a *weak* hydrogen peroxide (H_2O_2) solution. Note that hydrogen peroxide is acid-stabilized when purchased and consequently it will dissolve some carbonates if they are present; this problem can be overcome by adjusting the pH of the hydrogen peroxide solution to 8.2–8.5, but this adjustment will make the reaction between the sample and the now-destabilized hydrogen peroxide much more vigorous. Wick (1947, in Mueller 1967) suggested the following procedure: Reduce the dry sample to fragments 3–6 mm in size; pour 15% H_2O_2 solution (dilute 30% commercial solution with water in 1:1 ratio) over the fragments in a large beaker under a hood (just immerse the fragments). Boiling will occur within a few minutes and water vapor plus oxygen will be generated by fluid dissociation. (If boiling has not begun within 10 min, heat the sample and add a few millilitres of NaOH or KOH.) Repeat the process for strongly indurated rocks. Wet-sieve and recycle the remaining aggregates. Stand all beakers containing hydrogen peroxide and samples on a drip tray and use a larger beaker than you expect to need; the mixture may seem unreactive for a while, but if an oxidizable component is present, the reaction will generate heat and become self-accelerating, with considerable frothing. If you observe the mixture until the reaction peak has passed, there is a good chance that you will be able to break up the froth with a glass stirring rod and/or a small squirt of water before it overtops the beaker. If the reaction is still too vigorous, dilute the H_2O_2 further and try standing the beaker on ice water.

Caution: Concentrated H_2O_2 (30 vol or greater) can cause severe burns, fumes can ignite, and highly carbonaceous samples may react vigorously (explosively if concentrated solutions are used).

tive clothing and goggles, and work near a water supply for washing. When using strong acids or other dangerous chemicals, either do not work alone or request a reliable individual to check your welfare regularly. Before you start, ensure that you are familiar with the hazards of the reagents you propose to use and with the safety facilities in your laboratory (see Chapter 1).

Check whether the reagents you intend to use will remove or change the properties of both the target mineral and the nontarget minerals; also check whether undesirable reaction products are likely to form. In our experience, there is no such thing as a totally selective extractant (see later in this chapter [p. 79.]), not even pure water. Consequently, the best approach is to assume that nontarget minerals may be affected unless proved otherwise. There are various ways to

Table 5-5.   Chemical Methods for Removing Carbonates and Silica Cements from Sediment Samples

To remove carbonates:

For the least effect on other minerals, Jackson (1958) recommends sodium acetate treatment for calcium carbonate and gypsum (ineffective with large crystals or concentrations). Brewer (1964) suggests the following procedure: place 100 g fine sample in a large (e.g., 3 L) beaker, add 600 mL 2N sodium acetate solution (buffer for pH 5), and stir into suspension; allow to digest on water bath. Allow settling, decant the fluid, and wash with sodium acetate solution to remove excess calcium salts, then wash with distilled water and decant until clear. Repeat as necessary. (To avoid oxidizing organic matter, bubble SO_2 gas through the suspension overnight.)

When delicate components might be damaged by effervescence generated by a stronger acid, use dilute acetic acid (approximately 15%). The sample must generally remain for days in the solution until no reaction occurs when fresh acid is added; monochloracetic acid (50%) is much faster, but *avoid contact and exposure to fumes.*

Use warm dilute HCl (10–15%) to remove fine dolomite, magnesite, and siderite, as well as calcium carbonates. For the least harmful action on other minerals, use the smallest amount of HCl possible. Brewer (1964) suggests adding 100 mL 2N HCl to 1 L water plus 5 mL 2N HCl for each percentage of calcium carbonate above 2%. Expanding clays are markedly affected and some cryptocrystalline sulfides will dissolve.

Boil the sample gently in 20% phosphoric acid for 5 min. This treatment may not remove abundant calcite cement, but it may remove iron carbonates and some iron oxides. A finely crystalline calcium phosphate will precipitate (see Leith 1950). Oxalic, formic, citric, and sulfuric acids have also been used, and any one may be particularly effective for a sample suite.

Organic complexing agents that form soluble complexes with alkaline earths may be used (e.g., EDTA, ethylene-diamine-tetra-acetic acid or DTPA, diethylene-triamine-pentacetic acid and TEA, triethanolamine). A 10% EDTA solution is recommended, using 50 mL of solution for each 0.5–1 g limestone. Boil until the carbonate is dissolved (that is, until a small suspended sample does not react to HCl). pH levels are critical for iron removal (low pH) and to keep clays from altering (high pH; raise by adding NaOH). Other calcium, strontium, and barium minerals (such as apatite, barite, and gypsum) are totally or partially dissolved (see Glover 1961; Bodine and Fernald 1973).

Ion-exchange resins can be used as an effective means of dissolving carbonates (e.g., French, Warne, and Sheedy, 1984), but phosphates and gypsum are also attacked and smectites and chlorites may undergo some modification. Satisfactory results can be obtained by adding water and an excess of the hydrogen form of a cation exchange resin (e.g., Amberlite IR120 or IR50) to sample material, crushed to a size finer than the resin beads, and allowing the mixture to stand for 12 to 24 hours; after reaction, the sample material can be separated from the resin beads by wet sieving.

To remove silica cement:

Hot strong alkali solutions (KOH or NaOH) may dissolve opal, chalcedony, chert, or authigenic quartz overgrowths.

Hydrofluoric acid may be used to extract some heavy minerals. (Add crushed material to cold concentrated HF in a platinum or teflon crucible.)

Caution: HF is one of the most corrosive and dangerous acids in the laboratory! Ensure that you are familiar with all safety and first-aid procedures before starting work; a single drop on the skin can cause serious burns. Neutralize used HF with washing soda before disoposal.

---

check the extent to which nontarget minerals have been dissolved, including chemical analysis of the extractant solution after reaction with the sample, microscopic examination of the sample material before and after digestion, X-ray diffraction analysis of the sample before and after digestion, or chemical analysis of subsamples of the sample before and after digestion. Any of these tests will add to the work involved in studying the samples; hence their application to all samples in a suite may not be warranted. However, testing some of the samples will not only yield selectivity information but will also provide additional mineralogical and geochemical information that may be of value to the original study. For large studies, there are often advantages in quantitatively testing the selectivity of potential extractants on arti-

ficially prepared mixtures of the types of minerals known to be present in the samples. Quantitative measurement of sample losses from chemical action also can be achieved with proper experimental design. (Commonly it is sufficient to determine the weight lost by weighing before and after treatment, but reaction precipitates must be removed.)

If clays are to be retained, it may be necessary to flocculate them by adding NaCl to permit decantation of the chemical reaction products; an involved filtration procedure to clean the clays of electrolytes may be necessary as well. Note that after putting clays through many chemical treatments, size or mineralogical analysis is useless.

A preliminary mechanical disaggregation is necessary to produce rock fragments 1 cm or smaller before most chemi-

Table 5-6.   Chemical Methods for Removal of Sulfides and Sulfates, and Organic Matter from Sediment Samples

To remove sulfides and sulfates:

Pyrite and most other metal sulfides: use warm (to boiling) 15% nitric acid (which also removes phosphates), oxalic acid, or hydrogen peroxide (15%); for large recalcitrant crystals, adding strong (120 vol) hydrogen peroxide dropwise to the hot nitric acid works well.

Pyrrhotite: use dilute HCl.

Calcium sulfates: use warm concentrated HCl. Use dilute HCl if gypsum is fine (for drastic removal of gypsum, use a strong ammonical solution of ammonium sulfate).

Barium sulfate: use warm concentrated H_2SO_4 (which also removes carbonates, chlorite, and biotite and affects other minerals), or chelating agents such as EDTA or DTPA.

To remove organic matter (necessary before size analysis of modern sediments):

Use hydrogen peroxide (H_2O_2). Treat cold or warm, or boil in 10% solution, or treat cold in 30% H_2O_2 (e.g., Jackson 1958). *In all cases heat will be generated and use of a water bath is recommended; be particularly cautious with high concentrations.* Initially, just cover the sample with distilled water, then add small quantities of H_2O_2, stirring until any effervescence ceases. (Make sure the beaker is large enough that frothing will not cause overflow—e.g., a 1-L beaker for 50 g sediment. If excessive frothing occurs, spray lightly with an aerosol of ethanol, which will later fully evaporate—*however, do not use ethanol if substantial heat is being generated!*) A large watch glass cover may be necessary if frothing is excessive. Continue adding H_2O_2 until frothing ceases, then heat slowly to 60–70°C (H_2O_2 decomposes above 70°C). Observe for 10 min to ensure that any danger of a strong reaction has passed. Add H_2O_2 until no further reaction occurs; leave for several hours. Elemental carbon and paraffin-like compounds are unaffected; coarse fibrous organic material will have to be removed by hand. Wash and filter the residue. To avoid partial dissolution of carbonates, initially adjust the pH of the hydrogen peroxide up to 8.2–8.5 with NaOH (*this procedure will also destabilize the hydrogen peroxide and make the subsequent reaction more vigorous!*). pH-adjusted peroxide is very good at removing organic matter from carbonate sediments and corals and does virtually no damage to the carbonates (McConchie and Harriott 1992).

Use the sodium hypobromite method (e.g., Brewer 1964): for 100 g (of soil), add 400 mL fresh sodium hypobromite solution (mix equal volumes of bromine water and 40% NaOH solution). Stir and let the suspension stand 2 hr at room temperature with occasional agitation. Add another 400 mL of solution, stir, and let stand overnight (12 hr). Wash by decanting or filter.

Soak the sample in strong sodium hypochlorite (or household bleach) for 1 to 5 days, then decant, and wash the excess solution from the solid sample residue.

For bituminous matter (such as cements), use carbon disulfide, ether, or chloroform, washed with carbon tetrachloride, acetone, or alcohol. (Bituminous matter fluxes with hot Canada balsam; if thin sections are to be prepared of bituminous sediment, special precautions must be taken—see Milner 1962.) Solid asphalts and coal must be treated differently.

cal treatments are applied. Table 5-5 lists techniques for removing carbonates and silica; Table 5-6 lists methods for removing sulfides and sulfates, and organic matter. Iron oxides (and manganese oxides) are usually difficult to dissolve using common acids or bases, but tend to dissolve readily in mild acidic solutions containing a reducing agent (e.g., Table 5-7); Table 5-8 lists techniques that have been successfully employed in removing the various oxides.

Cleaning Solid Residues after Chemical Treatment of Samples

After any of the above chemical treatments the solid remnants of the sample should be washed free of reagents and any salts formed during reaction. If the solid residues are not too fine, washing can be readily achieved by repeated rinsing with water (ultrapure, if geochemical analysis is intended) and decantation; centrifuging prior to decantation is necessary for fine particles. When samples contain an abundance of clays and other very fine particles, even centrifuging may not cause sufficient settling and filtration is the only practical option. Filtration after each rinsing with (ultrapure) water is best achieved using a standard Buchner funnel and vacuum system; remember to apply the vacuum then moisten the filter paper to ensure a good seal before

Table 5-7.   Some Common Reducing and Oxidizing Agents Used in Various Chemical Treatments of Sediment Samples

Reducing Agents	Oxidizing Agents
ascorbic acid	. . . bromate
. . . dithionite	bromine
hydrazine	. . . dichromate
hydroxylamine	. . . hypochlorite
mercurous . . .	. . . iodate
. . . sulfite	perchloric acid
sulfur dioxide	. . . periodate
stannous . . .	. . . peroxide
. . . thiosulfate	. . . peroxydisulfate

Table 5-8.   Techniques for Removal of Iron Oxides from Grain Coatings

For the least harm to other sediments: add 300 mL distilled water to approximately 100 g of sample in a beaker. Add 24 g solid sodium citrate + 2.8 g solid sodium bicarbonate (the pH should remain at 7.0–7.5). Heat to 75–80°C while stirring to ensure complete solution. Add 7 g solid sodium hyposulfite (sodium dithionite, $Na_2S_2O_4 + H_2O$), stir constantly for 5 min and intermittently for 10 min. Wash by repeated sedimentation and decantation or filter with a Buchner funnel (Brewer 1964; see also Mitchell and MacKenzie 1954). Warming (with occasional shaking) for 20–30 min may work equally well after adding 20% sodium dithionite solution to a disperse suspension with a little water. The sample should turn gray. Wash it several times with 1% neutralizing NaCl solution, in this case before washing with distilled water. (*Note:* Dithionite solutions turn brown and lose their effect rapidly. Always use fresh solutions.)

Gently warm the sample in a solution of 20% HCl containing 10 g ascorbic acid per 100 mL HCl (make a fresh acid mixture each time it is to be used) for about half an hour, then filter and wash the solid residue.

Warm with 15–20% HCl + 10% stannous chloride (or 5 g stannous chloride in 10 mL HCl) until colorless (Drosdorff and Truog 1935).

Treat in oxalic acid and sodium sulfide (for the procedure, see Truog et al. 1937). This treatment also aids in removing other cementing materials. Alternatively, shake (30 min) or boil (10 min) in a solution of 31.5 g oxalic acid and 62.1 g ammonium oxalate in 2.5 L H_2O.

Boil gently for 5 min in 20% phosphoric acid; this procedure will also remove phosphates and carbonate cement.

Place the sample (approximately 20 g) in a 500 mL beaker. Add 300 mL distilled water then 15 g powdered oxalic acid and a cylinder of sheet aluminum (with a diameter slightly smaller than that of the beaker, extending approximately 2 cm above the fluid from the base). Boil gently for 20 min. Remove the cylinder and decant the liquid. Wash the sample and decant the wash water until it becomes clear (see Leith 1950). This procedure has a very slight effect on most minerals, but it probably affects clays. A yellow ferrous oxalate should precipitate on the aluminum cylinder.

Remove any carbonates. Place 2 g sample in a 150 mL beaker, add 40 mL potassium oxalate solution (103.7 g/L), and heat to 80°C. Add 10 mL oxalic acid solution (85 g/L), stir, and heat to 90°C. Add approximately 0.2 g of magnesium ribbon (wrapped around a rod) and stir for 3–5 min at 90–95°C . Add 5 mL oxalic acid solution, and continue heating for 3 min or until the material turns white or gray. Wash by repeated sedimentation and decantation, or by filtering. For first washing, use N/10 HCl (Brewer 1964).

Note: Try ultrasonic treatment first.

adding the suspension to be filtered. After the first filtration, further washing of the solids can be achieved by keeping the vacuum on and adding water to the Buchner funnel. When the suspension to be filtered contains abundant clays, filtration may be slowed by clays blocking the pores in the filter; the choice of appropriate filter papers can speed up this process, but ensure that the use of a faster (more porous) filter paper does not result in some of the finer sample material passing through.

DRYING

Virtually all sediment samples will need to be dried at some stage (e.g., prior to crushing and grinding or to determine weights required for chemical analytical or grain size distribution calculations). Drying is most commonly achieved by placing the samples in a standard laboratory drying oven at 105°C for 24 hr, but at this temperature some exchangeable ions may become fixed to their host minerals (they are then no longer exchangeable), or some aggregates (bricks) may form that cannot be easily dispersed for later grain size analyses; there may be other physical or chemical changes. When any side effect of drying at a high temperature is likely to pose a problem in subsequent studies, it is better to dry for longer at a lower temperature. Different laboratories have their own preferences for low-temperature drying, but all operate in the range of 40–60°C; samples should be kept at this temperature for between 48 hr and 1 week. Low-temperature drying ovens must have fan-forced circulation and a vent to allow water vapor to escape. It is also important to ensure that the samples will not be exposed to direct radiation from the heating element, because then they may be affected by short-term exposure to much higher temperatures than will be registered by the oven thermostat. If the oven does have an exposed heating element, place a metal tray between the element and any samples.

EXTRACTION OF MICROFOSSILS

General

Plant and animal microfossils may comprise substantive quantities of the sand or mud fraction of sediments, and can be important indicators of paleoenvironment (even though the vast majority have been transported from their position in life); in ancient sediments, they are the predominant indicators of the time of sedimentation.

Contamination is always the greatest problem, and maximum precautions must be taken to ensure that all equipment

Table 5-9.   Technique for the Extraction of Calcareous Microfossils from Sediment Samples

1. Disaggregate the sample in the gentlest way possible (see text discussion); some microfossils are fragile and can even be damaged by operations such as wet sieving. Simple soaking in water, or water with detergent such as "Quaternary O," is generally adequate with friable sediment. Ultrasonic disaggregation in an aqueous medium can also damage microfossils; test a subsample under various power settings to determine optimal conditions. Addition of about 1 tablespoon of washing soda, followed by boiling (and then cooling), is another favored technique that ensures dispersion and facilitates later wet sieving.

An alternative method, recommended only as a "last resort" with very-well indurated rocks, is to dissolve c. 1 cm^3 fragments of the sample in hydrofluoric acid (HF). *Use extreme caution!* The reaction commonly is vigorous; therefore, set the polyethylene beaker in a cold water bath. *Gradually* fill with 5% HF, decant when reaction ceases, and renew acid until dissolution is complete; note that other compounds will precipitate during this process. *Neutralize used HF with washing soda before disposal!* Wash sample with hot water, then repeatedly with cold water.

2. Wet sieve through 40 mesh (0.425 mm) *and/or* 120 mesh (0.125 mm) *and* 200 mesh (0.075 mm) sieves (or sieves close to these values, as long as the bottom sieve is no larger than 200 mesh). Ensure samples are sufficiently small so that clogging of the mesh does not occur; if the sieve fills and overflows, delicate fossils that can float may be lost. Drainage can be enhanced by tapping with fingers on the edge of the sieve. As a final step, stack the sieves and flow water through remaining sample until outflow is clear.

3. Air-dry fractions in the sieves; a photographic drying cupboard or infrared lamp may be used to speed up drying, but caking can occur if there is abundant clay present and brick is difficult to process further.

4. Remove dry sediment from the sieves by tapping onto large clean sheet of paper, then inverting sieve onto the paper and rapping the sieve *evenly* on the benchtop. Fold paper into funnel and pour into labeled jar. Bottle each fraction separately.

5. If necessary, concentrate the microfossil fraction either: (a) by repeated passes through a magnetic separator (Chapter 8; examine each fraction to ensure that filled microfossils do not separate into the discard fraction!); or (b) by flotation in carbon tetrachloride (CCl$_4$, S.G. = 1.59), using a device and procedure similar to that described for heavy mineral separations (Chapter 8). The float caught in filter paper will have most microfossils; dry and wash.

Caution: CCl$_4$ **is dangerous and volatile; cap bottles when not in use and carry out operations in a fume cupboard.**

6. Use *clean* microsplitter(s) to obtain a working sample for picking.

7. Send sample to the zoologist/paleontologist; they prefer to see the whole separate! (Or pour sample into tray, examine under binocular microscope, and pick fossils with fine wet brush; place each specimen carefully into special biological tray that has a water-soluble adhesive coating and outlined squares.)

Source: Method of K. M. Swanson, University of Canterbury.

Note: Hydrogen peroxide (H$_2$O$_2$) is a useful reagent for removing organic matter from calcareous fossils (both macro and micro) but when purchased the peroxide is acid-stabilized (usually sulfuric or phosphoric acid) and may dissolve or damage small fossils and delicate ornamentation on larger fossils. This problem can be overcome by raising the pH of the H$_2$O$_2$ to between 8.2 and 8.5 by adding sodium hydroxide (NaOH) before use (McConchie and Harriott 1992); because this procedure destabilizes the peroxide and reactions are more vigorous, more dilute peroxide solutions can be used. An alternative procedure for cleaning organic matter from calcareous fossils is to soak them in a solution of sodium hypochlorite (common bleach).

(and users thereof!) are clean. Sieves are a particularly likely source of contamination: brush all diagonally to the mesh, and apply water jets from the underside on those used for wet sieving. A wise procedure when extracting calcareous microfossils is to soak the mesh in methyl blue dye in an intermediate stage of the cleaning process; any calcareous contaminants in the final sample will show a deep blue color. *Ensure that the sieves are thoroughly washed before subsequent sieving, because any dye left in the mesh could then stain the sample!* Another underrated problem is labeling—confusion between samples is easy in a laboratory where many samples are being treated at the same time. Ensure that containers at all levels of the operation are clearly and consistently labeled. Table 5-9 lists a procedure for extraction of calcareous microfossils.

Many siliceous microfossils are small (diatoms range between 5 and 500 μm) and care is required to avoid loss of sample material during processing. Because siliceous microfossils (particularly diatoms) can be found in a wide variety of rock types, no one method for extraction is generally applicable, but fortunately these fossils are resistant to most standard chemical treatments (except HF). Possible chemical treatments include most of those described in the disaggregation and dispersion section of this chapter, but the best combination to be used for any particular sample will largely need to be determined by experimenting with alternatives. If using detergents containing abrasives, be particularly wary of contamination because some abrasive detergents contain diatomite! One widely applicable extraction procedure is listed in Table 5-10.

Table 5-10.   Technique for the Extraction of Siliceous Microfossils from Sediment Samples

1. Add about 1–2 g of dry sediment to 100–200 mL of 20% HCl and heat gently for few hours without boiling; if the sample is heavily stained by iron oxides or iron hydroxides, add about 0.5 g of ascorbic acid.

2. Decant and rinse five to seven times in distilled water at 2-hour intervals to eliminate HCl residues, salts, and the clayey fraction. The rinsing-decanting process is tedious, but it has the advantage over centrifugal methods of avoiding frustule breakdown.

3. Add 100–200 mL of hydrogen peroxide (try dilute peroxide first if organic matter is abundant) and, when frothing subsides, heat to boiling point for an hour to eliminate the organic matter.

4. Decant and rinse five to seven times in distilled water at two hour intervals to eliminate HCl residues, salts, and the clayey fraction. If the H$_2$O$_2$ treatment has not completely removed the organic matter, the sample should be retreated with stronger peroxide.

5. If the clay content is still high, try deflocculation with sodium hexametaphosphate and a few drops of ammonia; if this treatment is unsuccessful it may be necessary to boil the sample for 15 min in a 20% sodium hydroxide solution. Repeat step 2 if either of these treatments has been used.

6. Use a smear mount to check the extract for the presence of siliceous microfossils, then transfer all fossils with a little distilled water to a labeled vial and store for subsequent study.

Source: Method of K. M. Swanson, University of Canterbury.

Pollen

Special procedures are necessary with this group of micro-fossils. Contamination problems involving air, water, and the chemicals used for processing are most acute! Stirring rods should be thoroughly cleansed (e.g., by flaming) between steps in the procedure, samples should always be sealed, and caps of reagent bottles should always be placed on fresh clean paper. A clean room is generally necessary; do not attempt extractions in a general-purpose laboratory. All operations should be in a fume cupboard. Description is given in Table 5-11 for unconsolidated sediment samples; peat/coal and rock samples require other treatment (e.g., Dettman 1963; Barss and Williams 1973; Phipps and Playford 1984).

Table 5-11.   Method for the Extraction of Pollen Microfossils (Palynomorphs)

1. Disaggregate and disperse as (1) in Table 5-10, but ensure gentle treatments and avoid use of H_2O_2, which will damage the palynomorphs.

2. Wet sieve at 100 mesh (0.15 mm); wash finer fraction into 50 mL polypropylene centrifuge tube, centrifuge *(ensure equal balance of tube weights in opposite positions in centrifuge—even small differences can cause major problems for the centrifuge!),* and decant liquid.

3. Add 10 mL 10% KOH and heat in water bath for about 10 min to denature or mobilize unwanted organic matter. Stir continuously with polypropylene rod.

4. Fill remainder of tube with distilled water and centrifuge; repeat until supernatant liquid is clear.

5. Acidify with 10% HCl, centrifuge, and decant. This step neutralizes residual KOH and removes calcium carbonate, which would combine with HF in the next step to form CaF_2, which creates an insoluble clog.

6. Add 10–15 mL HF *(caution!),* cover with plastic dish, and heat in boiling water bath for 3–4 hours.

7. Add 5 mL 10% HCl to prevent precipitation of CaF_2 from any residual $CaCO_3$ that may have been released during the HF treatment; cool.

8. Fill remainder of tube with 10% HCl to ensure an acid solution, stir, centrifuge, and decant liquid. *Neutralize supernatant liquid with washing soda before disposal.*

9. Wash and centrifuge repeatedly to remove all traces of HF (test pH).

10. Add a few mL distilled water to disperse the residue thoroughly; withdraw a small amount of the suspension with a disposable micropipette, and add to a drop of distilled water on a microscope slide. Emplace cover glass and view with microscope.

Even though the sample has been treated with HF, some mineral content commonly remains (e.g., zircons and other resistant species); particularly if the concentration of palynomorphs is low, the residual mineral content can be reduced at this stage by floating off the light fraction and other treatments:

   (a) Wash sample with 10% HCl, centrifuge, and decant. If the solution is not acidic, during the next step a gel will form (irreversibly) and further extraction is impossible.
   (b) Add 15 mL of zinc bromide solution. (Solution is prepared by dissolving 250 g 98% $ZnBr_2$ in slightly less than 100 mL 10% HCl; measure specific gravity with S.G. bottle or glass disks and adjust to 2.0 ±0.05 by adding more acid. Use Pyrex beakers because the reaction is exothermic. Filter and store in well-stoppered bottle.) Stir for full suspension of sediment and centrifuge for 5 min at 2000 rpm.
   (c) Pipette off, or decant, the light fraction; alternatively the bottom two-thirds of the tube can be frozen with liquid nitrogen and the supernatant liquid poured off. Thoroughly wash light fraction with 10% HCl, centrifuge, and decant. Discard heavy fractions and associated $ZnBr_2$ solution.

11. Add 5 mL glacial acetic acid, stir, centrifuge, and decant.

12. *Optional:* Add 10 mL solution of acetic anhydride and concentrated sulfuric acid in the proportion 9:1, adding the acid slowly because the reaction can be vigorous. Stir and heat in boiling water bath for 3 min. Centrifuge, decant; flood with glacial acetic acid, stir, and centrifuge again. Decant glacial acetic acid and wash sample residue thoroughly with distilled water at least three times.

13. *Optional:* Remove excessive organic matter by treating with concentrated nitric acid. *Caution!* Wash thoroughly and centrifuge.

14. Stain residue with basic fuschin and mount on slide with gelatin adhesive (see Table 5-13). Seal edges of cover slip by painting with clear nail polish.

Source: Method of K. M. Swanson, University of Canterbury.

MOUNTING LOOSE GRAINS FOR MICROSCOPY

Once grains are dispersed and clean, most geologists will wish to study grain textures and identify mineral components under a microscope. Several procedures for mounting grains

Table 5-12.   Procedures for Mounting Loose Sand Grains on Slides for Microscope Study

1. Disaggregate and clean the grains; sieve them into the fractions to be studied.

2. Warm Lakeside Cement to 88–99°C and apply a thin coating to a microscope slide over an area about 1.5 times the width of the cover glass.

3. Cool the slide, and then reheat to 101°C to eliminate bubbles.

4. Scatter grains evenly across the cement; take care that a representative suite of grains is sampled (see earlier discussion and Chapter 4). If grains are to be studied after this stage, ensure their upper surfaces are not covered by the cement. Cool the slide.

If grain thicknesses are too uneven for easy study, or if a thin section is to be prepared:

5. (a) Grind the mounted grains with 400- and then 600-grade carborundum paste on a glass plate (ensure finer pastes are on plates never used for coarser abrasives).

(b) Reheat, apply a thin layer of cement, and carefully press down a coverslip (using, for instance, the end of a pencil eraser) until bubbles are eliminated and excess cement is forced out around the edges (creating an air seal). (Application of a coverslip after step 4 is difficult because of the common differences in thickness of particles.)

6. Cool and label the slide.

If etching and staining procedures are desired to assist identification of minerals (e.g., Chapter 8), etch/stain before 5b (a coverslip may not be desirable, depending on the brittle state of the stain coatings).

or

2. Apply a very thin coating of Canada Balsam to the slide (one drop spread over half the area of the slide).

3. Heat 3–4 min until the balsam is cooked.

4. Cool 1–3 min, and then spread the grains evenly on the slide.

5. Place the slide on a hot plate at 65°C and remove it when the grains begin to sink (no grain should be covered by balsam).

6. Cool and label. (Excess balsam can be removed with alcohol or acetone.)

or

2. Prepare an epoxy compound, and allow it to stand until bubbles disappear.

3. Spread liquid over part of the glass slide (if spread too thinly, the epoxy will blister when heated).

4. Dry for 12 hr at 70–80°C.

5. Mix more epoxy (for <3φ grains, use one part plastic to three parts thinner; use less thinner with larger grains).

6. Using a fine brush, paint the liquid across the hardened epoxy, then sprinkle grains on the slide, invert the slide to remove loose grains, and dry it for 12 hr at 90°C.

7. Cool and label.

for such studies are outlined below, because requirements vary depending on the materials and requirements of the workers (also see Chapter 7 for special techniques used to prepare grain mounts for textural analysis by SEM).

The first step in many routine analytical procedures, particularly for cores of loose sediment, is to prepare a *smear slide* by scraping a small amount of sediment in its natural state and smearing it on a previously labeled glass slide. The sediment is thinned with a wet brush and dried; then a drop of liquid adhesive (e.g., Canada Balsam) is applied and covered with a coverslip; the slide is heated (at 80°C for balsam) until the edges are hardened. The slide can then be examined with a binocular or petrographic microscope, either qualitatively or quantitatively with point counting techniques (see later discussion). Enhancements are desirable if adequate time is available, grains have coatings, or more precise identification of components is necessary.

Table 5-12 lists common procedures used in preparing grains for microscope studies. These methods at the most only grind one side of grains and leave uneven thicknesses of rounded grains for viewing, and most geologists are used to examining thin sections where parallel-sided slabs have been cut. If the convenience warrants the extra work involved, a thick slurry of grains and cold-setting epoxy resin can be prepared, let set into a block, and then treated as solid rock specimen (alternatively, loose grains may be tamped with cement in a vial, then cut after solidification—see Middleton and Kraus 1980). If a temporary mount is desired for immersion oil methods of mineral identification (Chapter 8), gelatin mounts can be prepared (see Table 5-13).

Table 5-13.   Two Methods of Mounting Grains in Gelatin for Study with Immersion Oil Techniques for Mineral Identification

1. Prepare solution A: 0.1 gelatin solution (0.25 g/50 mL H$_2$O). Prepare solution B: 10 mL distilled water, 5 mL acetone, 2 mL 2% formalin. Cork both well to avoid evaporation.

2. Place drops of solution A on a glass slide; dry at 80°C. (For large grains, use more gelatin.)

3. Place one or two drops of solution B on the gelatin film; spread the grains evenly on the soft surface with a needle. Dry at 80°C.

4. Place refractive index oils on the slide as necessary; between oil applications remove the oils with acetone (xylol, alcohol, or carbon tetrachloride—*caution!*) and dry.

An alternative gel preparation that is used to mount pollen is as follows:

1. Boil 19 mL distilled water with 33 g glycerine. Do not overboil—excessive evaporation will alter the character of the gel.

2. Add 7 g gelatin *very slowly* and stir. Finger tap the gelatin from a spatula, ensuring an even covering on the water-glycerine mix.

3. Add 1 g phenol. Let cool.

4. To use, heat gel in water bath at 80°C, then scatter grains on the surface.

Source: The first method is after Marshall and Jeffries (1946), the second from K. Swanson, University of Canterbury.

Note: See Chapter 8.

IMPREGNATION PROCEDURES

Samples of natural sediments collected in blocks or of friable sedimentary rocks may be artificially cemented preparatory to specialized studies of slabs or thin sections. The impregnation resin can be stained to enhance the resolution of porosity (see discussion of pore stains in this chapter).

Comprehensive reviews of impregnating procedures and materials can be found in Bouma (1969) and Stanley (in Carver 1971). The fundamental laboratory requirements are a vacuum chamber (low pressure; c. 10–15 mm Hg) and a nonreactive (to prevent grain corrosion), isotropic (for viewing under the microscope), low-viscosity cement (for minimizing air bubbles as well as maximum penetration) that cures at relatively low temperatures (to avoid changing temperature-sensitive sediment properties). Many cementing materials are available and experimentation may be necessary to discover the best for a particular sediment (e.g., Franklin 1969; Minoura and Lonley 1971; Crevello, Rine, and Lonesky 1981; Conway 1982; Fitzpatrick 1984; Palmer and Barton 1986; Miller in Tucker 1988). Awadallah (1991) illustrates an ingenious method utilizing plasticine molds to hold and impregnate multiple small samples; Jordan and Roady (1987) describe a different multisample technique for well cuttings. A water-miscible resin and/or special techniques are necessary if the sediment is moist (e.g., Chiou et al. 1983; Jim 1985); alternatively, the sample may be dried before placing in resin either by vacuum desiccation, air-drying, or treating with progressively more concentrated ethanol or acetone and water mixtures, then pure acetone (*caution*). Low viscosity is necessary for greatest penetration and can be achieved with most resins by mixing them in a high-volatile solvent. The cement should polymerize slowly so that as it solidifies it will not expand and shrink to develop cracks and strains or produce a birefringent solid; it should also have a refractive index of about 1.54. Naturally, a strong final product is essential. (Note that carborundum powder used in grinding processes will be collected by the resin—do not be fooled by those clusters of little black grains seen under the microscope in the intergranular pores!)

EMBEDDING PROCEDURES

Instead of impregnating samples of unconsolidated sediment—and particularly if the necessary preliminary desiccation procedure causes excessive cracking—the entire sample while still moist may be preserved by embedding it in a clear medium such as Plexiglas, Lucite, or epoxy resin. Bouma (1969) provides a thorough discussion of the various procedures. A wide range of commercial colorless unsaturated polyester resins suitable for embedding exists, and most of the solvents used to accelerate the hardening process are at least mildly dangerous, so procedures should be carried out in a fume cupboard (some preaccelerated resins can be obtained that do not require the storage of separate solvents). Heating may or may not be required, depending on the compounds used.

Molds can be made with glass, wood, metal, or plaster of paris, with edges sealed with silicone rubber compounds, and if rough/porous surfaces are involved (e.g., most woods and all plasters), with a coating of paraffin, petroleum jelly, or other pore sealants; in all cases a coating of mold-release agent must be applied immediately prior to embedding. Initially a base layer is generally prepared (and solidified), coated with a resin layer that will act as a glue, then the mold-with-base-layer inverted and placed on top of the sample. The sample must be prepared with care to avoid loose grains or smeared structures, then sprayed several times with a clear plastic aerosol (e.g., Krylon™) to seal the surface and prevent loss of moisture; impregnation with preaccelerated resin under vacuum may be necessary with some highly porous samples to avoid excessive air bubbles. The mold-plus-sample is then inverted, and depending on the sample thickness, multiple thin layers of bubble-free resin (mix carefully and allow to stand before use) should be applied around the sides (never on top—any air in the sample must be allowed to escape) to avoid excessive heat buildup and consequent sample cracking. Pour steadily in one position to minimize air bubbles. If possible, apply vacuum for a few minutes after each pouring to remove bubbles. *Take care to cover the mold at all stages between applications to prevent dust accumulation.* The top of virtually all samples will become irregular during the final curing process; grinding and polishing (finishing with a felt polishing disk with aluminum, tin, or magnesium oxide abrasive solutions . . . or even Brasso™) or a special preparation for a final top layer (see Bouma 1969) are necessary for maximum effect. Experimentation is generally necessary before the optimal procedure is found for the various sediments.

THIN-SECTION PREPARATION

Thin sections of rock or impregnated sediment (c. 0.03 mm thick) are routinely studied by geologists and pedologists under the transmitted-light polarizing petrographic microscope (see Chapter 8). Preparation procedures differ depending on the equipment available and sediment characteristics (e.g., Reed and Merguer 1953; Allman and Lawrence 1972, Hutchison 1974; Miller in Tucker 1988). Semi-automatic machines speed up and simplify thin-section manufacture, but are expensive and have a few drawbacks. For example, the glass slides on which samples are mounted must be ground perfectly flat, which results in frosting (also necessary for bonding with the epoxy cements used by these machines) that under the microscope gives a speckled effect and detracts from the clarity of the view. Nonetheless, these machines are necessary for the high-precision lapping and polishing of double-polished sections in which maximum detail can be resolved (see also Nentwich and Yole 1991 for additional suggestions on the preparation of these high-quality slides). Table 5-14 lists a procedure for a laboratory equipped with the basic diamond saw, lap, and hot plate.

Table 5-14. Procedure for Making a Thin Section from a Rock or Block of Embedded Sediment

Ensure that initially the specimens, and later the glass slides, are labeled so that there is no chance of confusion between the multiple samples usually present in the thin-section preparation room! Also ensure that the thin-sectioning room and equipment are kept clean; a grain of coarse grinding powder on a fine lap can result in substantial scratches on sections, and dust can be trapped between the section and the glass slide or coverslip.

1. Cut the sample with a diamond saw to produce a 3-mm-thick slice with parallel sides. Cut slowly to minimize the risk of shattering and blade chatter.

2. Grind the slice on one side, on a coarse grinding lap using 180-grade and then 320-grade carborundum powder with water as a lubricant to remove saw marks. Alternatively, a diamond lap may be used. If speed is adjustable, use a low speed (c. 30 rpm) for maximum control and to keep the slices cool. Move the rock slice across the full width of the lap during grinding to avoid uneven wear of the lap. Rinse thoroughly after each grinding operation, and avoid grinding with a finer abrasive on a lap used previously with a coarser one (an old toothbrush will help to remove grains of abrasive caught in microfractures or pores of the section, but even that treatment may not remove all coarser grains from the lap).

3. Continue grinding on a glass plate using 400-grade carborundum powder with water, and after rinsing, on a separate plate with 600-grade paste. Use a rotary action to produce a polished flat surface.

4. Scrub with running water to remove waste carborundum and then dry the slice on a hot plate. Examination with a binocular microscope of the wet surface is often worthwhile to see textural relationships and any sedimentary structures that may be present.

5. Mount the slice on a clean microscope slide (prelabeled with a diamond scribe) with "Lakeside 70" thermoplastic cement (or thermosetting epoxy cements or ultraviolet-curing adhesives), using a hot plate set at 100°C. Apply the cement to the polished face of the hot rock slice and press the slide down until all excess cement and any bubbles are squeezed out.

6. Grind the upper surface of the rock slice either on a diamond lap or on a coarse grinding lap using 180- and then 320-grade carborundum with water. Continue until you can see light through the rock.

7. Grind the surface on a glass plate using 400- and then another with 600-grade carborundum paste (do not use the same plate with different paste sizes). Use a rotary action and keep the surface parallel to the surface of the microscope slide. Check the thickness regularly, using a polarizing microscope under crossed polars (CPL, see Chapter 8), until a final thickness of 0.03 mm is attained (quartz will show a pale, first-order gray color CPL).

8. Wash the thin slice and then dry it thoroughly. Coat it with cellulose acetate to prevent it from breaking up, and trim it to a convenient size with a razor blade.

9. Heat the thin rock slice on a hot plate set at 85°C, then transfer it from the old slide to a new microscope slide with Canada Balsam on it (*Note:* Such transfers cannot be attempted if epoxy adhesives are used). The balsam should be heated initially for about 3 min or until a bead, picked up on the points of tweezers, forms a brittle thread that snaps upon cooling when the tweezers are opened. Place more Canada Balsam on the upper surface of the thin slice, and then cover it with a coverslip. Then clean the completed thin section with alcohol (or acetone—caution!). Store in warm place for 24 hours to cure the balsam.

10. Label the specimen both by scratching the slide with a diamond or carborundum pen and with a paper label (paper labels and coverslips cannot be used if the section is to be subjected to cathodoluminescence, scanning electron microscopy, or microprobe analysis).

Notes: Canada Balsam is expensive, Lakeside Cement less so but nonetheless not cheap, and some laboratories use other and cheaper mounting media, particularly various epoxy resins (various types and names). *(Solvents for these resins are generally xylol or toluene, both of which emit potentially toxic fumes, hence all work should be carried out in fume cupboards or in well-ventilated areas.)* Ultraviolet-curing adhesives also have been found to be useful in thin section preparation (e.g., Yangual and Paxton 1986). If another medium is used, mounting methods may differ from those stated above. Before standardizing on any other medium, ensure that its refractive index is known and lies very close to the standard 1.54 (the RI of quartz) before analysis with the petrographic microscope; otherwise relative relief studies are greatly restricted.

Unconsolidated but coherent materials may be cut with a hacksaw, spatula, or piano wire and initially ground on a paper-backed abrasive like emery paper. Water-soluble rocks and those with swelling components require kerosene or glycol to be used as a lubricant when grinding instead of water, or they may be ground dry (e.g., Martin, Litz, and Huff 1979; Socci 1980).

Initial lapping pastes must be much finer where soft minerals such as evaporites are to be sectioned.

POLISHED-SECTION PREPARATION

Polished sections of samples are required for many scanning electron microscopy or electron microprobe analyses, as well as for samples to be examined by reflected light microscopy (see Chapter 8). Opaque minerals, such as those that comprise most of the economic metal ores, and coal require polished sections for routine analyses. Polished thin sections are particularly useful if the sample contains both opaque minerals requiring examination by reflected light microscopy and nonopaque minerals requiring examination with a standard petrographic microscope, or if a sample is to be examined with a standard petrographic microscope and by electron microprobe.

To prepare a polished thin section, proceed as for the preparation of a normal thin section up to step 8 in Table 5-14, but stop just before reaching the final 0.03-mm target thickness (i.e., when quartz has a very pale first-order yellow), then continue as follows:

8. Wash the section thoroughly and inspect it with a binocular microscope to ensure that no grains of grinding powder remain lodged in fractures or in the mounting adhesive.
9. Polish the section with 1 µm diamond paste on a polishing lap used only with this grade of abrasive and continue until a final thickness of 0.03 mm is achieved. Inspect the section regularly to ensure that no scratches are appearing that might indicate contamination of the lap by coarse abrasive (transferred from an operator's fingernails perhaps?) or by a hard grain plucked from the section during polishing.
10. Clean the section thoroughly and carry out final polishing with 0.25 µm diamond paste on a polishing lap used only with this grade of abrasive. Check the section regularly to ensure that no scratches are developing and that the frosting associated with the 1 µm polishing is disappearing.

Polished sections also are commonly prepared using a piece of the rock sample embedded in a solid plastic block of an appropriate size to fit in the sample holder of the electron microscope or probe to be used in subsequent studies. To prepare these polished sections, proceed as shown in Table 5-15.

Polished Sections of Coal

Preparation of polished coal specimens differs significantly from that used for rocks and minerals. Coal samples are first encapsulated in synthetic resin (thermoplastic, thermosetting, or cold setting), either as blocks to reveal gross maceral structure, or as ground particles statistically representative of the sampled material. Block preparation may require the construction of special molds from synthetic rubber, metal foil, etc., to accommodate the size and shape of the surface to be polished. Removal of air bubbles and prevention of differential settling within the unhardened resin are particularly important. Small samples may be conveniently mounted in either thermosetting or thermoplastic resin by using a heat press, but if this is unavailable, or if high temperatures are to be avoided, cold-setting resins are satisfactory. If differential settling cannot be prevented, a polished section must be prepared parallel to the settling direction and the analyses performed in full traverses parallel to this direction.

Coal polishing may be carried out either by hand on a rotating lap, or by machine wherein the mounts are clamped to a turret. In both cases, the coal mounts are first wet-ground on a series of silicon carbide papers, usually ending with grit

Table 5-15. Procedure for Making a Polished Section from Rock or a Block of Embedded Sediment

1. Select a section of metal tube (or similar means of confining liquid plastic resin) about 1 cm long and with an inside diameter roughly equal to the outside diameter required for the mounted specimen. Smear the inside of the ring lightly with silicone grease to prevent the plastic from sticking to the ring. Stretch a piece of "mylar" (or similar plastic film) over one end of the ring and tape it to the outside of the ring to keep it in position.

2. Cut the rock sample with a diamond saw for a flat surface where the final section is desired, and cut away excess pieces of rock so that it will fit into the ring with the flat surface against the covered end. Clean the rock sample thoroughly and place it on a hot plate until dry.

3. Place the ring, covered end down, on a flat surface (because many plastics may generate heat when setting, use a surface that will not be damaged if it gets a little hot). Place the specimen in the ring with the flat side down, and slowly pour in the appropriate mixture of plastic and accelerator until the sample is covered. Allow the plastic to set (the time will depend on the type of plastic used).

4. Carefully remove the mounted sample from the ring, wipe off any grease adhering to the sides, then grind the section surface with 320-grade carborundum powder or an appropriate diamond lap. Clean the specimen thoroughly after each grinding operation.

5. Cut the plastic from the back of the mounted section (a parallel diamond saw and some light grinding will do nicely) so that the thickness of the section is compatible with the sample holder in any instrument to be used to examine it. Ensure that the section surface and the back of the section are as nearly parallel as possible.

6. Grind the section surface first on 400-grade then on 600-grade carborundum powder, using a rotary action, until the surface of the rock is flat and free from saw marks or other deep scratches. Scrub the sample thoroughly after using the 400-grade abrasive and again after using the 600-grade abrasive.

7. Polish the section surface as in steps 8–10 for preparation of polished thin sections.

Notes: Although diamond polishing paste is ideal for most samples, it is not the best polishing material for all minerals (particularly many soft minerals), and it may be necessary to try alternative abrasives such as tin oxide, alumina, chromic oxide or colloidal silica (e.g., Teague 1989) at step 10.

Because a polish can be rapidly ruined by even a single grain of coarse abrasive, absolute cleanliness is essential at all stages and the work should be carried out in a room dedicated to polishing (i.e., do not just set up polishing operations in a quiet corner of a laboratory).

Preparing polished sections is an art that will benefit greatly from experience and patience.

600. Hand polishing methods commonly use a "soft" or napped cloth lap and thick slurries of alumina, chrome oxide, or magnesium oxide. Machine polishing may use similar laps and polishing agents, or alternatively, "hard" or napless synthetic cloths and thin slurries or embedded diamond powder. In all cases, at least two polishing stages are used. Colloidal silica is increasingly popular as a final polishing agent.

Many different in-house methods of preparing polished coal specimens have been developed at various research centers, possibly due to the wide range of polishing behavior exhibited by coals from different deposits (e.g., ICCP 1971; Stach et al. 1982; Ward 1984; and the standard ASTM D2797-72). Coal samples are mounted as either blocks, allowing sectioning perpendicular to bedding and examination of gross maceral structures, or grains, prepared to represent a much larger sample such as a whole-seam intersection or a stockpile.

Coal blocks can be of any size or shape, limited only by the capacity of the polishing apparatus. Molds for blocks may be individually constructed from cardboard, metal foil, or synthetic rubber, but the use of irregular size molds may prevent automatic polishing. Preparation of grain mounts, as specified by standard methods, requires grinding of a representative sample to 1 mm (maximum) with minimum overbreakage, followed by embedding and preparation of a polished surface so as to minimize the effect of gravitational separation of coal components. Differential settling is largely overcome by mixing a viscous coal/resin slurry; remove air bubbles either by vacuum or by repeated compression/decompression cycles in a press. If these techniques cannot be applied, a low-viscosity (resin-rich) mixture must be prepared and air bubbles must be allowed to rise naturally. This method may require sectioning of the mount parallel to the settling direction, and analytical scans over the full "depth" of the mount. The shape of such low-viscosity mounts may preclude automatic polishing. Encapsulated coal mounts are usually prepared for polishing by wet grinding on a series of SiC papers, commonly ending with grit 600 (P1200).

Polishing procedures can be generally subdivided into two classes: "soft" napped laps and thick polishing slurries, and low-nap or nap-free laps, normally used with thin slurries or embedded polishing agents. Methods of the first class produce rapid removal of material and are best suited to manual polishing in which the coal mount is hand-held on a rotary wheel. These methods often produce a high-relief polish that is not desirable for high-power microscopy. Typical arrangements employ a rough polish stage with OC alumina or chrome oxide slurries, followed by a final polish with fine alumina or magnesium oxide slurries. The second class of procedures is more ideal for automated polishing, using a variety of polishing agents including alumina, cerium oxide, diamond, and colloidal silica. Optimum wheel speeds, specimen pressures, and polishing times may have to be determined by experiment, because of the many lap/polishing agent combinations possible. Some arrangements may prove more applicable for coals of a particular character. The method listed in Table 5-16 is suited to automatic polishing machines in which the mounts are fixed (with clamps or grub screws) to a rotating head. Excellent results are obtained for a wide variety of coals.

PORE STAINS

In many thin-section studies, particularly for work associated with the petroleum industry, examination of the nature and extent of porosity in the sediment may be as important as (or more important than) examination of the mineral components in the sample. However, microporosity in thin sections is difficult to examine in plane polarized light, and under crossed polars the presence of grains in extinction make estimation of porosity difficult; pores that are smaller than the

Table 5-16.   Procedure for Making a Polished Section from Coal Mounts on an Automatic Polishing Machine

1. Pregrind at grit 60–240 to level and flatten specimens. If too much material is removed to allow further grinding, the specimens must be remounted, and briefly reground to ensure level surfaces.

2. Grind successively on 400, 600, and 1200 grit papers for 1–3 min, using a wheel speed of 150 rpm and 0.2–0.6 kg·cm² pressure. Use abundant water feed. "Soft" bituminous coals may require only .05 minutes on the final paper, which may otherwise clog with smeared coally material.

3. Wash holder and specimens between stages, particularly when commencing the polishing step (4). Use of an ultrasonic cleaner may help remove trapped abrasive grains.

4. Polish for 2–3 min on a napless synthetic cloth (e.g., "Pellon"™ cloths—Struers, Ltd; Pan-W™ or Pan-K™—Leco Corporation; Texmet™—Buehler Ltd; Hyprocel Pad-K™—Engis Ltd) lightly charged with 1 μm diamond in aqueous slurry. Use 150 rpm 0.2–0.3 kg/cm² pressure, and approximately 1 drop water feed/second, after the first 15 seconds of polishing. Flush liberally to clean the lap during the last 15 seconds. This step may be repeated for coals, which are resistant to polishing.

5. Final polish for 2–3 min using the same lap type as (4), or a napless silk or nylon cloth that is first dampened thoroughly, then charged with colloidal silica (10 mL on an 8-in. wheel is adequate). Use 150 rpm, 0.05–0.1 kg/cm² pressure, and a water feed of around 1 drop/second commencing after 15 seconds. Flush liberally at the end of the polishing time as above. Repeat this step if fine scratches remain.

thickness of the thin section are even more difficult to examine. Use of colored and/or fluorescent pore stains (e.g., Yanguas and Dravis 1985; Ruzyla and Jezek 1987) makes it easy to examine the pores under plain polarized light or under UV illumination (fluorescent dyes), and to obtain rapid estimations of porosity. The basic pore-staining procedure involves using standard impregnation techniques (see above) to fill the pores with an epoxy resin that has been mixed with a colored fluorescent dye; nonfluorescent dyes can be used, but they will not allow UV fluorescent imaging of the pores. Fluorescent dyes can also be added to sections that have been previously impregnated; Ruzyla and Jezek (1987) found the following dyes to be successful with "Araldite 506" epoxy resin: "Poly Supra Royal Blue," nonfluorescent blue; "Poly Supra Hi Brite Yellow Concentrate," fluorescent yellow; "Poly Supra Hi Brite Lime Green Concentrate," fluorescent green; "Poly Supra Hi Brite Yellow Orange Concentrate," fluorescent yellow orange; "Poly Supra Hi Brite Red Concentrate," fluorescent red.

STREAK PRINTS

The streak (the color of the finely powdered mineral) produced by minerals is one of the well-established techniques used in mineral identification and it can be used to obtain a permanent 1:1 record (a streak print) of the mineralogy and texture of flat surfaces of rock samples that contain a high proportion of minerals that produce a distinctive streak

Table 5-17.   Procedure for Preparing a Streak Print for Metallic Minerals in Sedimentary Rock Samples

1. Grind a smooth surface on the core or rock to be printed; the surface need not be flat (the curved surface of a core can be used) but it must be smooth (i.e., all marks from saws or drilling bits should be removed).

2. Wash the core or rock sample to remove any dirt or coarse rock particles that remain; dry.

3. Rub the smooth dry surface briefly with waterproof abrasive paper (about 320 mesh for softer minerals such as sulfides or 220 mesh for harder rocks such as banded iron formations) and rinse the surface thoroughly in a stream of water (do not rub the surface; also avoid touching it). Allow the surface to dry by evaporation.

4. Stick transparent adhesive tape (any transparent pressure-sensitive adhesive tape will do, but the "magic" or "invisible" tape gives the best results) over the surface to be logged and press the tape down gently and evenly.

5. Carefully pull the tape off the rock and stick it onto a piece of white paper or card. You now have a permanent record of the sample at a 1:1 scale; the streak is recorded because fine particles of the minerals adhere to the tape. If required, the mineral particles adhering to the tape can be stained before the tape is stuck to the paper/card. The stain can be applied by carefully pressing the tape onto a piece of soft porous paper moistened with the staining reagents, then removing the tape and allowing it to dry before sticking it to the white paper/card.

(Morris and Ewers 1978). Streak prints are particularly useful for logging cores that have a high sulfide or metal oxide mineral content (e.g., banded iron-formation cores) and can also be stained for phosphate (Morris and Ewers 1978) or carbonates and feldspars (see procedures in Chapter 8). Well-prepared streak prints can record fine textural and mineralogical detail of the rock material, are easily photographed, and provide a means of recording very detailed core logs in a small folder. Table 5-17 lists steps in preparing a streak print.

ACETATE PEELS

Etching of polished rock surfaces in acid produces a differential relief because constituents dissolve at different rates depending on their density, crystal size, and crystal orientation. The peels mold themselves to the irregularities and form castings that can be studied with a microscope and/or used to prepare photographic contact prints (use a high-contrast paper and set the lens at f/11 or higher for maximum depth of field). The method has several advantages over thin-section studies: analysis is easier and more rapid; the cost is lower; greater surface area is available for study; and many details are generally seen as well as, or better than, in thin section. Serial sections can be prepared easily. A variation on the theme—preparation of a Plexiglas replica (Frank 1965)—provides especially fine detail for high-power microscope studies. Peels may also be made of surfaces stained for mineral identification (e.g., Katz and Friedman 1965 and see Chapter 4). Details of peel preparation vary; experimentation

is necessary for particular rock types (e.g., see Lane 1962; McCrone 1963; Bouma 1969). Best results are obtained with nonporous samples; if porosity is too great, impregnate the sample (see above). The procedure described in Table 5-18 is for limestones, with which acetate peels are most commonly used. However, peel techniques are also applicable to etched surfaces of any kind of rock, including silicate rocks, for which the technique is as described for limestones except that etching is done with hydrofluoric acid (HF) fumes or even 5–40% HF solutions (*caution!* e.g., Bjorlykke 1966). Such techniques can provide information on composition (by the difference in material dissolved) as well as crystal/grain shape and fabric, and have been applied to archeological studies of tool-making materials such as chert and petrified wood (e.g., Young and Syms 1980).

See also Chapter 4, where all citations in this section are listed.

Table 5-18.   Procedure for Preparing an Acetate Peel of a Carbonate Rock in the Laboratory

1. Polish the surface of the sample with sandpaper and emery paper or carborundum powder down to 800 grade or even finer, depending on the rock and the desired degree of detail.

2. Etch the surface uniformly for 10–45 sec with 20% vol. acetic acid or 1%, 5%, 10%, or 15% HCl, avoiding excessive effervescence, which gives unselective and uneven etches. Disodium ethylene diamine tetraacetic acid (diNaEDTA) provides an even more subtle etch and is effective with single particles. The easiest method is to dip the surface into a bowl of acid at an angle to avoid trapping of bubbles. Etching time may vary from seconds to an hour (with the weakest acids), depending on the solution and the rock.

3. If desired, apply stains for mineralogical discrimination (see Chapter 8).

4. Wash *gently* with distilled water, and dry thoroughly.

5. Arrange the etched surface to face upward and to lie horizontally (embed the base in a bowl of sand or shaped plasticine).

6. Cover the surface with acetone, allow excess fluid to evaporate, and then while the entire surface is glistening, apply acetate sheet matte surface down (medium thickness, c. 0.1 mm, is usually best). Remove electrostatic charge on the acetate sheet before application (e.g., with an antistatic gun). The sheet should overlap the etched surface by about 1 cm. It is best applied by pinching the edges inward to form a U-shape, then applying the base of the U to the center of the surface and flattening the sheet out from the center to eliminate all bubbles. Application of an even weight (such as a sandbag) over the peel for about 10 sec ensures even contact.

Caution: Acetone fumes can damage cell membranes. Do not inhale too much vapor or permit excessive contact with skin (natural oils are rapidly dissolved and dermatitis may result).

7. Allow the peel to dry fully; circulating warm air or an infrared lamp may be used. Remove the peel carefully. If it sticks, the peel is probably not yet dry.

8. Trim the acetate sheet and mount it onto glass, either by sandwiching between glass plates or by sealing the edges of the sheet to glass using a brush soaked in acetone or with invisible mending tape (ordinary transparent tape splits and becomes brittle with age).

9. Study with binocular or petrographic microscope, or insert into photographic enlarger and print the areas of interest. Do not leave specimens in the open: they attract dust and can be easily scratched.

ROCK CRUSHING AND GRINDING

Rocks that are to be chemically analyzed, by methods other than those based on electron microscopy, will generally need to be crushed and ground prior to analysis. In most laboratories, crushing is achieved using a powered jaw-crusher or a hand-held hammer, and grinding is achieved using a ring mill, or for very small samples, a vibro-mill. Physically, these devices usually perform very well, but chemically there are traps for the unwary. All crushing and grinding equipment is a source of both primary and secondary sample contamination that must be anticipated and allowed for in analytical work.

The type and degree of primary contamination will depend on the type of alloy used to construct the jaws of the crusher and the material (commonly tungsten carbide, various steel alloys, partially stabilized zirconia, or agate) used to make the grinding bowl and rings for the ring mill. Most manufacturers will provide information on what contaminants to expect from their particular products; this information should be checked before purchasing new equipment. If there is any uncertainty, it is useful to run some blanks (clean acid-washed quartz crystals are commonly used for this purpose) through the equipment and subsequent analytical procedures. (*Note:* The amount of material transferred from grinding equipment to the sample will depend on the hardness of the minerals in the sample, grinding time, particle size, and the quantity of the rock placed in the grinding head.)

Secondary contamination, or cross-contamination, results from inadequate cleaning of the crushing and grinding equipment between samples. Secondary contamination is avoided only by carefully cleaning the equipment between samples; useful aids to cleaning include a compressed air gun, a brush, acetone (dries quickly whereas water not only dries slowly but may rust metal parts), tissues, and a good vacuum cleaner to keep the sample preparation room as dust-free as possible. When grinding soft samples, such as those containing carbonates or clay minerals, patches of ground sediment may adhere to the grinding bowl or rings and become difficult to remove during normal cleaning; when this happens, the best approach is to clean the equipment as well as possible, grind some clean quartz, then clean the equipment again before grinding the next sample.

PREPARATION FOR CHEMICAL ANALYSIS

Some quantitative analytical techniques (e.g., Mössbauer, pressed powder X-ray fluorescence, X-ray diffraction, neutron-activation analysis, determination of C and S by inductive combustion and arc-emission spectrometry; see Chapters 8 and 9) can be used directly on samples that have been dried, crushed, and ground and little additional pretreatment of the powdered sample is necessary. However, even for whole rock analyses, many routine analytical techniques require one of the additional sample pretreatments (usually some form of digestion) described below. If analyses are to be carried out on specific grain size fractions (often a very informative approach) or mineral separates from the samples, the separations will need to be made before the sample material is ground.

Before selecting the analytical technique to be used, it is important to decide exactly why you need to perform the analysis and what you want to discover; only then can decisions be made on the most suitable analytical technique and whether whole rock analyses, selective extractions, or individual grain (microprobe methods) analyses will best provide the required data. For example, whole rock analyses are not of much use in most pollution studies, when the requirement is the concentration of particular elements in a bioavailable form. In a recent study of metals in river sediments (by D. McConchie), bulk sample analyses revealed local concentrations of chromium and nickel that suggested point-source anthropogenic pollution; follow-up work showed that pollution was not significant because most of the Cr and Ni were present in naturally derived fuchsite mica that selectively accumulated in low-energy parts of the river.

Preparation of Samples for X-Ray Fluorescence Analysis

Major element compositions are conventionally determined using fused glass discs formed by heating powdered sample material and a flux to 1000–1100°C in a Pt/Au or Pt/Rh/Au alloy crucible for 10–15 min (with a forced-air burner or in a furnace), pouring the fused liquid into a duralumin casting device (e.g., Kerrigan, 1971), then annealing the resulting disc at 200–250°C. Probably the most widely used procedure is that of Norrish and Hutton (1969), in which a flux consisting of $Li_2B_4O_7$ and Li_2CO_3, plus La_2O_3 (added as a heavy absorber) and $LiNO_3$ (decomposes during fusion; added as an oxidant), is combined at a flux:sample ratio of 6:1 to form a glass disc about 1 mm thick. Another fusion recipe that is gaining popularity is the low-dilution fusion procedure of Lee and McConchie (1982), which uses 1.5 g sample in 3.0 g flux (56% $Li_2B_4O_7$, 44% Li_2CO_3), with 0.2 g $LiNO_3$, and no heavy absorber, to form a disc 2.5–3.0 mm thick; in the absence of the heavy absorber, the thicker disc is needed to meet infinite thickness requirements when shorter fluorescent X-ray wavelengths (e.g., $Zr_{K\alpha}$) are examined. Other flux recipes and sample:flux ratios are described by Thomas and Haukka (1978) and Hutton and Elliott (1980). The low-dilution (e.g., sample:flux ratio of 1:2) fused discs are more difficult to prepare, but they have the advantage that both major and trace elements can be analyzed on a single disc without significant loss of accuracy, precision, or sensitivity relative to analyses that employ higher-dilution fusions for major elements and pressed powder discs for trace elements.

When high-dilution fused discs (e.g., Norrish and Hutton, 1969) are used for major element analysis, pressed powder pellets are normally used for determination of trace elements. Pressed powder pellets are prepared by tamping 6–10 g of powdered sample into a small disc (about 5 mm thick), surrounding this disc with boric acid powder in a die, and compressing both sample and boric acid into a pellet at a pressure of about 10 tons; some analysts add an organic binding agent to the pellet to increase its strength (e.g., Fabbi 1978; Harvey and Atkin 1982). Because the intensity of fluorescent X-rays from the sample will be affected by the size of particles in the sample powder and the pressure applied when forming the pellet, it is important to standardize preparation procedures. The influence of particle size and pelletization pressure becomes small at particle sizes ≤25 μm and pressures ≥10 tons.

Whole Rock Digestions

For certain geochemical analyses (see Chapter 9) it is necessary to dissolve an entire rock subsample (e.g., for trace elements bound in several minerals). There are two alternative techniques that can be used to dissolve whole rock samples: (1) digestion using strong acids, or (2) digestion by fusion with alkali salts. Acid digestion involves the use of more hazardous reagents and is more tedious to perform, but it has the advantage that it does not involve addition of extra salts to the digest. Digestions involving strong acids can be carried out in Teflon or Pt crucibles, but fusions with alkali salts must be carried out in Pt or Pt-alloy crucibles.

Slightly different acid digestion procedures are employed in different laboratories, but a typical procedure would be:

1. Place about 0.2 g (weighed accurately) of finely ground sample in a 100-mL Teflon or platinum crucible and add about 1 mL of concentrated nitric acid or aqua regia.
2. Add 5 mL HF (40%)—caution!—and heat on a water bath for 30 min.
3. If a clear solution is obtained, add 50 mL saturated boric acid solution (the boric acid will neutralize any remaining HF), and dilute to 100 mL. If a precipitate remained after step 2, add the boric acid solution, then reheat in a closed-reaction vessel for 30 min before diluting to 100 mL.

Unfortunately, several minerals (e.g., zircon, chromite, tourmaline, rutile, and other heavy minerals) are resistant to this digestion procedure. Some resistant minerals can be dissolved by heating the digestion mixture for several hours, but better results can be obtained by heating the mixture in a Teflon bomb to around 150°C (e.g., Bernas 1968; Van Eenbergen and Bruninx 1978). Alternative digestion procedures can be employed when silicon does not need to be retained in the digest.

A variety of alkali salts can be used to achieve successful rock decomposition by fusion with flux (e.g., see Ingamells, 1970; Cremer and Schlocker 1976), but the favored fluxes are lithium metaborate (LiBO$_3$; melting point 845°C), lithium tetraborate (Li$_2$B$_4$O$_7$; melting point 930°C), and sodium carbonate (Na$_2$CO$_3$; melting point 851°C). Other fluxes and combinations of fluxes are widely used. Digestions usually involve sample:flux ratios of 1:3 to 1:7 and are carried out in a Pt/Au or Pt/Rh/Au alloy crucible by heating for 10–15 min with a forced-air burner or in a furnace. The molten fusion mixture is normally poured directly into a beaker of dilute nitric acid (1 M); this process shatters the sample into small fragments that dissolve in the acid more readily than would the large glass bead that would remain if the melt was allowed to cool. As with acid digestions, some minerals are resistant to decomposition in the molten flux (see Cremer and Schlocker, 1976); hence, after dissolving the fused material in the nitric acid the solution should be checked for solid residues.

Selective Extractions

For many studies it is not necessary to digest the entire sample and it may be more informative to selectively digest particular components in the samples. In both pollution monitoring and geochemical exploration programs, the most common selective extraction procedure is physical rather than chemical and involves the selective extraction and analysis of the mud fraction in the sample. Because trace metals and many organic pollutants are normally much more concentrated in the finer (clay and fine silt) grain size fractions than in sand size and coarser fractions, the concentration in a particular bulk sample may rather reflect the texture of the sample than the site where the sample was collected (e.g., Tessier et al. 1982; Förstner and Wittman 1981; Förstner, 1989; Luoma 1990). By analyzing only the mud fraction (< 62.5 μm because this is the lowest convenient size for wet sieving, but other size limits are used in some studies), the influence of sediment texture on metal or pollutant load is largely overcome, and the measured concentrations more likely reflect the proximity of the sample site to the source of metals or pollutants, their concentration at the source, and other factors of interest.

Whether or not grain size separations have been carried out, selective chemical extractions can be used to determine the amount of an element, or compound, present in a particular chemical form (i.e., as a particular chemical species). For environmental and geochemical exploration studies, in particular, determination of chemical speciation is extremely important (e.g., Förstner and Wittmann 1981; Chao 1984; Campbell and Tessier 1987; and numerous papers in Bernhard, Brinckman, and Sadler 1986), because the chemical form of an element or compound will have a major influence on its mobility, dispersion, and bioavailability.

Individual selective extractions (e.g., for exchangeable metals or sulfides or oxides) are appropriate in some studies, but in others a series of selective extractions (sequential extractions) are of more value. Sequential extractions provide a more clear definition of chemical processes operating within the sediment, but although some procedures do have a high selectivity for specific chemical species, no procedure that we know of is totally selective for the intended target chemical species. Numerous sequential extraction procedures have been described in the literature (e.g., Engler et al. 1977; Deurer, Förstner, and Schmoll 1978; Tessier, Campbell, and Bisson 1979; Rapin and Förstner 1983; Rapin et al. 1986; Kersten and Förstner 1986); they all have their advantages and disadvantages. Some useful selective extraction procedures and the chemical species they extract are listed in Table 5-19. It is not possible to describe a single universally applicable sequential extraction procedure because the best procedure for each study will be governed by the information requirements of the study and the nature of the sediments involved, but typical examples are shown in Figs. 5-2 and 5-3. As a general rule, the sequential extraction procedure should be kept as simple as possible to minimize analytical errors and to save time. As Rapin and Förstner (1983) have shown, it is relatively easy to devise a sequential extraction procedure for oxic sediments, but there are significant difficulties when sulfidic material is present (especially hydrated iron monosulfides, which can decompose in ordinary oxygenated water). Insoluble residues remaining after a single selective extraction, or after a sequential series of extractions, can be analyzed by one of the procedures used for whole rock analysis described above.

In environmental and soil fertility studies, selective extractions have the added importance of providing information on the concentrations of chemical species that are in a form that makes it possible for them to be taken up by biota. Adsorbed or exchangeable species are readily bioavailable, and some other chemical forms (e.g., elements bound to organic compounds and sulfides, which may decompose if conditions become oxidizing) are moderately bioavailable; in contrast, the chemical constituents of silicates and most carbonates and oxides are effectively unavailable for uptake by most biota. Several selective extraction procedures have been developed to provide concentration data that can be shown to correlate well with biological uptake rates in tested plants and animals. In agriculture, the method of Bray and Kurtz (1945) provides a good guide to phosphate availability; the method of Gillman and Sumpter (1986) provides a good guide to the availability of Na, K, Mg, and Ca; and the DTPA + TEA extraction of Lindsay and Norvell (1978) provides a good indication of the availability of trace metals (both biologically necessary and biologically unnecessary) to plants. In pollution studies, either the Gillman and Sumpter (1986) method or the Lindsay and Norvell (1978) method are often used to indicate the concentration of readily bioavailable chemical species; digestion in hot 20% HCl (commonly after ashing or oxidization with hydrogen peroxide) is used to indicate the concentration of potentially bioavailable chemical species.

Table 5-19.   Examples of Methods of Extraction of Metals from Major Chemical Phases in Sediments

Chemical Extractant	Reference
Adsorbed and exchangeable species	
Barium chloride plus triethanolamine (pH 8.1)	Jackson 1958
Ammonium acetate (pH 7)	Engler et al. 1977
Barium chloride plus ammonium chloride	Gillman and Sumpter 1986
Cation exchange resins	Duff et al. 1989
Diethylenetriaminepentacetic acid (DTPA) plus triethanolamine (pH 7)	Lindsay and Norvell 1978
Carbonate phases	
Acidic cation exchange	Deurer, Förstner, and Schmoll 1978
$NaOAc/HOAc$—buffer (pH 5)	Tessier, Campbell, and Bisson 1979
Dilute HCl or HNO_3 (low selectivity)	
Reducible phases (approximate order of iron release)	
Hydroxylamine—acetic acid	Chester and Hughes 1967
Dithionite—citric acid buffer	Holmgren 1967
Acidified hydroxylamine (+0.01 M nitric)	
Ascorbic acid (with or without added HCl or HNO_3)	
Organic phases and sulfides	
Peroxide and ammonium acetate (pH 2.5)	Engler et al. 1977
Peroxide and nitric acid	Gupta and Chen 1975
Organic solvents	Cooper and Harris 1974
0.1 M NaOH and sulfuric acid	Volkov and Formina 1974
Na—hypochlorite	Gibbs 1977
(DPTA)—NaOAc (pH 7)	Khalid, Grambrell, and Patrick 1981

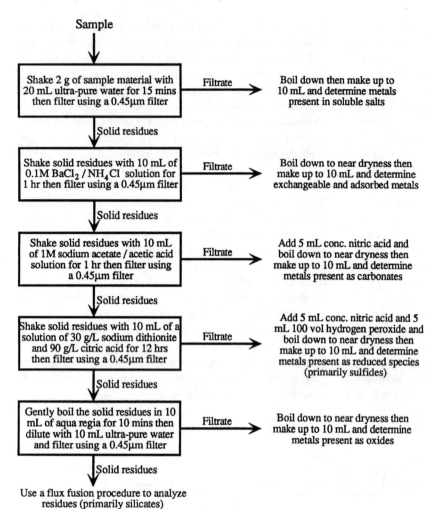

Figure 5-2.   Example of a sequential chemical extraction procedure for metals that can be applied to samples in the laboratory. *Caution:* Concentrated solutions of the acids used are dangerous. Exercise extreme care when handling them.

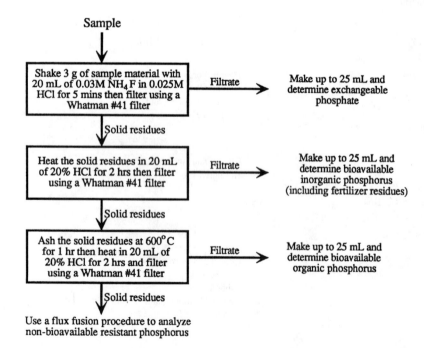

Figure 5-3.   Example of a sequential extraction procedure for phosphates in sediment samples.

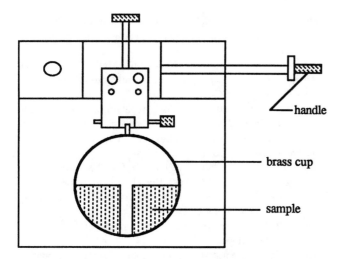

Figure 5-4. Sketch of the liquid limit device. (*Note:* Construction of a precise instrument is described in various standards' books.)

Table 5-20.   Procedure for Determining the Liquid Limit of Sediment Samples

1. Prepare the mechanical liquid limit device (Fig. 5-4): it must be clean and dry, and the cup should fall freely with a 1-cm drop of the cup from where it is released by the cam. Adjust if necessary to minimize side play at the hinge of the cup. Check that the grooving tool is 2 mm wide at the tip; reshape it if it is over 1 mm wider. Ensure that the grooving tool has not scratched a groove in the cup itself. If it is necessary to adjust the height of drop: (a) Place a 1-cm gauge between the cup and the base of the device and rotate the handle slowly. The cam should just touch the follower attached to the cup. (b) Loosen the clamping screws on the top of the device and adjust the screw at the rear of the machine until the follower just touches the cam. Tighten the clamping screws and repeat.

2. Mix c. 250 g samples of soil at natural water content to a stiff consistency that is slightly over the liquid limit (experience will tell!); if necessary, moisten with successive small increments of distilled water to that state. Stand overnight in sealed container at constant temperature so that equilibrium is attained.

3. Remix sample for 5–10 min. Spread subsample at the bottom of the brass cup of the liquid limit device; squeeze and level it with the top of the cup. Retain excess sediment in sealed container. Separate the subsample into two parts with the grooving tool, ensuring that the sides of the groove are smooth.

4. Using the eccentric crank, lift and drop the cup at a rate of 2 rps, counting the number of blows until the two sides of the sample come into contact along the bottom of the groove over a distance of c. 1 cm. The total should be no more than 15 to 25 blows; if less or more, remix sample after drying or with addition of distilled water. Repeat on the same subsample c. four times or until the difference between the number of blows for closure in two consecutive determinations is not greater than one. Record the last number of blows.

5. Take a 5–10-g slice of soil, weigh, dry at 110°C for 24 hours, and weigh again to discover the percentage water loss (100 × wt. water/wt. dry sample).

6. Repeat for at least four additional subsamples in more or less fluid conditions; ideally two determinations should occur in each of the ranges 15–25 and 25–35 blows. Plot the water content (abscissa) vs. number of shocks on semi-log graph paper and plot a best-fit straight line through the points. The intersection of the 25 shock ordinate (nearest whole number) with the curve is the liquid limit value.

LIQUID AND PLASTIC LIMITS

Liquid and plastic limits, called the *Atterberg limits,* are commonly measured by engineers working with plastic soils (these do not apply to fully granular materials). The liquid limit is the percentage water content at which a sediment passes from the plastic to the liquid state, whereas the plastic limit is the lowest percentage water content at which the soil becomes plastic. The *plasticity index* (liquid limit minus the plastic limit) also is commonly specified. Because moisture content is critical, samples should remain sealed until complete processing can be performed rapidly. Whole soil samples should begin at natural water content, and distilled water is added in small increments as necessary. *Civil Engineering Soil Testing Standards* are published in most countries and provide full descriptions in precise detail of the operations required to determine these values; the description below provides the essentials but may differ in some details.

Whole soil samples (from single soil layers) should have all or nearly all sediment finer than 425 μm. Remove any particles coarser than 2 mm by hand sieving, and if the sample remains too coarse, sieve for the less than 425-μm fraction. Coarse organic matter may also have to be removed by hand. The procedure for the liquid limit test is given in Table 5-20; the mechanical liquid limit device is shown in Fig. 5-4 (note that the construction of both this and the grooving tool are critical for precise measurements). The procedure for determining the plastic limit is listed in Table 5-21.

Table 5-21.   Procedure for Determining the Plastic Limit of Sediment Samples

1. Thoroughly mix a subsample of at least 30 g of the soil remaining from the sample used for the liquid limit test, then air dry until the material becomes sufficiently plastic to be shaped easily into a ball without sticking to the fingers when squeezed. Roll between hands until sufficiently dry that slight cracks appear on the surface. Take two subsamples for separate determinations and keep in sealed container until used.

2. Divide subsample into four parts and apply the following procedure to each part.

3. Roll a part between finger and thumb to form a thread c. 6 mm diameter. Roll this thread mass on a ground glass plate using the palm of the hand or the heel of the thumb, applying pressure uniformly to reduce the diameter of the thread to 3 mm. As soon as a diameter of 3 mm is reached with the thread still coherent, fold and knead the soil into a ball to reduce the moisture content further.

4. Repeat the rolling and kneading process until the soil has dried to the condition where the rolled thread just breaks into a number of pieces at the 3-mm diameter. The plastic limit has been reached!

5. Place soil in weighing bottle and seal.

6. Repeat steps 3 to 5 using each of the other parts of the subsample, adding the sediment each time to the same sealed container.

7. Repeat all the above for the second subsample.

8. Determine the water content of the soil in each weighing bottle (as for liquid limit test). If the water contents determined for the two subsamples differ by more than 5%, repeat the test. The plastic limit is the average of the water contents rounded to the nearest whole number, and is written without a percentage symbol.

SELECTED BIBLIOGRAPHY

Pretreatment of Sediment Samples (see also general laboratory books listed in Chapter 1)

Anderson, J. U., 1963, Effects of pretreatments on soil dispersion. *New Mexico State College Agricultural Experimental Station Research Report* 78, 14p.

Bodine, M. W., Jr., and T. H. Fernald, 1973, EDTA dissolution of gypsum, anhydrite, and Ca-Mg carbonates. *Journal of Sedimentary Petrology* 43:1152–6.

Bouma, A. H., 1969, *Methods for the Study of Sedimentary Structures.* Wiley-Interscience, New York, 458p.

Brewer, R., 1964, *Fabric and Mineral Analysis of Soils.* Wiley, New York.

Charm, W. B., 1967, Freeze drying as a rapid method of disaggregating silts and clays for dry particle size analysis. *Journal of Sedimentary Petrology* 37:970–1.

Drosdorff, M., and E. Truog, 1935, A method for removing iron oxide from minerals. *American Mineralogist* 20:669–73.

Edwards, A. P., and J. M. Bremner, 1965, Dispersion of soil particles by sonic vibrations. *Journal of Soil Science* 18:47–63.

Felix, C. J., 1963, Mechanical sample disaggregation in palynology. *Micropaleontology* 9:337–9.

French, D. H., S. S. J. Warne, and M. T. Sheedy, 1984, The use of ion-exchange resins for the dissolution of carbonates. *Journal of Sedimentary Petrology* 54:641–3.

Funkhouser, J. W., and W. R. Evitt, 1959, Preparation techniques for acid-insoluble microfossils. *Micropaleontology* 5:369–75.

Gipson, M. J., 1963, Ultrasonic disaggregation of shale. *Journal of Sedimentary Petrology* 33:955–8.

Glover, E. D., 1961, Method of solution of calcareous material using the complexing agent EDTA. *Journal of Sedimentary Petrology* 31:622–26.

Kerrigan, G. C., 1971, A sample preparation device for X-ray fluorescence. *Journal of Physical and Earth Sciences Instruments* 4:544–5.

Kravitz, J. H., 1966, Using an ultrasonic disruptor as an aid to wet sieving. *Journal of Sedimentary Petrology* 36:811–2.

Krumbein, W. C., 1933, Dispersion of fine grained sediments for mechanical analysis. *Journal of Sedimentary Petrology* 3:121–35.

Kunze, G. W., 1965, Pretreatment for mineralogical analysis. in C. A. Black (ed.), *Methods of Soil Analysis,* vol. 1, American Society of Agronomy, Inc., Madison, Wisc. pp. 568–77.

Leith, C. J., 1950, Removal of iron oxide coatings from mineral grains. *Journal of Sedimentary Petrology* 20:174–6.

Milner, H. B., 1962, *Methods in Sedimentary Petrography,* vol. 1. Allen & Unwin, London, 643p.

Mitchell, B. D., and R. C. MacKenzie, 1954, Removal of free iron oxide from clays. *Soil Science* 77:173–84.

Moston, R. P., and A. I. Johnson, 1964, Ultrasonic dispersion of samples of sedimentary deposits. *United States Geological Survey Professional Paper 501-C,* pp. C159–60.

Mueller, G., 1967, *Methods in Sedimentary Petrology,* H.-U. Schmincke (trans.). Hafner, New York.

Nelsen, T. A., 1983, Time- and method-dependent size distributions of fine-grained sediments. *Sedimentology* 30:249–59.

Overbey, W. K., Jr., and B. R. Henniger, 1970, Disaggregation of sandstones by ultrasonic energy. *Journal of Sedimentary Petrology* 40:465–72.

Percival, S. F., Jr., E. D. Glover, and L. B. Gibson, 1963, Carbonate rocks: Cleaning with suspensions of hydrogen ion exchange resins. *Science* 142:1456–7.

Prokopovich, N. P., and C. D. Nish, 1967, Methodology of mechanical analysis of subaqueous sediments. *Journal of Sedimentary Petrology* 37:96–101.

Savage, E. L., 1969, Ultrasonic disaggregation of sandstones and siltstones. *Journal of Sedimentary Petrology* 39:375–8.

Suczek, C. A., 1983, Disaggregation of quartzite. *Journal of Sedimentary Petrology* 53:672–3.

Sulcek, Z., and P. Povondra, 1989, *Methods of Decomposition in Inorganic Analysis.* CRC Press, Boca Raton, Fla., 368p.

Tchillingarian, G., 1952, Study of the dispersing agents. *Journal of Sedimentary Petrology* 22:229–33.

Tolmachoff, I., 1932, Crystallization of certain salts used for the disintegration of shales. *Science* 76:147–8.

Troell, E., 1931, The use of sodium hypobromite for the oxidation of organic matter in mechanical analyses of soils. *Journal of Agricultural Science* 21:476–84.

Truog, E., J. R. Taylor, R. W. Pearson, M. E. Weeks, and R. W. Simonson, 1937, Procedure for special type of mechanical and mineralogical soil analysis. *Soil Science Society of America Proceedings* 1:101–12.

Walker, P. H., and J. Hutka, 1973, Grain fragmentation in preparing samples for particle size analysis. *Soil Science Society of America Proceedings* 37:278–80.

Zingula, R. P., 1968, A new breakthrough in sample washing. *Journal of Paleontology* 42:1092.

Grain Mount and Thin-Section Preparation

Ashley, G. M., 1973, Impregnation of fine-grained sediments with a polyester resin: A modification of Altemuller's method. *Journal of Sedimentary Petrology* 43:298–301.

Awadallah, S. A., 1991, A simple technique for vacuum impregnation of unconsolidated, fine-grained sediments. *Journal of Sedimentary Petrology* 61:632–3.

Carver, R. E., 1971, *Procedures in Sedimentary Petrology.* Wiley-Interscience, New York, 653p.

Chiou, W. A., L. E. Shephard, W. R. Bryant, and M. A. Looney III, 1983, A technique for preparing highwater content clayey sediments for thin and ultrathin section study. *Sedimentology* 30:295–9.

Conway, J. S., 1982, A simplified method for impregnation of soils and similar clay-rich sediments. *Journal of Sedimentary Petrology* 52:650–1.

Crevello, P. D., J. M. Rine, and D. E. Lanesky, 1981, A method for impregnating unconsolidated cores and slabs of calcareous and terrigenous muds. *Journal of Sedimentary Petrology* 51:658–60.

Franklin, J. A., 1969, Rock impregnation using monomers, epoxide, and unsaturated polyester resins. *Journal of Sedimentary Petrology* 39:1251–3.

Gardner, K. L., 1980, Impregnation technique using colored epoxy to define porosity in petrographic thin sections. *Canadian Journal of Earth Sciences* 17:1104–7.

Ginsburg, R. N., H. A. Bernard, R. A. Moody, and E. E. Daigle, 1966, The Shell method of impregnating cores of unconsolidated sediments. *Journal of Sedimentary Petrology* 36:1118–25.

Jim, C. Y., 1985, Impregnation of moist and dry unconsolidated clay samples using Spurr resin for microstructural studies. *Journal of Sedimentary Petrology* 55:597–9.

Jordan, C. F., Jr., and C. D. Roady, 1987, A practical method for preparing thin sections of well cuttings. *Journal of Sedimentary Petrology* 57:759–60.

Lindholm, R. C., and D. A. Dean, 1973, Ultra-thin sections in carbonate petrology: A valuable tool. *Journal of Sedimentary Petrology* 43:295–7.

Lumsden, D. N., 1979, Discrepancy between thin-section and X-ray estimates of dolomite in limestone. *Journal of Sedimentary Petrology* 49:429–36.

Marshall, C. E., and C. D. Jeffries, 1946, Mineralogical methods in soil research: I. The correlation of soil types and parent materials with supplementary information on weathering processes. *Soil Science Society of America Proceedings* 10:397–405.

Martin, R., P. E. Litz, and W. D. Huff, 1979, A new technique for making thin sections of clayey sediments. *Journal of Sedimentary Petrology* 49:641–3.

Middleton, L. T., and M. J. Kraus, 1980, Simple technique for thin-section preparation of unconsolidated materials. *Journal of Sedimentary Petrology* 50:622–3.

Minoura, N., and C. D. Lonley, 1971, Technique for impregnating porous rock samples with low-viscosity epoxy resin. *Journal of Sedimentary Petrology* 41:858–61.

Nentwich, F. W., and R. W. Yole, 1991, Polished thin section preparation of fine-grained siliciclastic rocks. *Journal of Sedimentary Petrology* 61:624–26.

Orlansky, R., 1968, Method for making slides of fine-grained unconsolidated sediment and ooze. *Journal of Sedimentary Petrology* 38:1378.

Palmer, S. N., and M. E. Barton, 1986, Avoiding microfabric disruption during the impregnation of friable, uncemented sands with dyed epoxy. *Journal of Sedimentary Petrology* 56:556–7.

Reed, F. S., and J. L. Mergner, 1953, Preparation of rock thin sections. *American Mineralogist* 38:1184–1203.

Socci, A., 1980, A method for dry and semi-dry thin-sectioning of certain water-sensitive rocks. *Journal of Sedimentary Petrology* 50:621–2.

Teague, T., 1989, An improved technique for polishing difficult geological materials using a colloidal silica suspension. *Journal of Sedimentary Petrology* 59:635.

Tucker, M. (ed.), 1988, *Techniques in Sedimentology.* Blackwell Scientific Publications, Oxford, 394p.

Waldo, A. W., and S. T. Yuster, 1922, Method of impregnating porous materials to facilitate pore studies. *American Association of Petroleum Geologists Bulletin* 21:259–67.

Yanguas, J. E., and S. T. Paxton, 1986, A new technique for preparation of petrographic thin sections using ultraviolet-curing adhesive. *Journal of Sedimentary Petrology* 56:539–40.

Microfossil Extraction

Barss, M. S., and G. L. Williams, 1973, Palynology and nannofossil processing techniques, *Geological Survey of Canada Paper 73-26,* pp. 1–25.

Dettman, M. E., 1963, Upper Mesozoic microfloras from south-eastern Australia. *Royal Society of Victoria Proceedings* 77:11–2.

Phipps, D., and G. Playford, 1984, Laboratory techniques for extraction of palynomorphs from sediments. *Department of Geology, University of Queensland, Papers* 11:1–23.

Wolf, K. H., A. J. Easton, and S. Warne, 1967, Techniques of examining and analyzing carbonate skeletons, minerals and rocks. In G. Chilingar, H. J. Bissell, and R. W. Fairbridge (eds.), *Carbon-*

ate Rocks, Developments in Sedimentology 9B, Elsevier, New York, pp. 253–341.

Coal Sample Treatment

ASTM, 1987, *Annual Book of ASTM Standards,* vol. 05.05. Gaseous Fuels; Coal and Coke. American Society for Testing and Materials, Philadelphia.

International Committee for Coal Petrography Handbook, 1971, Centre Nationale de la Recherche Scientifique, Paris.

Stach, E., M.-Th. Mackowsky, M. Teichmuller, R. Teichmuller, G. H. Taylor, and D. Chandra, 1982, *Coal Petrology.* Gebruder Borntraeger, Berlin.

Ward, C. R. (ed.), 1984, *Coal Geology and Coal Technology.* Blackwell, Melbourne, 345p.

Chemical Analysis Methodology

Bernas, B., 1968, A new method for decomposition and comprehensive analysis of silicates by atomic absorption spectrometry. *Analytical Chemistry* 42:1682–6.

Bernhard, M., F. E. Brinckman, and P. J. Sadler (eds.), 1986, *The Importance of Chemical "Speciation" in Environmental Processes.* Life Sciences Research Report 33, Springer-Verlag, Berlin, 762p.

Bray, R. M., and L. T. Kurtz, 1945, Determination of total organic and available forms of phosphorus in soils. *Soil Science* 59:39–45.

Campbell, P. G. C, and A. Tessier, 1987, Current status of metal speciation studies. In J. W. Patterson and R. Pussino (eds.), *Metal Speciation, Separation and Recovery.* Lewis Publishers Inc., Chelsea, Mich., pp. 201–24.

Chao, T. T., 1984, Use of partial dissolution techniques in geochemical exploration. *Journal of Geochemical Exploration* 20:101–35.

Chester, R., and M. J. Hughes, 1967, A chemical technique for the separation of ferro-manganese minerals, carbonate minerals, and adsorbed trace elements from pelagic sediments. *Chemical Geology* 2:233–48.

Cooper, B. S., and R. C. Harris, 1974, Heavy metals in organic phases of river and estuarine sediments. *Marine Pollution Bulletin* 5:24–6.

Cremer, M., and J. Schlocker, 1976, Lithium borate decomposition of rocks, minerals and ores. *American Mineralogist* 61:318–21.

Deurer, R., U. Förstner, and G. Schmoll, 1978, Selective chemical extraction of carbonate-associated metals from recent lacustrine sediments. *Geochimica Cosmochimica Acta* 42:425–7.

Duff, S. J., G. W. Hay, R. K. Micklethwaite, and G. W. Vanloon, 1989, Distribution and classification of metal species in soil leachates. *Science of the Total Environment* 87/88:189–97.

Engler, R. M., J. M. Brannon, J. Rose, and G. Bigham, 1977, A practical selective extraction procedure for sediment characterization. In T. F. Yen (ed.), *Chemistry of Marine Sediments.* Science Publishers, Ann Arbor, pp. 163–80.

Fabbi, B. P., 1978, *Geology.* In H. K. Herglotz and L. S. Birks (eds.), *X-ray Spectrometry.* Marcel Dekker, New York, pp. 297–353.

Förstner, U., 1989, *Contaminated Sediments.* Lecture Notes in Earth Science, No. 21. Springer-Verlag, Berlin. 157p.

Förstner, U., and G. T. W. Wittmann, 1981, *Metal Pollution in the Aquatic Environment.* Springer-Verlag, Berlin, 486p.

Gibbs, R. J., 1977, Transport phases of transition metals in the

Amazon and Yukon rivers. *Geological Society of America Bulletin* 88:829–43.

Gillman, G. P., and E. A. Sumpter, 1986, Modification to the compulsive exchange method for measuring exchange characteristics of soils. *Australian Journal of Soil Research* 24:61–6.

Gupta, S. K., and K. Y. Chen, 1975, Partitioning of trace metals in selective chemical fractions of near-shore sediments. *Environmental Letters* 10:129–58.

Harvey, P. K., and B. P. Atkin, 1982, Automated X-ray fluorescence analysis. In *Sampling and Analysis for the Mining Industry*. Institute of Mining and Metallurgy, London, pp. 17–26.

Holmgren, G. G. S., 1967, A rapid citrate-dithionite extractable iron procedure. *Soil Science Society of America Proceedings* 31:210–1.

Hutton, J. T., and S. M. Elliott, 1980, An accurate XRF method for the analysis of geochemical exploration samples for major and trace elements using one glass disc. *Chemical Geology* 29:1–11.

Ingamells, C. O., 1970, Lithium metaborate flux in silicate analysis. *Anales Chimica Acta* 52:323–34.

Jackson, M. L., 1958, *Soil Chemical Analysis*. Prentice Hall, Englewood Cliffs, N.J., 498p.

Kersten, M., and U. Förstner, 1986, Chemical fractionation of heavy metals in anoxic estuarine and coastal sediments. *Water Science and Technology* 18:121–30.

Khalid, R. A., R. P. Grambrell, and W. H. Patrick, 1981, Chemical availability of cadmium in Mississippi River sediment. *Journal of Environmental Quality* 10:523–8.

Lee, R. F., and D. M. McConchie, 1982, Comprehensive major and trace element analysis of geological material by X-ray fluorescence using low dilution fusion. *X-ray Spectrometry* 11:55–63.

Lindsay, W. L., and W. A. Norvell, 1978, Development of a DTPA soil test for zinc, iron, manganese, and copper. *Soil Science Society of America Journal* 42:421–8.

Lord, C. J., 1982, A selective and precise method for pyrite determination in sedimentary materials. *Journal of Sedimentary Petrology* 52:664–6.

Louma, S. N., 1990, Processes affecting metal concentrations in estuarine and coastal marine sediments. In R. W. Furness and P. S. Rainbow (eds.), *Heavy Metals in the Marine Environment*. C.R.C. Press, Boca Raton, Fla., pp. 51–65.

McConchie, D. M., and V. J. Harriott, 1992, The partitioning of metals between tissue and skeletal parts of corals: Application in pollution monitoring. *Proceedings of the International Coral Reef Symposium,* Guam.

Morris, R. C., and W. E. Ewers, 1978, A simple streak print technique for mapping mineral distributions in ores and other rocks. *Economic Geology* 73:562–6.

Norrish, K., and J. T. Hutton, 1969, An accurate X-ray spectrographic method for the analysis of a wide range of geological samples. *Geochimica Cosmochimica Acta* 33:431–53.

Rapin, R., and U. Förstner, 1983, Sequential leaching techniques for particulate metal speciation: The selectivity of various extractants. In *Proceedings 4th International Conference on Heavy Metals in the Environment,* Heidelberg, pp. 1074–7.

Rapin, F., A. Tessier, P. G. C. Campbell, and R. Carigan, 1986, Potential artifacts in the determination of metal partitioning in sediments by a sequential extraction procedure. *Environmental Science and Technology* 20:826–40.

Ruzyla, K., and D. I. Jezek, 1987, Staining method for recognition of pore space in thin and polished sections. *Journal of Sedimentary Petrology* 57:777–8.

Tessier, A., P. G. C. Campbell, and M. Bisson, 1979, Sequential extraction procedure for the speciation of particulate trace metals. *Analytical Chemistry* 51:844–50.

Tessier, A., P. G. C. Campbell, and M. Bisson, 1982, Particulate trace metal speciation in stream sediments and relationships with grain size: Implications for geochemical exploration. *Journal of Geochemical Exploration* 16:77–104.

Thomas, I. L., and M. T. Haukka, 1978, XRF determination of trace and major elements using a single-fused disc. *Chemical Geology* 21:39–50.

Van Eenbergen, A., and E. Bruninx, 1978, Losses of elements during sample decomposition in an acid-digestion bomb. *Analytica Chimica Acta* 98:405–6.

Volkov, I. I., and L. S. Formina, 1974, Influence of organic material and processes of sulphide formation on distribution of some trace elements in deep-water sediments of Black Sea. *American Association of Petroleum Geologists Memoir* 20:456–76.

Yanguas, J. E., and J. J. Dravis, 1985, Blue fluorescent dye technique for recognition of microporosity in sedimentary rocks. *Journal of Sedimentary Petrology* 55:600–2.

6
Analysis of Sedimentary Structures

Sedimentary structures are generally studied in the field, where sketches or photographs are taken to record characteristics and for later analysis, and where a compass and clinometer are utilized to quantitatively measure the three-dimensional attitude of the elements composing the structures (see Chapter 3 and **PS** Chapter 4). Devices such as the disk shown in Fig. 6-1 can prove convenient in the field for measuring the strike and dip of cross-bedding that is not exposed along the bed surfaces. Video cameras are utilized increasingly in both the laboratory and the field to obtain quantitative as well as qualitative data on the development of sedimentary structures. Peels (see Chapter 3) and surface replicas (see below) are also obtained in the field to permit laboratory studies of small-scale structures. Obscure small-scale internal structures can be artificially enhanced or must be studied in the laboratory utilizing special photographic equipment, X-ray radiography, or stains (e.g., Chapter 8). In the case of laboratory specimens, if any directional data are desired, it is essential that their spatial orientation be recorded when they are collected. A thorough discussion of field and laboratory procedures and both principles and applications of paleocurrent analysis are given in Potter and Pettijohn (1977), as well as a number of the journal papers listed in the Selected Bibliography.

ENHANCEMENT OF ORIGINAL STRUCTURES

Surface Replicas

Preparation of various kinds of peels in the field has been discussed in Chapter 3; these show both structures and textures of the sediment in vertical cross-section. Occasionally in the field (or in the laboratory when dealing with the results of experiments) it may be desirable to prepare a replica of the structures on the surface of a bed (e.g., ripplemark, flutes).

Most commonly, plaster of paris is used to make casts in the field, whereas in the laboratory the same material, dental plaster, silicone rubber, or various plastics can be used to make positive replicas (e.g., see Bouma 1969).

In the field, a frame is placed around the selected area and pushed into the sediment sufficiently firmly that it will not move and the plaster will not flow out. An initial thin mixture of water and plaster is necessary for even and gentle pouring (avoid creation of new structures!). Start pouring near a corner and do not pour directly onto the sediment (use intermediary hand or other object); pour a small quantity into a separate hole in the sediment to use as a monitor for the hardening process. A thick mixture of plaster can be poured onto the first layer, after it has begun to harden, to create a slab at least 2 cm thick over the thinnest spots (to avoid breakage upon removal). The thin mix will require more time to harden than the thick; hence check carefully before removing the plaster by inserting a knife between the frame and the plaster. Upon removal, clean the surface of loose material and mark a cardinal compass direction indelibly on the side, bottom or top.

When preparing a positive replica in the laboratory, ensure that the surface is smooth and clean, and treat it with a mold-release agent such as a nitrocellulose lacquer. Place a high-sided frame about the cast and pour in the casting medium evenly to avoid air bubbles. Some touching up of the resultant replica may be necessary.

Sand-Blasting

With poorly consolidated sedimentary rocks, sandblasting can etch structures into relief (e.g., Hamblin 1962a). Success depends on the degree of differential cementation in different structural units within the sample. A small multipurpose air-cleaning spray-gun or sandblasting unit that operates on air pressures as low as 75 lb/in^2 can be used. The sand used in the

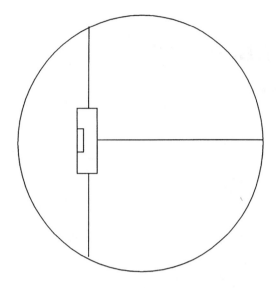

Figure 6-1.   Sketch of a simple device for measuring strike and dip of cross-bedding in the field. A disk of any convenient size is cut from aluminum sheeting and a straight line is scribed along a chord; a carpenter's level is then attached to the disk along the scribed line. In the field, the disk is aligned parallel to the cross-bed and the bubble is centered; the scribed line is now a strike line, and the dip can be measured by placing a clinometer on the disk at right angles to the scribed line.

spray should be clean and unsorted, with a particle size smaller than the average grain size of the sample. Sand and particles blasted from the sample can cause respiratory problems; use a mask or work under a laboratory hood.

Photographic Enhancement

Even without selective staining, subtle structures in sediments may be enhanced in photographs by careful lighting techniques, control of exposure time and depth of focus, and/ or selection of film speed and grain (see Chapter 3). Experimentation is always necessary with particular suites of rocks; there is no universally applicable "best" method. When dealing with rocks, in general a cut-and-polished surface of rock is prepared, then submerged in alcohol (just to the level of the surface) or coated with oil or glycerol and placed on a matte black surface to prevent reflection from the sample sides. Photographs taken on all sides of a single block can be mounted on cardboard and glued together to provide three-dimensional representations, which themselves can be photographed from any angle (e.g., Bockelie 1973). The final prints can be treated with a solution (e.g., 1 part iodine:10 parts potassium iodine in 100 parts water) that removes the emulsion but leaves any india-inked outlines made earlier.

Infrared (or other restricted wavelength) photographs of thin slabs of loose sands and muds or fine limestones illuminated by a light source from the underside can provide considerable enhancement of subtle structures that do not appear using stains or X-ray radiographic techniques (Ali and Weiss 1968). Organic matter and clay minerals are particularly prone to absorb infrared light, but all differences in sediment grain size and composition are influential. Suitable thicknesses of slabs are determined qualitatively by detecting by eye whether any light is transmitted through them using a 100-watt bulb. A single 500-watt tungsten bulb produces sufficient infrared intensity; a filter (e.g., Wratten 25A, 29F, 70, 87, 88A, 98B, and 15G; Bouma 1969) should be used to remove blue light. A light box must be constructed that allows only that light transmitted through the sample to reach the camera lens, and photography must be carried out in a darkroom. A fan may be necessary to cool the light box on which the specimens are placed. Care must be taken to focus the lens, since infrared rays will focus on a different point than that for visible light (c. 0.05 focal length extension), and exposure times must be determined empirically (2–100 sec at 50 cm)—coarse-grained samples are more transparent than fine-grained samples.

Note: It is commonly worthwhile when photographing structures to insert a scale that indicates a cardinal direction (e.g., chalked onto a black-painted scale).

X-Ray Radiography

Since its introduction to sedimentological studies by Hamblin (1962b), X-ray radiography has become a standard tool for determining internal structures in sediments and rocks with obscure structures or apparently homogeneous fabric. Cores collected from both recent and ancient sediments are now routinely subjected to these procedures, using filters that compensate for the difference in sediment thickness penetrated by the X-rays and scales that appear in the radiographs (e.g., Baker and Friedman 1969). Box samples of both soils and subaqueous sediments as well as sawed slabs of rock or impregnated sediment blocks are commonly treated, and time-lapse radiography of burrowing operations of organisms in laboratory tanks has produced remarkable results (e.g., Howard 1968); other ingenious uses of the technique have been found. Thorough descriptions of principles, techniques, and equipment, together with examples of applications, are provided in Hamblin (1962b), Bouma (1969), and Krinitsky (1970). Analyses are carried out in commercially available, generally small, and reasonably lightweight, lead-lined radiography boxes that contain both X-ray source and sample chamber; some medical and industrial X-ray machines may also be used, and even X-ray diffraction equipment has been adapted (Clifton 1966). Voltage used and time of exposure are critical elements of the procedure, and they must be determined by experimentation with the particular sediments and sample thicknesses (which for rocks may range from 2 mm up to 2 cm); film type can also be important. Guidelines for optimal combinations of these variables are available in Hamblin (1967) and Bhargava (1971). X-rays are injurious to health! Routinely use a scintillometer or other approved device to detect whether any X-rays are escaping from the theoretically sealed containers!

The short-wavelength, high-energy radiation penetrates materials that absorb or reflect visible light; as it passes through a sediment, an X-ray beam is attenuated by absorption and is scattered, and the fabric and composition of the sediment thus dictate the resultant image. The film is blackened most where the emulsion has received and absorbed most radiation; hence the darkest lines reflect the least concentration of particles (e.g., joints, fractures) or a concentration of the least absorbent particles (e.g., bedding planes, organic matter); light colors on the print reflect the most opaque portions of the sample. Rocks are denser than loose sediment; hence they require longer times of exposure and/or greater intensity of bombardment. Because concentrations of different grain size and composition or grain fabrics comprise discrete sedimentation units or layers, the structural arrangement of these units is revealed by the different intensities of X-rays that the film emulsion absorbs. Irregular thicknesses of samples can obscure internal fabric differences; hence surfaces should be smooth. Consistent thickness changes, e.g., toward the sides of cores, provide consistent changes in intensity; therefore, internal differences are still apparent, but it is desirable to study uniformly developed images, and various means of compensating for the thickness changes are commonly attempted. If mineralogy, grain size, and grain packing are homogeneous, no structures will be detected. Because of the low absorption level of carbonate minerals and the small difference in absorption between carbonate grains, this technique is usually of little use with limestones.

COLLECTION OF PALEOCURRENT DATA

Structures that show paleocurrent orientation or direction are measured by taking the strike and dip of planar structures (e.g., planar cross-beds; see DeCelles, Langford, and Schwartz 1983 for methods of dealing with trough cross-beds) or the trend and plunge of linear structures (Chapter 3); vector (rather than arithmetic) analysis is generally performed on the data. If the structures are abundant in an area, it may be necessary to make statistically representative sample measurements rather than measure all those available (e.g., Chapter 4). The number of measurements required varies with the objectives of the project as well as the type and variability of the structures present—between 10 and 20 of all kinds may be sufficient to indicate a regional mean direction if samples are measured over a dispersed area and current patterns are consistent, whereas many more of each kind are needed if the goal is to determine variability of process in the deposits of a complex sedimentary environment (see Rao and Sengupta 1970; Freeman and Price 1979 for discussions of means of evaluating requirements in the field). In most situations where structures are abundant, cost effectiveness dictates the number of measurements taken! Sole marks typically have a low variance and require fewer measurements than ripple-mark or cross-bedding to obtain regional patterns. If directional structures are abundant and the area is readily accessible, it is generally desirable to sample in two stages: the first making a few measurements distributed over the area to determine the regional pattern(s), the second to refine

the data, particularly concentrating where there is marked variability in direction or at boundaries between areas showing divergent patterns. Hierarchical or nested sampling may be necessary (see Chapter 4) of variability within beds, between beds in a single outcrop, between outcrops in an area, and between areas. Vertical variability should be investigated: patterns of flow may change during the deposition of a thick sedimentary unit.

The number of paleocurrent measurements necessary for reliable estimation of the mean flow vector depends on the variability of the values recorded. The variance (V) is a function of the standard deviation (S) and the number of measurements (n, which should represent the mean of the measures from individual beds, but since variability is low within most single beds, only one measure is taken per bed and n generally is the total number of measures; n may also represent the means of individual localities in regional studies):

$$V = \frac{S}{\sqrt{n}}$$

Where n has been increased to the point beyond which the variance does not change significantly, sufficient measurements have been made.

CORRECTION OF DIRECTIONAL DATA FOR TECTONIC DEFORMATION

If the sedimentary rocks from which the field data have been collected have been deformed significantly by folding or faulting, that data must be corrected to restore the original attitude of the measured structures. For linear structures (e.g., current lineation, ripple-crest orientations), only where principal bedding surfaces dip more than about 25° is correction essential. When planar structures are measured (e.g., strike and dip of cross-beds), however, tectonic dips of greater than 5° often require corrections to be applied. Both the attitude of the principal bedding and the attitude of the structures themselves must be measured in the field; care must be taken at that stage to ensure the full three-dimensional geometry is known. (*Note:* If deformation is caused by folds that plunge—i.e., with axes that are not horizontal—or if substantial compression—flattening—or shear has been involved, then further structural data must be recorded in the field and a more complex correction procedure must be followed—see discussion by Graham in Tucker 1988.) A computer program that corrects for both plunge and tilt has been presented by Cooper and Marshall (1981). If structures have been rotated about a vertical axis, only the difference between directions in structurally-bound areas is likely to indicate the tectonic modifications, and a regional structural map will be required to indicate which areas have been least affected.

Field Correction

The relationship between the sedimentary structure to be measured and the principal planes of stratification is generally independent of the amount of structural deformation. Hence, immediate correction for simple tilt deformation can be made in the field. Mark the strike (line formed by the intersection of a horizontal plane with the bedding surface) on a principal bedding plane. For a linear structure, measure the angle between this strike line and the structure (e.g., with a protractor or compass clinometer). To determine the original orientation of the structure, if this angle opens clockwise relative to the strike line, add the angle to the value of the strike angle; if this angle opens counterclockwise, subtract the angle from the strike value.

Even more simply, adapt a carpenter's rule by inserting a leveling bubble in one length, orient that length along the strike of the principal bedding, align the other length of the rule along the structure, and record the spacial orientation of that length. Okada and Arita (1970) describe the construction

of a simple perspex instrument from which the corrected direction can be read directly from an attached protractor.

Laboratory Correction

The simplest correction procedures for tilted sequences involve the use of a stereonet (see Hoyt in Carver 1971 for detailed treatment of the trigonometric method). Solutions using stereonets are discussed by Ramsay (1961) and Graham (in Tucker 1988). Essentially, a stereonet such as Fig. 6-2 represents a two-dimensional vertical cross-section of a sphere; the sphere is cut by planes oriented at 10° intervals from 0–360° that appear as curved lines (*great circles*) in the two-dimensional cross-section. In that sphere (i.e., on that cross-section), the spacial position of both planar (e.g., bedding planes, cross-bed foresets) and linear features (e.g., cur-

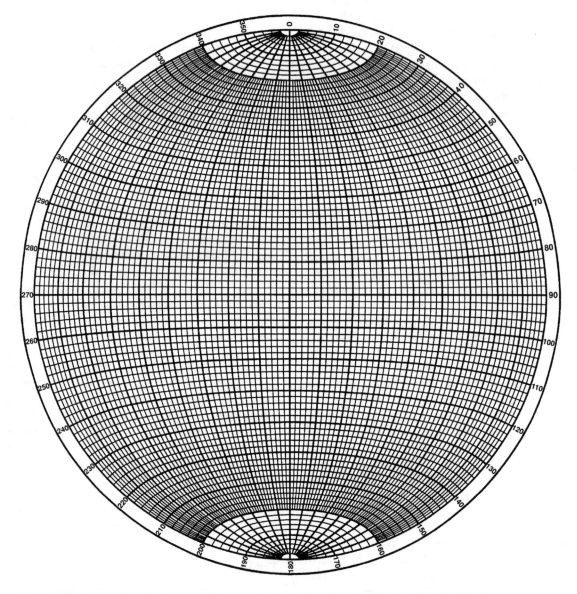

Figure 6-2.   Equal-area stereonet (equal-angle stereonets could also be used), as used for plotting paleocurrent data and for correcting measurements for any tectonic deformation of original bed attitudes.

rent lineation, dip direction of asymmetric ripplemark, crestal trends of ripplemark) can be depicted; then all can be rotated to remove the amount of tilt imposed by the tectonic deformation.

EVALUATION OF DIRECTIONAL DATA

In most cases, all directional structures encountered are measured and plotted on a rose or a composite-ray (or spoke) diagram (**PS** Fig. 4-3) for a locality (or from small areas). Where data are few, visual inspection serves to estimate variance and provides sufficient information for interpretation. Where statistical procedures are followed in the plotting of rose diagram data (see below and e.g., Ballantyne and Cornish 1979; Krause and Geiger 1987), the essential calculations have been made and the data can be considered reliable.

To detect whether two or more samples are drawn from a single population, High and Picard (in Carver 1971) present cogent reasons why the common student's t test and the chi-square test are not appropriate for most paleocurrent data. They recommend the Kolmogorov-Smirnov test, which is a graphic method of measuring the spread between two or more frequency curves and estimating the probability that the two curves are part of a homogeneous population. For a given sample size, the probability that they are part of a homogeneous population increases as the spread decreases. The null hypothesis is that the separate curves represent different populations, and this hypothesis is tested by looking at the maximum spread of the two curves and comparing it with

$$d_{.05} = 136 \sqrt{\frac{1}{n_1} + \frac{1}{n_2}}$$

where n_1 and n_2 are the number of measurements for each curve and the .05 subscript refers to a selected 0.05 significance level. If the maximum spread is less than the expected value ($d_{.05}$), the null hypothesis is rejected and it is concluded that the samples come from a homogeneous population.

Where mathematical computations are involved, summary statistical measurements must be adjusted for the circular nature of azimuthal data. Arithmetic means can be calculated meaningfully only where the data lie within 180°. Vector means thus are generally computed for directional data; the resultant vector of a set of data must be calculated, then the magnitude of that vector, using formulae such as that given by Potter and Pettijohn (1977):

R (magnitude or length of vector) $= (V^2 + W^2)^{1/2}$

L (percent magnitude of resultant vector) $= 100 \ (R/n)$

X_v (azimuth of resultant vector) $=$ arc tan W/V

$$V = \sum n_i \cos x_i$$
$$W = \sum n_i \sin x_i$$

where x_i is the midpoint of the ith class interval and n_i is the number of observations in each class. The greater the magni-

tude of the vector (L), the greater the concentration of the azimuths.

Testing of the calculated vector against random distributions will give a measure of the central tendency of the population of data. If the statistical chi-square test indicates that the observed distribution does not depart significantly from a uniform distribution, then the vector mean is not significant. These and further steps in vector analysis are well discussed by High and Picard (in Carver 1971) and by Graham (in Tucker 1988), as well as by Potter and Pettijohn (1977).

PRESENTATION OF RESULTS

Quantitative paleocurrent data can be represented by a variety of two-dimensional azimuthal graphs (e.g., **PS** Fig. 4-3) or by plotting (and contouring) three-dimensional data such as poles of cross-beds or orientation of grain axes in a stereonet such as Fig. 6-2 (all data points appear, in contrast to the grouping that occurs with histogram or vectorially averaged plots; see Graham, in Tucker 1988, for examples). In most cases, azimuth is the indicative angle for paleocurrent direction or orientation, and only the two-dimensional distributions are presented. Graphical presentation of data is essential even if summary statistical values are calculated, because the statistics only summarize, mostly deal with directional as opposed to orientational data, are usually calculated assuming unimodal distributions (which is commonly false with paleocurrent data), and do not show the full range of population values. It is important that only the same *kind* of data be plotted in the same graph (or band, in the composite-ray diagram)—different kinds of structures may be produced by different processes or events. It is also important to relate the paleocurrent patterns to the depositional facies as delimited by *all* data, particularly textures and nondirectional sedimentary structures.

Data from regional studies should be shown initially on map projections, with clear distinction between factual (initial raw measures) and interpretive (variously summarized and mathematically treated) data. Statements should be made on the sampling procedure as well as on the later methods of manipulating the data and on the statistical significance of any summary values displayed. Moving averages and trend surfaces can be constructed to show two-dimensional *areal* variation of scalar properties such as orientation and directional data. Any arbitrary grid can be used to create moving averages (e.g., see below for the method of Wise in compiling data for rose diagrams); resultant vectors are plotted within squares on a map overlay to show regional trends and local anomalies (e.g., see Potter and Pettijohn 1977). Quantitative analysis can be performed utilizing least squares regression procedures to fit different surfaces (linear, quadratic, or other) to the observed data; the difference between the computed and actual data defines the *residual* and calls attention to anomalies.

The most common way of summarizing abundant quantitative paleocurrent data of the same kind for an area is in rose

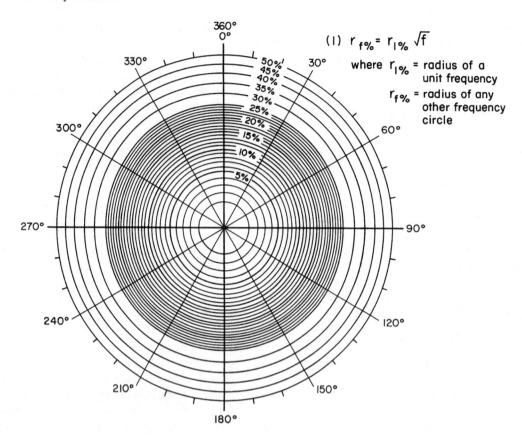

Figure 6-3.   A frequency net to facilitate rapid construction of accurate circular histograms of vectorial directional and orientational data from paleocurrent measurements. The frequency circles can be extended to higher values, if needed, by calculating from the formula provided; $r_{1\%}$ = 10 mm in the example drawn. The % symbols must be omitted if pure number frequencies (rather than percentages) are plotted.   (After Nemec 1988.)

diagrams, which are essentially histograms laid out in a 360° (azimuthal) radial graph (e.g., Andreassen 1990). The majority of those published to date are constructed simplistically and incorrectly, for the same reasons that histograms in general are misleading (i.e., choice of azimuthal intervals for the "bars"; see Fig. 7-15 for normal bar graph effects), because frequency linear scales are used (see discussion by Nemec 1988 and Fig. 6-3). Data concentration factors and average smoothing techniques must be applied (Wise, pers. comm., 1991). The data concentration factor (DCF) used by Wise is:

$$DCF = NFW \times \frac{180 \vee 360}{TND \times W}$$

where NFW = number of points in the window at that location (see below)

TND = total number of data points in the set

W = width in degrees of the counting window

180 or 360 = range of azimuths in the data set

The average value of DCF for the full azimuthal range is equal to 1. The smoothing procedure can be performed manually:

raw azimuthal data are plotted along a linear x axis through the azimuthal range (and with data repeated at either end to avoid edge effects), an (odd) number of degrees in width is moved across the graph in 1° increments, and numbers of points falling within the window at each location are recorded above the window. For double smoothing, the same procedure is repeated (reversing the direction of smoothing) on the numerical results of the first smoothing. Final plots can be linear; one vertical scale division represents one DCF unit and is always plotted with a length equal to 10° of azimuth on the x axis. Rather than perform these tedious measurements and plots, it is far wiser to use a computer program that does all the calculations and also plots the rose (contact Don Wise, University of Massachusetts, for the program).

SELECTED BIBLIOGRAPHY

Ali, S. A., and M. P. Weiss, 1968, Fluorescent dye penetrant technique for displaying obscure structures in limestone. *Journal of Sedimentary Petrology* 38:681–2.

Andreassen, C., 1990, A suggested standard procedure for the construction of unimodal current rose-diagrams. *Journal of Sedimentary Petrology* 60:628–9.

Baker, S. R., and G. M. Friedman, 1969, a Non-destructive core analysis technique using x-rays. *Journal of Sedimentary Petrology* 39:1371–83.

Ballantyne, C. K., and R. Cornish, 1979, Use of the chi-square test for the analysis of orientation data. *Journal of Sedimentary Petrology* 49:773–6.

Bhargava, J., 1971, Choice of voltage and film in radiography of thin rock slabs. *Journal of Sedimentary Petrology* 41:1141–2.

Bockelie, T. G., 1973, A method for displaying sedimentary structures in micritic limestones. *Journal of Sedimentary Petrology* 43:537–9.

Bouma, A. H., 1969, *Methods for the Study of Sedimentary Structures.* Wiley-Interscience, New York, 458p.

Carver, R. E., 1971, *Procedures in Sedimentary Petrology,* Wiley-Interscience, New York, 653p.

Cheel, R. J., and G.V. Middleton, 1986, Measurement of small-scale laminae in sand-sized sediments. *Journal of Sedimentary Petrology* 56:547–8.

Clifton, H. E., 1966, X-ray radiography with x-ray diffraction equipment. *Journal of Sedimentary Petrology* 36:620–35.

Cooper, M. A., and J. D. Marshall, 1981, ORIENT: a computer program for the resolution and rotation of paleocurrent data. *Computers in Geoscience* 7:153–65.

Curray, J. R., 1956, Analysis of two-dimensional orientation data. *Journal of Geology* 64:117–31.

DeCelles, P. G., R. P. Langford, and R. K. Schwartz, 1983, Two new methods of paleocurrent determination from trough cross-stratification. *Journal of Sedimentary Petrology* 53:629–42.

Dott, R. H., Jr., 1974, Paleocurrent analysis of severely deformed flysch-type strata—a case study from South Georgia Island. *Journal of Sedimentary Petrology* 44:1166–73.

Freeman, T., and K. Price, 1979, Field statistical assessment of cross-bed data. *Journal of Sedimentary Petrology* 49:624–5.

Hamblin, W. K., 1962a, Staining and etching techniques for studying obscure structures in clastic rocks. *Journal of Sedimentary Petrology* 32:530–3.

Hamblin, W. K., 1962b, X-ray radiography in the study of structures in homogeneous sediments. *Journal of Sedimentary Petrology* 32:201–10.

Howard, J. D., 1968, X-ray radiography for examination of burrowing in sediments by marine invertebrate organisms. *Sedimentology* 11:249–58.

Jipa, D., 1967, Cross-stratification as a criterion of palaeocurrent direction in flysch deposits. *Scottish Journal of Geology* 3:227–34.

Krause, R. G. F., and T. A. M. Geiger, 1987, An improved method for calculating the standard deviation and variance of paleocurrent data. *Journal of Sedimentary Petrology* 57:779–80.

Krinitsky, E. L., 1970, *Radiography in the Earth Sciences and Soil Mechanics.* Plenum Press, New York, 103p.

Miller, R. L., and J. S. Kahn, 1962, *Statistical Analysis in the Geological Sciences.* Wiley and Sons, New York, 483p.

Nemec, W., 1988, The shape of the rose. *Sedimentary Geology* 59:149–52.

Okada, H., and M. Arita, 1970, An instrument for measuring pretectonic current directions. *Journal of Sedimentary Petrology* 40:1048–51.

Potter, P. E., and F. J. Pettijohn, 1977, *Paleocurrents and Basin Analysis,* 2d ed., Springer-Verlag, New York, 425p.

Ramsay, J.G., 1961, The effects of folding upon the orientation of sedimentation structures. *Journal of Geology* 69:84–100.

Rao, J. S., and S. Sengupta, 1970, An optimum hierarchical sampling procedure for cross-bedding data. *Journal of Geology* 78:533–44.

Risk, M. J., and R. B. Szczuczko, 1977, A method for staining trace fossils. *Journal of Sedimentary Petrology* 47:855–9.

Shrock, R. R., 1948, *Sequence in Layered Rocks.* McGraw-Hill, New York, 507p.

Tucker, M. (ed.), 1988, *Techniques in Sedimentology.* Blackwell Scientific Publications, Oxford, 394p.

7

Textures

Textural attributes of grains and relationships between grains are examined by a variety of procedures ranging in scale from naked-eye examination (coarse particles) to electron microscopy (magnifications to more than 300,000×). Salient procedures for examining texture are largely treated here, but essential preparation procedures are discussed in Chapter 5. Microscopy essential for mineralogical identification also provides powerful means of textural analysis (Chapter 8). Qualitative interpretations from studies of texture are reviewed in **PS** Chapter 5 and a few of the salient papers are provided in the Selected Bibliography of this chapter; interpretations from quantitative studies of size distributions are briefly discussed here following the review of quantitative methodology.

SHAPE

Flemming (1965), Barock (1974), Barrett (1980), and Illenberger (1991) have presented thorough discussion and evaluation of the relative theoretical and practical merits of the wide variety of available shape measures. Interpretive values of quantitative analyses of shape data are illustrated by Patro and Sahu (1974) and Spalletti (1976). Here we focus on common measures of sphericity and roundness.

Sphericity

True sphericity has been defined as the surface area of a grain divided into the surface area of a sphere of the same volume—a rather impractical property to measure! *Operational sphericity* (Wadell 1932) is:

$$(V_p/V_{cs})^{1/3}$$

where V_p = volume of particle and V_{cs} = volume of smallest sphere that would enclose the particle. Volume measures of individual particles are not generally practicable (but see Moussa 1973), and V_{cs} is approximated (Krumbein 1941) by

$$(LIS/L^3)^{1/3} = (IS/L^2)^{1/3}$$

where I = intermediate axis, S = short axis, and L = long axis (mutually perpendicular). The numerical values that result from these calculations do not fall into visually distinct categories, but may cross a variety of shape fields as defined for example in the Zingg diagram (Fig. 7-1a; see also Perez 1987 for another method of plotting shape on triangular diagrams).

Another quantitative sphericity measure commonly applied is that proposed by Sneed and Folk (1958; Fig. 7-1b). As in the case of the relationship of operational sphericity measures to visual shapes, their numerical values of "effective settling sphericity" cross "form" categories. These authors argued that their measure is more appropriate than many others because it will be the maximum projection area (determined in the plane containing the long and intermediate axes, which their quantitative measure emphasizes), which primarily determines the way grains behave while settling. However, there is little merit in selecting this method on those grounds: most sand (and all gravel) particles do not "settle" passively during sedimentation, but are deposited in turbulent traction or active suspension, where other factors are more influential.

In these and other sphericity procedures such as measuring Cailleux "flatness" (L + I/ 2S; see Cailleux 1945), only in the case of gravels and friable conglomerates are three-dimensional measures sometimes attempted. Most workers qualitatively estimate sphericity of particles from visual-comparison charts (e.g., **PS** Fig. 5-7 for a simplified version; Rittenhouse 1943; Catacosinos 1965; Crofts 1974); the estimates can be reliably precise once the worker gains some experience (Rosenfeld and Griffiths 1953). In any kind of

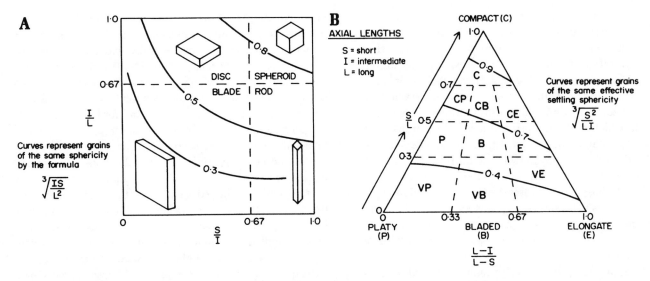

Figure 7-1. Two methods for quantitative measurement of particle shape. *A:* Sphericity; representative solids within the box represent minimal roundness values for each shape. Curves within the box represent values of the quantitative formula for sphericity (shown on the left). Note that the same numeric value spans a variety of shapes. *B:* Sphericity-form diagram of Sneed and Folk (1958). (*A:* After Zingg 1935, in Pettijohn 1975.)

measurement, the inherent limitations of any two-dimensional techniques used should be borne in mind.

Whereas measurement of particle sphericity is usually contemplated only with some gravels, time-consuming determinations of sand particle shapes also may be necessary for fundamental studies on sediment behavior (e.g., Moss 1962).

Roundness

Quantitatively, true roundness is generally expressed by the Wadell formula:

roundness = $\Sigma\,(r/R)/N$

where r = the radius of curvature of grain corners, R = the radius of the largest inscribed circle, and N = the number of corners.

Because it is not practical to determine r values from three-dimensional particles, quantitative measures are generally taken from two-dimensional projected images (direct projections or photographic images expanded to scale). Wherever possible a consistent orientation of the particles is attempted—grains are agitated so that they lie on their surfaces of maximum cross-section (i.e., maximum projection sphericities are viewed). Swan (1974) reviews techniques applied to roundness measures. Sources of experimental error are discussed by Folk (1972). **PS** Chapter 5 discusses and figures some simplified visual comparison charts used for semi-quantitative estimations (see also Harrell 1984).

As in the case of sphericity, there are inherent limitations involved in the two-dimensional measurement techniques generally applied. And other measures, such as Cailleux roundness ($2R/L \times 1000$, where R = minimal radius of curva-

ture seen in the plane of maximum projection sphericity), have been applied by various workers, particularly with gravels for which a variety of arcs with known radii can be constructed for rapid comparison in the field (e.g., King and Buckley 1968). Hence it is important to state the technique utilized when presenting a report on investigations.

Pivotability

Pivotability is an expression of the ability of particles to roll down an inclined plane; it reflects a combination of the sphericity and roundness attributes of grains (e.g., Shepard and Young 1961; Pryor in Carver 1971; Li and Komar 1986; also cf. the "rollability" measure of Winkelmolen 1971). Because size also influences pivotability, analyses must be carried out on specific sieve size fractions. Although not widely determined, the various categories of grains that can be distinguished by this method can almost certainly be profitably related to detailed interpretation of grain behavior in traction transport.

METHODS FOR DETERMINING THE SIZE OF DETRITAL SEDIMENTS

The Udden-Wentworth grain size scale and its phi scale equivalents (a copy of which should be posted on the laboratory wall) used by most sedimentologists has been introduced in **PS** (Chapter 5; see also Wentworth 1922; Tanner 1969; addendum to this chapter, page 129). Quantitative analysis of the size distributions of sediments is necessary for detailed descriptions of sediments, which may differ subtly as well as dramatically in their textural characteristics. It is also necessary for comparison between samples and to discover significant relationships between sediment properties and geologic processes or settings (e.g., see Krumbein 1968; Klovan 1966). Commonly it is the subtle differences, overlooked in a purely qualitative description, that can show trends or refine interpretation of depositional conditions. Methods of quantitative size analysis emphasize precision (reproducibility) rather than accuracy (approach to a measure of "true size," e.g., see Griffiths 1967) because of the various ways in which *size* is defined and measured (consider how the concept cannot be separated from that of *shape*). Standardization of procedure is crucial: comparison of size distributions obtained by different methods is fraught with problems! Thus, in many cases the technique used for a particular analysis will depend on the method used to obtain previous data with which results are to be compared. It is also important to specify the procedure used when reporting results of analyses.

All grain size analyses involve indirect methods of measuring size and are biased by the variable of shape (e.g., the length of intermediate axis determines which grains pass through a sieve mesh), and in settling-tube analyses, by particle density as well. Thus, care must be taken to compare only samples that do not differ markedly in grain shape and composition, and to ensure that samples are fully disaggregated and dispersed (see Chapter 5). Since most gravels are inequant, sieve analysis is not generally a very useful method for determining precise size distributions (but, e.g., see Boggs 1969 for an example applied to pebble size study); for quantitative analysis, sphericity is measured (and can give close agreement with weight or volume measures of size; e.g., Cui and Komar 1984). Also, sieving is usually not warranted for shell fragments because they are inequant, nor should the settling properties of shells (or heavy minerals) be compared with data for quartz/ feldspar and common rock fragment particles (e.g., Maiklem 1968); hence pretreatments to remove carbonates (as well as organic material) may be necessary (e.g., Ellingboe and Wilson 1964; Gibbons 1967). Where diagenetic modification to grain texture has taken place, quantitative size analysis is generally not warranted. Measurement of areas from views of grains' maximum projection sphericity can provide consistent results when dealing with loose grain mounts, but both they and measures of apparent long axes in grains that have been cut randomly (e.g., as in thin section) are of very limited value in size determinations (e.g., Connor and Ferm 1966).

Sample Parameters

The most important preliminary step to quantitative analysis is to ensure representative samples from the sediments, and then appropriate subsamples from those samples (Chapter 4). It is at this stage that the potential applicability of any results is decided.

Samples must be large enough to give statistically meaningful results; hence they must be "large" relative to the largest particle size present; Fig. 7-2 provides a guide to the

minimum sample required. Conversely, sieves should not be overloaded with sediment to avoid damaging the mesh, nor should 1-L settling cylinders be loaded with much more than 10–25 g for pipette analysis, or 30–40 g for hydrometer analysis, to avoid interference effects. The inset in Fig. 7-2 shows total subsample weights for sand/mud mixtures that will provide suitable quantities of mud for pipette analysis. Keep a record of the weight of each subsample and the initial total sample. Grain size segregation commonly occurs during transport of sample bags—ensure that your subsample is representative of the whole by thorough mixing prior to subsampling. Dry the sediment, mix it thoroughly, then pour it evenly into the hopper of a sample splitter, repeating until a sufficiently small subsample is obtained. Equally satisfactory and applicable with wet sediments is successive quartering of cones on a piece of glazed or waxed paper: flatten the cone, quarter with a spatula, reject two opposite corners, mix the other two quarters, and repeat as necessary. For most homogeneous sandy or muddy sediments, a tablespoon blindly dipped into the mixture is an adequate subsampling technique, as long as differential segregation has not occurred.

Rapid Analysis for (Gravel)/Sand/Silt/Clay Ratio

In sediments of mixed grain size, a simplified but precise sieve plus pipette analysis can be made rapidly when ratios of (gravel to) sand to silt to clay are all that is necessary, which is commonly the case. If gravel sizes are sparsely present, they should be removed and weighed by sieving through a 2-mm sieve.

1. Prepare a representative subsample that contains 15–20 g of mud.
2. Disaggregate and disperse the subsample in a solution of distilled water plus (known quantity of) dispersant.
3. Wet-sieve the subsample, dry, and weigh the sand fraction (as in steps 3 and 4 of Table 7-3). Return the pan fraction (finer than 4ϕ) to the mud fraction.
4. Perform pipette analysis as described below, but only for two subsamples: 4ϕ (for total mud) and 8ϕ (for total clay).
5. Subtract weight % of 8ϕ fraction from that of the 4ϕ fraction to obtain total silt weight %. Calculate weight % of sand fraction.

Individual Particle Analysis

Direct measurement of particle size is generally attempted only with gravels, although some analysis of projected images of sands has been performed. A multitude of methods have been applied; none is considered a standard. Callipers can be used to measure the *L, I,* and *S* axes of individual particles with appropriate conventions (e.g., axes mutually perpendicular, *L-* or *I*-axis determinant), but there are commonly several alternatives to the axial lengths even with these con-

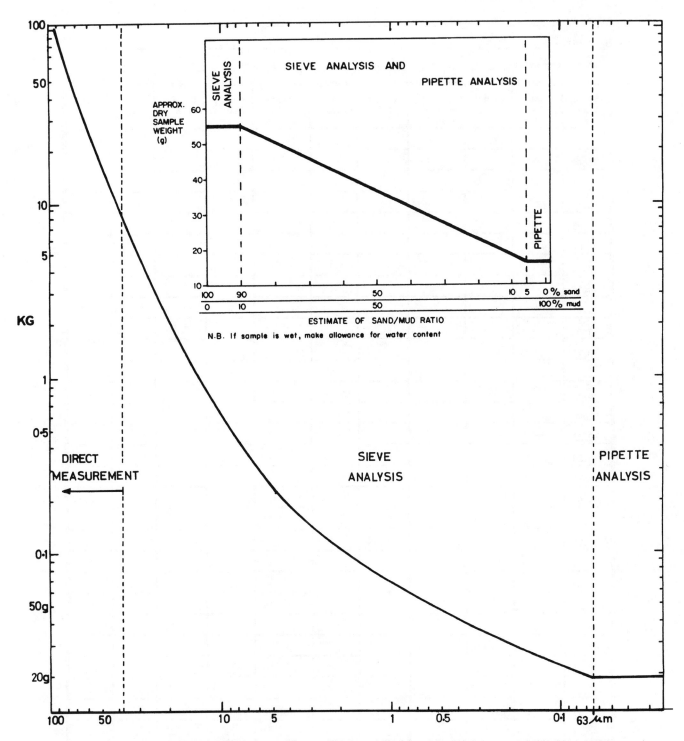

Figure 7-2.   Guide to the relationship between minimum sample weight (kg) for standard grain size analysis and size of the largest particle present in substantial proportion (abcissa scale in mm and μm). Inset shows approximate subsample weights from sediment mixtures that will provide suitable quantities of mud for pipette analysis.   (Diagram by G. Coates.)

ventions, and the lengths determined are treated in various ways (e.g., add all lengths and divide by 3 for a mean diameter). Templates through which gravels can be pushed have been used (e.g., Billi 1984), or the dimensions as seen in maximum projection sphericity can be measured on photographs or an overhead projector (e.g., Burke and Freeth 1969); neither of these methods adequately reflect the three-dimensional shape of these generally inequant particles. Nor does the method of immersion to determine volume, and calculation of a nominal diameter. Experimentation is necessary with any particular suite to determine what consistent measuring procedure will provide useful results. Generally, 100–

(Text continues on page 101.)

Sample No. _____ Analysed by _____ Treatment _____ Date _____

Particulars _____

Weights : dry sample _____sand _____mud _____sand & mud _____

Sieve Diam. Φ	Exact Diam. Φ	Weight Beaker	Weight Beaker + Spl.	Weight Spl.	% Aggs.	Corrected Weight	Cumul. Weight	Cumul. %	% Shell	Notes
-5.00										
-4.00										
-3.00										
-2.50										
-2.25										
-2.00										
-1.75										
-1.50										
-1.25										
-1.00										
-0.75										
-0.50										
-0.25										
0.00										
+0.25										
+0.50										
+0.75										
+1.00										
+1.25										
+1.50										
+1.25										
+2.00										
+2.25										
+2.50										
+2.75										
+3.00										
+3.25										
+3.50										
+3.75										
+4.00										
+4.25										
+4.50										
+4.75										
pan										
Total										

Figure 7-3.   Example of a data sheet for recording the results of sieve analysis. Percentage error = [(original sample weight – total weight retained) × 100]/original weight.

Sample no. _____ Analyst _____ Date _____

Description of sample _____

Modal grade: _____ to _____ mm. Secondary modes: _____ to _____ mm.

_____ to _____ mm.

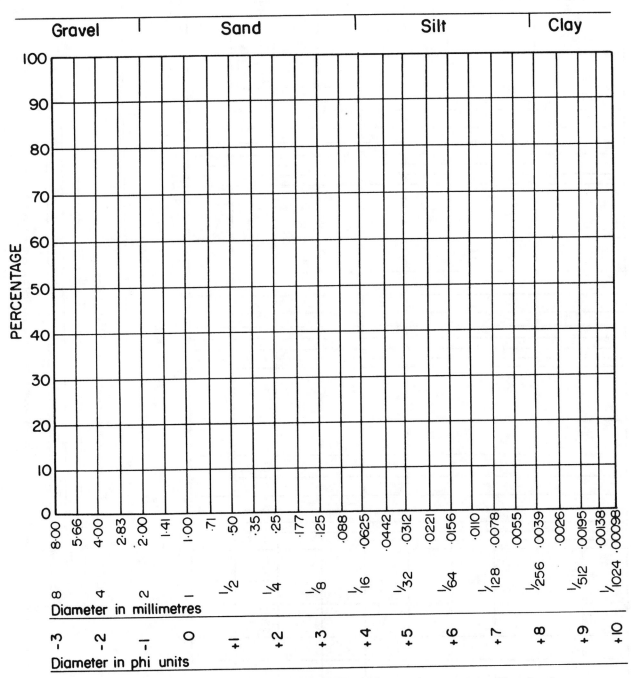

Figure 7-4.   Example of graph paper for plotting Udden-Wentworth grain sizes against weight % of size classes as histogram.

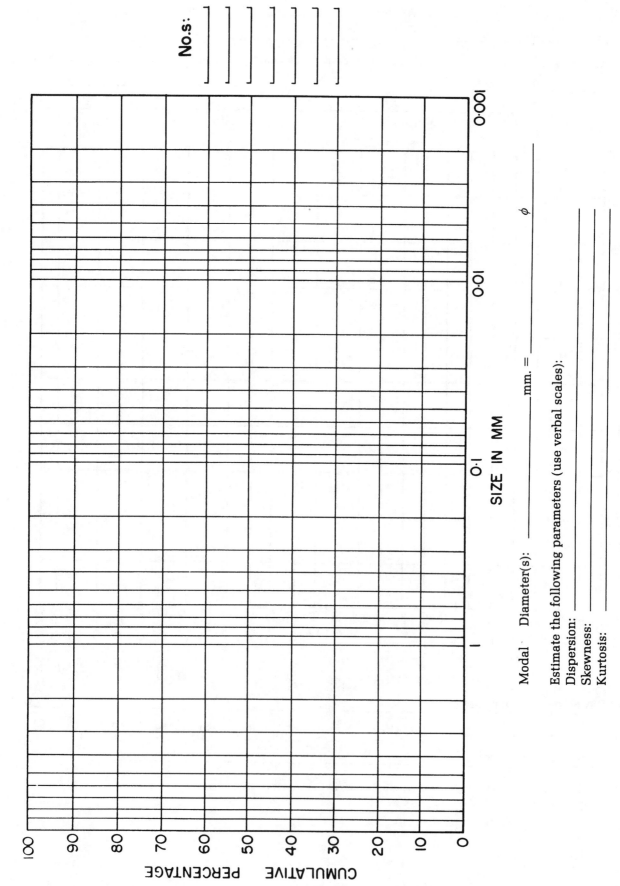

No.s: ⌐⌐⌐⌐⌐⌐

SIZE IN MM

CUMULATIVE PERCENTAGE

Modal Diameter(s): _____ mm. = _____ ϕ

Estimate the following parameters (use verbal scales):
Dispersion: _____
Skewness: _____
Kurtosis: _____

Figure 7-5. Semi-logarithmic (to the base 2) graph paper for plotting the Udden-Wentworth scale (in mm) against cumulative weight % sediment.

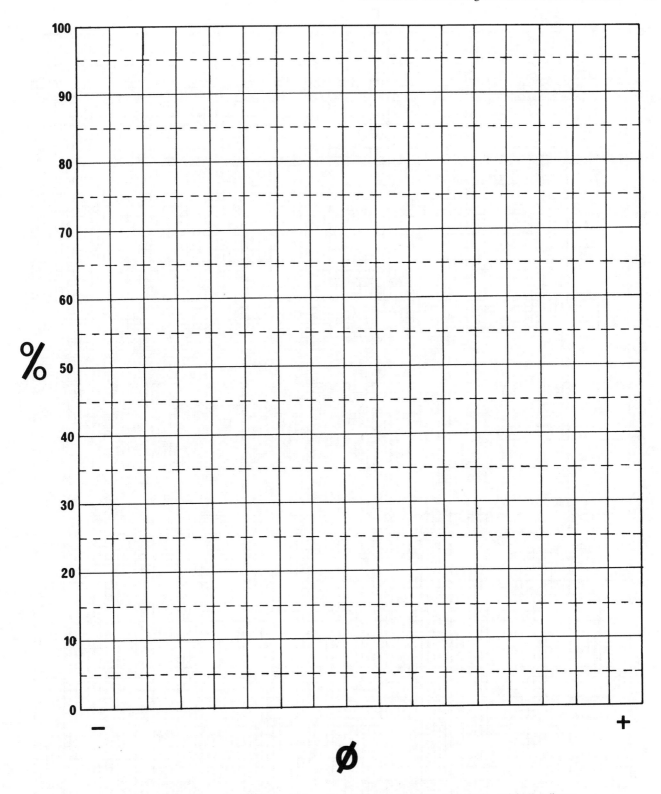

Figure 7-6.   Arithmetic graph paper for plotting the Udden-Wentworth scale (in phi units) against cumulative weight % sediment.

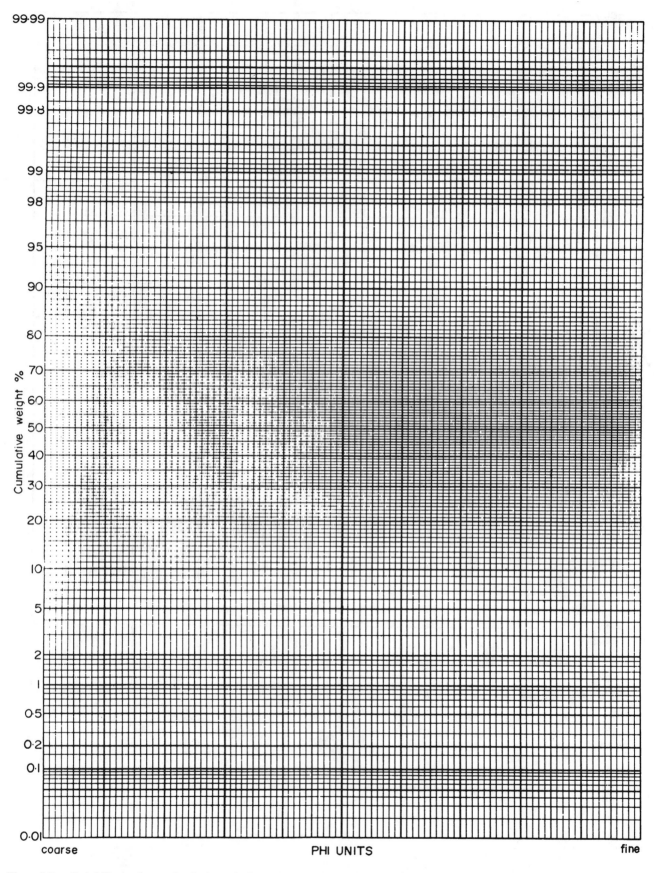

Figure 7-7.   Probability graph paper for plotting grain size (phi units) against cumulative weight % sediment.

400 pebbles should be counted as a representative sample, and in most cases it appears that workers use the intermediate axis as the index of mean particle size.

Sieve Analysis

Most previous data on sand and gravel fractions have been obtained from sieve analysis, which has marked theoretical and experimental limitations in the way it provides size data (see Ludwick and Henderson 1968). Sieving sorts on the basis of smallest cross-sectional diameter (the plane of the intermediate and short axes—the diagonal length between mesh corners determines the intermediate axis length). Since most sedimentary particles are not spherical, the number of particles that pass through a given sieve is time dependent; there will always be more grains that *could* pass through a given sieve if they were to land with just the right orientation on the mesh (e.g., Kennedy, Meloy, and Durney 1985). Hence, as well as other procedures, the time of sieving must be standardized, which is a major problem for field-based sieve analysis. Particularly difficult is field sieving of gravel, since most gravel particles are highly inequant, and yet large field box screens may be the only practical way of analyzing samples, which to be representative may weigh over 100 kg! A comprehensive "cookbook recipe" is provided below for sieving sand distributions in the laboratory. Slight variations may be necessary for particular sample suites (e.g., for heavy mineral sands where grains are commonly inequant; see Wang and Komar 1985) or may be suggested for use in your particular laboratory; if so, make sure you follow the same procedure with all samples. A field method for rapid determination of dry weight from wet sand, involving laboratory calibration with a small number of samples, is described by Kraus and Nakashima 1986.

It is commonly forgotten that sieves do not last forever and that the meshes are easily damaged. Routine testing should be carried out in the laboratory at intervals dependent on sieve usage; standard sets of glass spheres can be obtained from the U.S. Bureau of Standards, mesh openings can be measured with a microscope (at least 50 openings along a diagonal traverse), or a standard sample can be established from natural sediments. New sieves should be checked against the standards and/or those they are replacing in a stack; if necessary, calibration corrections may be applied to ensure consistent results. Table 7-1 lists the procedure for sieve analysis of a sand.

Settling Tubes

The principle is the same for virtually all settling-tube analyses: sizes are calculated on the basis of an ideal settling velocity (v) formula, usually Stokes' Law:

$$v = Cd^2$$

where d is the diameter in centimetres of an assumed sphere and

Table 7-1. Example of a Standard Procedure for Sieve Analysis of a Sand

Preparation

1. Disaggregate thoroughly, remove salts and organic matter (see Chapter 5). Select a representative subsample and label a data sheet for it (e.g., Fig. 7-3). Weigh to 0.001 g.

2. (a) For samples with less than about 10% mud and when analysis of the mud is not necessary, dry the subsample at no more than 65°C (to avoid baking clays). Leave to cool and equilibrate with the atmosphere for at least 1 hr before weighing. Then thoroughly disaggregate the sample—for most loose sands, a rubber bung on a piece of glazed paper is adequate.

(b) For samples with a mud fraction to be analyzed, wet-sieving is necessary (see Table 7-3, steps 3 and 4). It is wise to perform a wet-sieving operation on two subsamples—for one, dry both fractions and determine the proportion of mud to sand, whereas for the second only the sand fraction is dried and weighed and the wet mud fraction used for pipette or hydrometer analysis. After wet-sieving, dry the coarse fraction and weigh.

Analysis

3. Select a nest of sieves to cover the grain size range of the sample. If the sample has been wet-sieved, the finest sieve should be 4φ; otherwise sieves as fine as 4.75φ may be used. For detailed work and where polymodal distributions are present, use 0.25φ intervals.

4. Clean the sieves before using them: invert each sieve and tap it gently onto a flat surface or, using your hand, rap the side diagonally to the mesh to knock out any loose grains. Then brush the screen, again diagonally to the mesh, with a soft sieve brush. If any grains are trapped in the mesh, do not attempt to force them out—leave them there (or distortion of the mesh may result). Stack the sieves in order, with the pan at the bottom. If two nests are necessary, use the coarser set first, then transfer the contents of the pan to the finer stack (with another pan under it!).

5. Pour the sample into the top sieve and add the cover (the greatest load on a sieve should not exceed 5 grain-diameter thickness; otherwise mass-trapping effects or mesh distortion will occur). Secure the sieve nest firmly in the sieve shaker. Shake for a standardized time—usually 10 or 15 min.

6. After shaking, invert and clean each sieve as in step 4; retain each fraction on a large sheet of glazed paper, and transfer each to a labeled, preweighed beaker or envelope. If the sample has previously been wet-sieved and mud analysis is to follow, add sediment passing the 4φ sieve (pan fraction) to the mud fraction.

7. Weigh the beakers (or envelopes). Retain each fraction in a labeled envelope for future use.

8. Check each fraction for grain aggregates and other properties (e.g., compositional differences, shape properties) with hand lens or under a binocular microscope (there may be significant differences between fractions). If aggregates are common, either disaggregate and resieve, or carefully estimate the percentage of aggregates in each fraction and subtract this percentage of the weight of the fraction from both the weight of the fraction and the total weight of the subsample.

9. Compute the weight percentage of each fraction, then compute cumulative percentages. The weight percent of each sand fraction is:

$$100 \times \frac{\text{weight of sand on sieve}}{\text{total sample weight (sand plus mud)}}$$

Add these percentages incrementally to obtain cumulative weight percentages.

10. Plot the data on a histogram (if desired) and as a cumulative curve on graph paper (e.g., Figs. 7-4–7-7). Consistent "kicks" at the same size grade in cumulative curves for different samples may indicate a defective sieve.

$C = [(ds - df) \, g]/18\mu$

where ds = density of solid (generally invalidly assumed to be that of quartz = 2.65 g/cm^2)

    df = density of fluid at its particular temperature

    g = acceleration of gravity (980 cm/sec^2)

    μ = the viscosity of the fluid at the particular temperature

Some of the problems that arise in practical methods of measuring and applying settling velocity analyses are discussed by Lovell and Rose (1991).

Many laboratories use settling tubes for sand analysis (e.g., Gibbs 1972; Reed, leFever, and Moir 1975; Taira and Scholle 1979). The major advantage over sieving is rapidity and automation of analysis, although another strong argument is that the procedure takes into account not only size but also shape and density to provide a more realistic measure of hydraulic equivalent sizes. Analysis involves smaller samples (1–10 g) than those used in sieving (to avoid mass-trapping effects); therefore, special care must be taken to obtain representative subsamples from the initial sample. A great many instruments are available, with various tube diameters (greater for coarser particles) and lengths (most more than 2 m), and various devices for measuring the proportion of grains that reach a given point or accumulate on the bottom in relation to time (e.g., Colby and Christensen 1956). Each laboratory will detail specific analytical procedures for its particular instrument. A simple settling tube device that can be constructed relatively inexpensively is described by Felix (1969). In the most common *sedimentation balance* devices, a small pan is present near the base of the tube and the varying weight of sediment caught is automatically recorded as a function of time (Fig. 7-8); multiple samples can be run as long as the recorder can be recalibrated to zero after each run. Calibration is generally necessary (e.g., Cook 1969; Chapman

1981). The size distributions determined by all these techniques are not directly comparable with sieve size distributions, and size data should ideally be expressed in psi (ψ) (hydraulic equivalent size units) rather than phi (ϕ) values (same scale, different names); there can be a very close similarity in the results from the two procedures (Fig. 7-9 and see Komar and Cui 1984).

A variety of techniques may be used to analyze the size of silt and clay fractions (e.g., Swift, Schubel, and Sheldon 1972); the most common are the pipette and hydrometer procedures detailed below. Another technique used to measure size distributions in settling tubes is to pass a light beam through an aqueous suspension of the grains; as grains settle, the characteristics of the light received by the recorder vary systematically with the changing density of the suspension (e.g., Taira and Scholle 1977). The Coulter Counter is an expensive device for analyzing both muds and sands (0.0005–0.85 mm; see McCave and Jarvis 1973; Stein 1985) that is available in a variety of industrial (and medical) and a few earth-science laboratories. Particles are suspended in an electrolyte drawn through apertures of various diameters; automatic measures are made of current fluctuations as different volumes of the low-conductivity particles pass through. Precise results are attained very rapidly; hence large numbers of samples can be processed daily. Nonetheless, precision varies between the techniques and comparative results commonly differ significantly (e.g., Creager and Steinberg 1963; Steinberg and Creager 1961; Behrens 1978; Coates and Hulse 1985; Singer et al. 1988; Fig. 7-10). Regardless of the technique used, any disaggregation and dispersion tech-

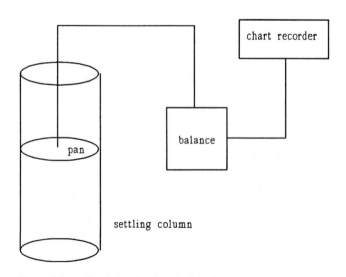

Figure 7-8. Sketch showing the principle of the settling tube procedures that utilize a pan to trap sediment.

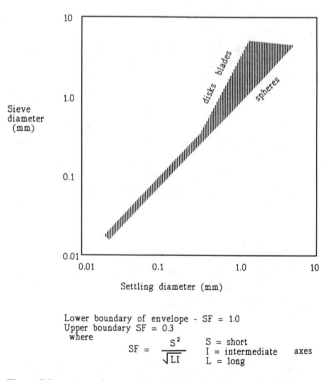

Lower boundary of envelope - SF = 1.0
Upper boundary SF = 0.3
   where

$$SF = \frac{S^2}{\sqrt{LI}}$$

    S = short
    I = intermediate   axes
    L = long

Figure 7-9. Approximate relationships between sieve and settling diameters for grains of different shape.   (After Fritz and Moore 1988.)

niques are likely to have modified the physical and chemical properties of the natural clay fraction (e.g., by deflocculating them, removing organic material, and changing the chemistry of the system). Natural-state studies must be carried out before bacterial or organic decomposition changes the properties, but commonly will not be feasible because natural flocculation may prevent settling. Swelling clays present a special problem, and it is doubtful whether useful data can be obtained from sediments that contain them or have been diagenetically modified.

Most data on the mud fraction have been obtained from pipette analysis. Precision is high—better than 0.1ϕ unit—but the method is time-consuming. Size analysis by hydrometer gives results that compare well with those of pipette analysis, and reproducibility is almost as good as that of pipette analysis; size analysis by other electronic instruments is even easier and faster, but precision of the particular instrument and comparability with results of other methods must be tested (e.g., Fig. 7-10, Kaddah 1974; Coates and Hulse 1985). Hydrometer analysis is considerably easier and

quicker to carry out than pipette analysis and is widely used by soil scientists. Elutriation techniques (where controlled upward-flowing fluid or air differentially suspends the grains) is an example of a relatively less-used method. Sonic sifters, which can sieve dry particles from c. 150 down to c. 5 microns in size within an oscillating column of air, are new devices that may prove a viable standard if many laboratories acquire them.

Pipette Analysis of Mud

Subsamples of a specific volume are extracted from a suspension of mud at specified times and depths; the weight of each dried subsample is representative of the proportion of the total mud fraction remaining in suspension above that specified depth at that specified time. Thus, each subsample measures the proportion of total mud that is finer than the size that will have settled to the specified depth in the specified time. For example, from Fig. 7-11, the maximum grain size that will have settled to the specified (10- or 20-cm) depth can be read as a function of time (e.g., after 2 hr, the

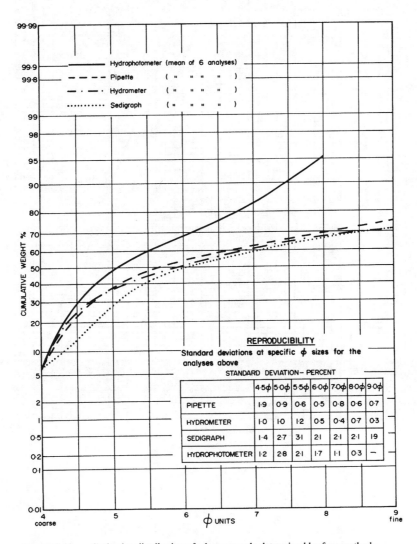

Figure 7-10. Grain size distribution of a loess sample determined by four methods. (Analysis and diagram by G. Coates.)

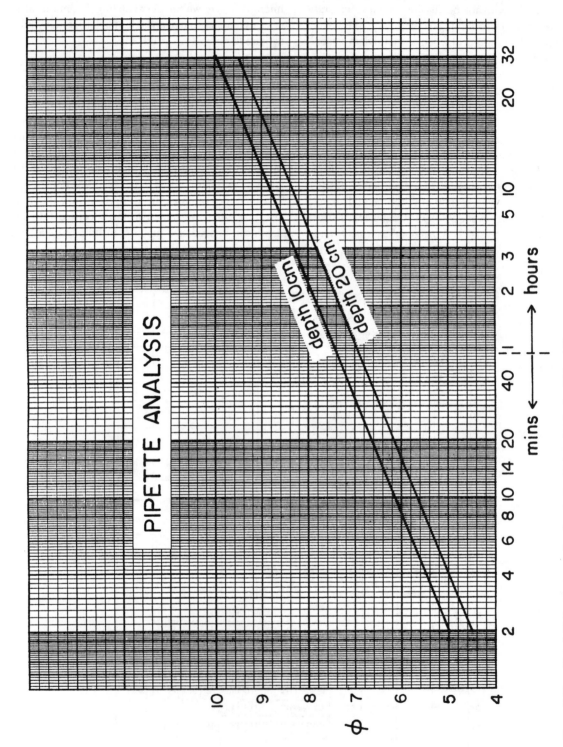

Figure 7-11. Plot of phi diameter versus time for pipette withdrawal at depths of 10 and 20 cm.

maximum grain size 10 cm below the surface is 8ϕ and a withdrawal at this depth will contain sediment finer than 8ϕ). Temperature affects the viscosity of water and therefore settling velocities. A correction for the temperature effect can be made indirectly by altering the depth of withdrawal (Table 7-2). Note the temperature of tap water in a separate column before and during any pipette analysis: change the sampling depth as necessary. The following equipment is needed:

1 data sheet
1 beaker or basin for disaggregation
1-L measuring cylinder(s)
eight or nine 50-mL beakers
1 watch glass for cylinder
4ϕ (0.063-mm) wet sieve (do not use 4ϕ dry sieve)
large evaporating basin
wash bottle with distilled water (may have dispersant)
large funnel (about 20 cm in diameter)
solution of dispersant (e.g., make up a stock solution of 50 g/L of "Calgon"; 20 mL of this solution—use a pipette—will have 1 g of dispersant)
brass stirring rod

thermometer (0–100°C)
20-mL pipette with depth graduations (mark depths with insoluble ink 10 cm and 20 cm from lower tip of pipette; a few additional marks at 0.5-cm intervals around those depths may also prove useful. Some laboratories use a 50-mL pipette and take 50-mL samples—we prefer to leave the maximum amount of fluid in the cylinder)
rubber pipette bulb for consistent suction
timepiece with hours, minutes, and seconds
pipette analysis schedule and temperature/withdrawal depth sheet
oven for drying mud fractions (60–65°C)

Allow up to one full day for preparing the samples, depending on how many are to be analyzed. Label a separate data sheet for each sample (e.g., Fig. 7-12).

The subsequent procedure for pipette analysis is listed in Table 7-3. Because settling time of the clay fraction is so slow, careful organization of the work schedule is necessary to minimize late-evening candle-burning and/or to avoid missing dinner! Table 7-4 presents a time schedule for pipette analyses of up to 21 samples in a day.

Table 7-2.   Depth (cm) of Pipette Insertion for a Given Temperature (assuming particles of S.G. = 2.65)

Temp. °C	Total Suspension 20 sec	4.5ϕ 44μ 2 min	5ϕ 31μ 4 min	5.5ϕ 22μ 8 min	6ϕ 16μ 15 min	7ϕ 8μ 30 min	8ϕ 4μ 2 hr	9ϕ 2μ 8 hr	10ϕ 1μ 32 hr
14.0	20.0	17.8	17.7	17.8	17.7	8.8	8.8	8.8	8.8
14.5	20.0	18.1	17.9	18.1	17.9	9.0	9.0	9.0	9.0
15.0	20.0	18.3	18.2	18.3	18.1	9.1	9.1	9.1	9.1
15.5	20.0	18.4	18.4	18.5	18.4	9.2	9.2	9.2	9.2
16.0	20.0	18.8	18.6	18.8	18.6	9.3	9.3	9.3	9.3
16.5	20.0	19.1	18.9	19.1	18.9	9.4	9.4	9.4	9.4
17.0	20.0	19.3	19.1	19.3	19.1	9.6	9.6	9.6	9.6
17.5	20.0	19.5	19.4	19.5	19.4	9.7	9.7	9.7	9.7
18.0	20.0	19.8	19.7	19.8	19.6	9.8	9.8	9.8	9.8
18.5	20.0	20.0	19.9	20.0	19.9	9.9	9.9	9.9	9.9
19.0	20.0	20.2	20.1	20.2	20.2	10.0	10.0	10.0	10.0
19.5	20.0	20.5	20.4	20.5	20.3	10.2	10.2	10.2	10.2
20.0	20.0	20.7	20.6	20.7	20.6	10.3	10.3	10.3	10.3
20.5	20.0	21.0	20.9	21.0	20.9	10.4	10.4	10.4	10.4
21.0	20.0	21.3	21.1	21.3	21.1	10.5	10.5	10.5	10.5
21.5	20.0	21.5	21.3	21.5	21.3	10.7	10.7	10.7	10.7
22.0	20.0	21.8	21.6	21.8	21.6	10.8	10.8	10.8	10.8
22.5	20.0	22.0	21.9	22.0	21.8	10.9	10.9	10.9	10.9
23.0	20.0	22.3	22.1	22.3	22.1	11.1	11.1	11.1	11.1
23.5	20.0	22.6	22.4	22.6	22.3	11.2	11.2	11.2	11.2
24.0	20.0	22.8	22.7	22.8	22.6	11.3	11.3	11.3	11.3
24.5	20.0	23.1	22.9	23.1	22.9	11.4	11.4	11.4	11.4
25.0	20.0	23.3	23.2	23.3	23.2	11.6	11.6	11.6	11.6
25.5	20.0	23.6	23.5	23.6	23.4	11.7	11.7	11.7	11.7
26.0	20.0	23.9	23.8	24.0	23.7	11.9	11.9	11.9	11.9
26.5	20.0	24.2	24.0	24.2	24.0	12.0	12.0	12.0	12.0
27.0	20.0	24.5	24.3	24.5	24.2	12.1	12.1	12.1	12.1
27.5	20.0	24.7	24.6	24.7	24.5	12.3	12.3	12.3	12.3

Note: *Time (min) = Depth of withdrawal (cm)/$(1500 \times A \times d^2)$, where A is a constant for a given water viscosity and particle density and d is the particle diameter.

Sample No. _____

Peptizer _____   Amount of Peptizer / 1000 ml. _____ gm.   Total mud _____ gms.

Diameter Φ	°C	Withdrawal depth (cm)	Time	Beaker No.	Wt. Sample + Beaker (gms)	Weight Beaker (gms)	Weight Sample (gms)	x50	Weight of Fraction	Cumulative Weight	Cumulative Percent
+4.0			20 sec.								
+4.5			2 min								
+5.0			4 min.								
+5.5			8 min.								
+6.0			15 min.								
+7.0			30 min.								
+8.0			2 hr.								
+9.0			8 hr.								
+10.0			32 hr.								

Notes: _____

Figure 7-12. Example of a sample data sheet for pipette analysis.

106

Table 7-3.   Example of a Standard Procedure for Pipette Analysis of a Mud

Preparation

1. Obtain a representative subsample that will yield no more than 15–20 g of mud.

2. Fully disaggregate the subsample (see Chapter 5). It may be adequate to cover the sample with a little distilled water plus dispersant (keep track of dispersant added) in a beaker and to use fingers in a rubber glove to break up the sample fully (rinse mud off glove back into the beaker). Alternatively, standardize on a time with an ultrasonic device.

3. Wet-sieve the sample with a *reserved-for-the-purpose* 4φ wet sieve. Place the sieve over a large evaporating basin and wash all its fines into the sieve using as little distilled water (usually the standard dispersant solution) as possible—end up with no more than 900 mL of water and mud! (After about 600 mL, let the silt settle out, then use the partly clear water for further wet-sieving; wash finally with clean water.)

4. Transfer all the sand fraction retained on the sieve to an evaporating basin or a beaker, using the wash bottle. Dry the sand fraction, leave to cool for 1 hr, and weigh to 0.001 g. (If there is a significant amount of sand, dry-sieve it before carrying out the pipette analysis, and extract any new mud fraction that may appear after dry sieving.)

5. Transfer all the mud collected in the basin to the 1-L measuring cylinder via a large funnel (label each cylinder).

6. Add 20 mL of prepared dispersant solution to the column if you have not previously used a solution with dispersant in your wash bottle or for disaggregation (see Chapter 5, "Dispersion of Clays"). Between about 0.5 and 1 g of sodium hexametaphosphate ("Calgon") is normally sufficient to prevent flocculation of clays, but this compound may dissolve fine carbonate grains such as foraminifers and may interfere with later X-ray analysis of clays. It is essential to know the exact amount of dispersant in each column for later calculations.

7. Top the column up to 1000 mL with distilled water. Thoroughly stir the column with a brass stirring rod (a disk with holes at the bottom of the tube is designed to generate maximum turbulence).

8. Label, and weigh to 0.001 g, eight (or nine) 50-mL beakers (one for each withdrawal on the pipette data sheet). Arrange the beakers in front of the column.

9. Cover the column with a watchglass and let it stand overnight to check for flocculation before running the pipette analysis. Fill a beaker with tap water and insert a thermometer (preparatory to the next step).

Analysis

Begin pipette analysis early in the morning, because the time between first and last withdrawals is at least 8 hours.

Before beginning, check that no columns have flocculated. Flocculation can be recognized by a curdling and rapid settling of clumps of particles, or by the presence of a thick, soupy layer on the bottom of the cylinder that passes abruptly into relatively clear water above. If flocculation is evident, try adding more dispersant solution or make up a new suspension with a smaller amount of sample. Using a mechanical stirrer for 5 minutes may assist dispersion.

10. Take the temperature of the water in the beaker of tap water and look up the corrected depths in Table 7-2. Note these depths on the pipette schedule, and monitor any temperature changes during the analysis (or ensure constant temperature by air conditioning). Viscosity changes with temperature and settling velocities will change significantly if there is variation.

11. Select a 20-mL pipette (one that empties quickly) with depth graduations. Connect a rubber pipette filler and check that the suction works efficiently. Have a large beaker of distilled water ready on the bench for rinsing.

12. Start the timepiece 1 min before the initial withdrawal (if using an electronic timepiece, set it at 11:59 P.M.). Immediately begin stirring column 1, using a brass stirrer. Start with short, quick strokes at the bottom and stir up all the settled mud, then work up the column with long, vigorous strokes, being careful not to mix air in with the suspension. Precisely at time zero (12:00:00 on the electronic timepiece), withdraw the stirrer. Lower the pipette to 20 cm. At exactly 20 sec, extract a 20-mL sample. Empty it into the respective 50-mL beaker and then rinse the pipette into the same beaker after sucking up 20 mL distilled water (also wash outer part with distilled water from the wash bottle).

This first withdrawal is particularly critical since it represents everything finer than 4φ (that is, total mud). Insertion of the pipette for subsequent withdrawals should be made with much more care to avoid creating turbulence.

13. The next withdrawal is for the fraction finer than 4.5φ. At exactly 2 min, withdraw 20 mL, empty it into the next beaker, and rinse as before.

Repeat the procedure for all subsequent withdrawals. Efficiency is essential, particularly where multiple samples are to be analyzed. Initially, a withdrawal must be made and the next column stirred within 1 min. Withdrawal and rinsing need to be completed in 30 sec, leaving 30 sec for stirring the next column. (To ensure thorough stirring of every column, carry out a preliminary stir in each one during an earlier spare moment.)

If withdrawal must be made at the wrong depth or time, make a note of the error and use Fig. 7-11 to find the grain size represented.

When there are long periods between withdrawals, cover each column with a watch glass. Any external source of vibration must be eliminated during the analysis.

14. When all withdrawals are completed, put beakers onto trays and oven dry them; it may take up to 48 hr to evaporate all the water. If further analysis of the clays is to follow, do not heat above 65°C.

15. Remove dry beakers from the oven and leave them to equilibrate with the atmosphere for at least 1 hr. Weigh to 0.001 g; record on data sheet.

16. Calculate cumulative weight percentages:

(a) Subtract beaker weights from beaker + sediment weights to get sediment weights.
(b) Multiply the weight of sediment from the 4φ sample by 50 and subtract the weight of dispersant in the column. This gives the total weight of mud, e.g., 0.405g (4φ sediment weight) × 50 – 1 g (wt. of Calgon in the procedure suggested) = 19.25 g (weight of mud, F).

This value, added to the weight of the sand fraction (S) determined from step 4, provides total sample weight. To test for experimental error, either (1) measure total sample dry weight initially (however, even low-temperature drying may cause problems in subsequent dispersion of the clay fraction); or (2) dry and weigh the suspension remaining in the cylinder after full analysis. If error has crept in to the above calculations, correct as necessary.

(c) Add the sand percentages cumulatively to obtain their cumulative percentages (step 9 of Table 7-1).
(d) Remember that each pipette sample represents material in the column finer than a certain grain size. To obtain cumulative percentages for mud intervals, multiply each mud weight by 50, subtract the weight of dispersant, divide by the total sample weight, and subtract from 100:

cum. % (mud range) =

$$\frac{100 - (50 \times (\text{pipette sample wt.})) - 1 \text{ (assuming 1g/L dispersant)}}{S + F}$$

A computer program can be constructed easily in standard spreadsheet software packages to process the raw data (all cells other than those for data entry should be "locked"; see also Slatt and Press 1976; Coates and Hulse 1985).

17. Plot results on graph paper as required (see Figs. 7-4–7-7) and proceed to graphical statistical analysis, or process by Method of Moments.

108   Textures

Table 7-4.   Schedule for 21 Pipette Analyses in One Day, 8:00 A.M. to 8:14 P.M. (Lunch: 45 minutes; Dinner: 1 hour)

Time (hr:min:sec)	Sample No	Approx. Depth (cm)	Size (φ)	Time (hr:min:sec)	Sample No	Approx. Depth (cm)	Size (φ)
0:00	1	—	Start	1:12	10	—	Start
0:20	1	20	4.0	1:12:20	10	20	4.0
2	1	20.	4.5	1:13	7	10	7.0
4	1	20	5.0	1:14	10	20	4.5
				1:16	10	20	5.0
5	2	—	Start	1:17	11	—	Start
5:20	2	20	4.0	1:17:20	11	20	4.0
7	2	20	4.5	1:18	8	10	7.0
8	1	10	5.5	1:19	11	20	4.5
9	2	20	5.0	1:20	10	20	5.5
10	3	—	Start	1:21	11	20	5.0
10:20	3	20	4.0	1:22	12	—	Start
12	3	20	4.5	1:22:20	12	20	4.0
13	2	20	5.5	1:23	9	10	7.0
14	3	20	5.0	1:24	12	20	4.5
15	1	20	6.0	1:25	11	20	5.5
18	3	20	5.5	1:26	12	20	5.0
19	4	—	Start	1:27	20	20	6.0
19:20	4	20	4.0	1:30	12	20	5.5
20	2	20	6.0	1:32	11	20	6.0
21	4	20	4.5	1:37	12	20	6.0
23	4	20	5.0	1:42	10	10	7.0
24	5	—	Start	1:46	13	—	Start
24:20	5	20	4.0	1:46:20	13	20	4.0
25	3	20	6.0	1:47	11	10	7.0
26	5	20	4.5	1:48	13	20	4.5
27	4	20	5.5	1:50	13	20	5.0
28	5	20	5.0	1:51	14	—	Start
29	6	—	Start	1:51:20	14	20	4.0
29:20	6	20	4.0	1:52	12	10	7.0
30	1	10	7.0	1:53	14	20	4.5
31	6	20	4.5	1:54	13	20	5.5
32	5	20	5.5	1:55	14	20	5.0
33	6	20	5.5	1:56	15	—	Start
34	4	20	6.0	1:56:20	15	20	4.0
35	2	10	7.0	1:58	15	20	4.5
37	6	20	5.5	1:59	14	20	5.5
39	6	20	5.5	2:00	15	20	5.0
40	3	10	7.0	2:00*	1	10	8.0
43	7	—	Start	2:01	13	20	6.0
43:20	7	20	4.0	2:04	15	20	5.5
44	6	20	6.0	2:05	2	10	8.0
45	7	20	4.5	2:06	14	20	6.0
47	7	20	5.0	2:10	3	10	8.0
48	8	—	Start	2:11	15	20	6.0
48:20	8	20	4.0	2:16	13	10	7.0
49	4	10	7.0	2:19	4	10	8.0
50	8	20	4.0	2:21	14	10	7.0
51	7	20	5.5	2:24	5	10	8.0
52	8	20	5.0	2:25	16	—	Start
53	9	—	Start	2:25:20	16	20	4.0
53:20	9	20	4.0	2:26	15	10	7.0
54	5	10	7.0	2:27	16	20	4.5
55	9	20	4.5	2:29	16	20	5.0
56	8	20	5.5	2:29*	6	10	8.0
57	9	20	5.0	2:3	16	20	5.5
58	7	20	6.0	2:40	16	20	5.5
59	6	10	7.0	2:43	7	10	8.0
1:01:00	9	20	5.5	2:48	8	10	8.0
1:03	8	20	5.5	2:53	9	10	8.0
1.08	9	20	6.0	2:55	16	10	7.0

Table 7-4.   (Continued)

Time (hr:min:sec)	Sample No	Approx. Depth (cm)	Size (φ)	Time (hr:min:sec)	Sample No	Approx. Depth (cm)	Size (φ)
3:12	10	10	8.0	4:22	21	20	5.5
3:17	11	10	8.0	4:24	20	20	6.0
3:22	12	10	8.0	4:25	16	10	8.0
3:25	17	—	Start	4:29	21	20	6.0
3:25:20	17	20	4.0	4:34	19	10	7.0
3:27	17	20	4.5	4:39	20	10	7.0
3:29	17	20	5.0	4:44	21	10	7.0
3:30	18	—	Start		LUNCH BREAK		
3:30:20	18	20	4.0	5:25	17	10	8.0
3:32	18	20	4.5	5:30	18	10	8.0
3:33	17	20	4.5	6:04	19	10	8.0
3:34	18	20	5.0	6:09	20	10	8.0
3:38	18	20	5.5	6:14	21	10	8.0
3:40	17	20	6.0	8:00	1	10	9.0
3:45	18	20	6.0	8:05	2	10	9.0
3:46	13	10	8.0	8:10	3	10	9.0
3:51	14	10	8.0	8:19	4	10	9.0
3:55	17	10	7.0	8:24	5	10	9.0
3:56	15	10	8.0	8:29	6	10	9.0
4:00	18	10	7.0	8:43	7	10	9.0
4:04	19	—	Start	8:48	8	10	9.0
4:04:20	19	20	4.0	8:53	9	10	9.0
4:06	19	20	4.5	9:12	10	10	9.0
4:08	19	20	5.0	9:17	11	10	9.0
4:09	20	—	Start	9:22	12	10	9.0
4:09:20	20	20	4.0	9:46	13	10	9.0
4:11	20	20	4.5	9:51	14	10	9.0
4:12	19	20	5.5	9:56	15	10	9.0
4:13	20	20	5.0	10:25	16	10	9.0
4:14	21	—	Start		DINNER		
4:14:20	21	20	4.0	11:25	17	10	9.0
4:16	21	20	4.5	11:30	18	10	9.0
4:17	20	20	5.5	12:04	19	10	9.0
4:18	21	20	5.0	12:09	20	10	9.0
4:19	19	20	6.0	12:14	21	10	9.0

Note: For 32-hr withdrawals (10φ measures), add 24 hr to the times for the 9φ withdrawals.

*Coincident withdrawal times: the 2-hr withdrawals will not be harmed by about 30 sec delay.

Hydrometer Analysis of Mud

The density of a sediment-water suspension is measured at intervals and the progressive decrease in density correlated with the size fractions that have settled past the measuring depth in the intervening times. Whereas the actual measurements are quickly and easily made, care must be taken to calibrate each hydrometer initially and to correct for any extraneous variables that may affect fluid density (such as temperature fluctuations). The following equipment is needed:

hydrometer (ASTM 152H)
mechanical shaker
thermometer (0–50°C)
4φ wet sieve
1000-mL graduated cylinder(s)

20-mL pipette
stirring plunger
analytical balance (0.001 g)
stopwatch
evaporating dish
500-mL shaking bottle
oven
wash bottle
250-mL measuring cylinder

The initial step is to calibrate the hydrometer (Table 7-5). Thereafter, to achieve greater accuracy or to compensate for samples of clearly differing composition, determine the specific gravity (S.G.) of the sample by the method listed in Table 7-6. For general purposes, an S.G. of 2.65 is assumed.

Table 7-5.   Procedure for Calibration of a Hydrometer

1. Immerse and steady the hydrometer in approximately 170 mL H₂O in a 250-mL cylinder; measure the water level before and after immersion.

2. Slowly withdraw hydrometer until the water level is halfway between the two previous levels; clamp in position. Record reading (R) on the hydrometer stem at the level of the cylinder top (one way is to lay a ruler across the cylinder for reference).

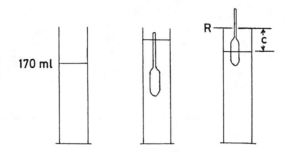

3. Measure distance from the cylinder top to the water level (nearest 0.5 mm); this is distance (C) from R to the center of gravity of the hydrometer bulb.

4. Remove hydrometer; measure (to the nearest 0.5 mm) and record distances (y) from R to each graduation on the hydrometer stem (+y above R, −y below R).

5. Fill a litre cylinder to the 1000-mL mark; immerse the hydrometer and record the change in level (L) to the nearest 0–3 mm.

6. Record the meniscus correction (Cm) as the difference between the top of the meniscus (R'h), determined from sighting on the cylinder wall, and the level water surface (Rh), read off the hydrometer stem. Hydrometers are normally read at the level of the liquid, but with opaque suspensions only the top of the meniscus is evident.

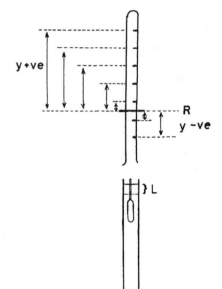

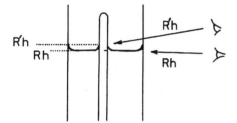

7. For each of the major graduations on the hydrometer stem, calculate:

   (a) the effective depth (Hr) by: $Hr = c + y - L/2$ (mm)
   (b) the hydrometer reading at the top of the meniscus by: $R'h = Rh - Cm$

Plot the calculated values of Hr and R'h on arithmetic graph paper; read the values of Hr from this graph for each reading of R'h you make with this hydrometer during subsequent analyses.

Table 7-6.   Procedure for the Determination of the Solid Density (S.G.) of a Sediment Sample (as is necessary with hydrometer analyses)

1. Dry 50–100 g of the sample in an oven; weigh to 0.1 g.

2. Place the weighed sample into a bottle half filled with water; insert a glass stopper and shake until dispersion is complete. Almost fill the bottle with water and allow the sample to settle for several minutes. Completely fill the bottle and insert the stopper. Dry the outside of the bottle and weigh it to 0.1 g.

3. Empty and wash the bottle; fill with water, insert stopper, dry bottle, and weigh to 0.1 g.

4. The solid density is calculated as follows:

$$\rho_s = \frac{\rho_w \times (\text{wt. of dry sample})}{(\text{wt. of bottle} + H_2O) + (\text{wt. of dry sample}) - (\text{wt. of bottle} + H_2O \text{ and sample})}$$

where (ρ_w) is the density of the water at the temperature of the test.

Table 7-7 lists the procedure for completing the hydrometer analysis up to the initiation of data recording; Fig 7-13 is a data sheet for recording the results. Table 7-8 shows a program for performing 21 analyses in one day.

Once the data have been compiled, transfer them (from Table 7-8) to a data sheet such as Fig. 7-13. The values for $R'h$ - Cc, diameter (D), and percent (P) must then be calculated according to the following formulae:

(1)   Corrected hydrometer reading (Rh) = $R'h - Cc$

where $R'h$ is the hydrometer reading at the top of the meniscus and Cc is the composite correction factor from the readings in the reference cylinder.

(2) d (diameter$_{mm}$) = $K (H_r/t)^{1/2}$

where K = a constant depending on the temperature of the suspension and the solid density of the particles (see Table 7-9)

H_r = effective depth (read from the graph that plots $R'h$ versus Hr)

t = elapsed time after stirring (in min)

(3)   % coarser than (P) = $100 - [(100 \times Rh \times a)/W]$

where Rh is the corrected hydrometer reading, a is the S.G. correction (see below), and and W is the weight of total sample.

Specific Gravity correction

S.G.	a
2.95	0.94
2.90	0.95
2.85	0.96
2.80	0.97
2.75	0.98
2.70	0.99
2.65	1.00
2.60	1.01
2.55	1.02
2.50	1.03
2.45	1.04

Table 7-7.   Example of a Standard Procedure for Hydrometer Analysis

Preparation

Obtain a representative subsample of 30–40 g of mud and treat as for pipette analysis (Table 7-3, steps 2 and 3). Dry in oven at 65° C or less, let cool for 1 hr, and weigh to 0.001 g to obtain weight (W). Prepare settling cylinders as for pipette analysis (steps 5–7) and make sure the samples are dispersed.

Analysis

1. Fill a reference cylinder (label separately to the sequence of multiple cylinders used for analysis) with water and dispersant solution as used for sample suspensions. Shortly before beginning the analysis, take a hydrometer reading in this cylinder at the top of the meniscus and record it as the composite correction (Cc), which in the calculations will correct for temperature, dispersing agent, and the meniscus. If the temperature varies by more than 1° during the analysis, remeasure Cc and note the time.

2. Stir each cylinder for 1 min (as for pipette analysis, step 11) and take hydrometer readings at 1 min, then at 2 min from withdrawal of the stirring rod (these times give approximately 4.5φ and 5.0φ size fractions).

To take readings: dry the hydrometer, then 20–30 sec before reading time, gently lower it into the suspension (allow 10 sec). Steady the hydrometer to take a reading at the appropriate time. Remove the hydrometer gently (allow 10 sec), rinse, and store in a beaker of clean water.

Repeat the 1- and 2-min readings at least once more and enter the averaged readings on your data sheet. (Repeated readings are necessary because the density of the suspension changes rapidly in the early stages of sedimentation.)

3. Start the timer 1 min ahead of zero, immediately begin stirring the column, and withdraw the rod at time zero. Take readings as appropriate (see example data sheet provided).

Cover the cylinders with watch glasses when there are any long periods between readings.

Table 7-8.   Schedule and Data Sheet for 21 Hydrometer Analyses in One Day

Time (hr.min)	Col. No.	R′h	Time (hr.min)	Col. No.	R′h	Time (hr.min)	Col. No.	R′h	Time (hr.min)	Col. No.	R′h
0.00	1	ST*	1.13	7		2.29**	16		4.19	4	
0.04	1		1.16	10					4.19**	21	
						2.30	17	ST	4.24	5	
0.05	2	ST	1.17	11	ST	2.33	16		4.25	16	
0.08	1		1.18	8		2.34	17		4.29	6	
0.09	2		1.19	4					4.30	17	
			1.20	10		2.35	18	ST	4.35	18	
0.10	3	ST	1.21	11		2.38	17		4.43	7	
0.13	2					2.39	18		4.48	8	
0.14	3		1.22	12	ST	2.40	16		4.53	9	
0.15	1		1.23	9		2.43	7		5.09	19	
0.18	3		1.24	5		2.43**	18		5.12	10	
			1.25	11		2.45	17		5.14	20	
0.19	4	ST	1.26	12		2.46	13		5.17	11	
0.20	2		1.27	10		2.48	8		5.19	21	
0.23	4		1.29	6		2.50	18		5.22	12	
			1.30	12		2.51	14		5.46	13	
0.24	5	ST	1.33	11		2.53	9		5.51	14	
0.25	3					2.55	16		5.56	15	
0.27	4		1.38	12		2.56	15		¹/₂-hr break		
0.28	5		1.42	10		3.00	17		6.25	16	
			1.43	7		3.05	18		6.30	17	
0.29	6	ST	1.46	13	ST				6.35	18	
0.30	1		1.47	11		3.09	19	ST	¹/₂-hr break		
0.32	5		1.48	8		3.12	10		7.09	19	
0.33	6		1.50	13		3.13	19		7.14	20	
0.34	4					3.14	20	ST	7.19	21	
0.35	2		1.51	14	ST	3.17	11		40-min break		
0.37	6		1.52	12		3.17**	19		8.00	1	
0.39	5		1.53	9		3.18	20		8.05	2	
0.40	3		1.54	13					8.10	3	
			1.55	14		3.19	21	ST	8.19	4	
0.43	7	ST				3.22	12		8.24	5	
0.44	6		1.56	15	ST	3.22**	20		8.29	6	
0.47	7		1.59	14		3.23	21		8.43	7	
			2.00	1		3.24	19		8.48	8	
0.48	8	ST	2.00**	15		3.25	16		8.53	9	
0.49	4		2.01	13		3.27	21		9.12	10	
0.51	7		2.04	15		3.29	20		9.17	11	
0.52	8		2.05	2		3.30	17		9.22	12	
			2.06	14					9.46	13	
0.53	9	ST	2.10	3		3.34	21		9.51	14	
0.54	5		2.11	15		3.35	18		9.56	15	
0.56	8		2.12	10		3.39	19		¹/₂-hr break		
0.57	9		2.16	13		3.44	20		10.25	16	
0.58	7		2.17	11		3.46	13		10.30	17	
0.59	6		2.19	4		3.49	21		10.35	18	
1.00	1		2.21	14		3.51	14		11.09	19	
1.01	9		2.22	12		3.56	15		11.14	20	
1.03	8		2.24	5		4.00	1		11.19	21	
1.05	2					4.05	2				
1.08	9		2.25	16	ST	4.09	19				
1.10	3		2.26	15		4.10	3				
1.12	10	ST	2.29	6		4.14	20				

Note: This program gives reading times for multiple columns at 4, 8, 15, and 30 min, 1, 2, 4, and 8 hr, corresponding to approximately 5.5, 6.0, 6.5, 7.0, 7.5, 8.0, 8.5 and 9.0 φ size fractions, respectively. For finer sizes, take one or two readings the next day; noting the time, record readings on this sheet at the specified times and any departure from these times.

*ST = start time (i.e., precise moment stirring ceases).

**Coincident withdrawal times: take the 2-hr (or 4-hr) readings as soon as convenient (an error of about 30 sec over several hours is not significant).

No.	ET	R'h	D	R'h - C_c	P	No.	ET	R'h	D	R'h - C_c	P	No.	ET	R'h	D	R'h - C_c	P
	.01						.01						.01				
	.02						.02						.02				
	.01						.01						.01				
	.02						.02						.02				
	.01						.01						.01				
	.02						.02						.02				

Figure 7-13.   Worksheet for hydrometer analysis of a sediment. ET = elapsed time.

Table 7-9.   Values of K for a Range of Temperatures and Solid Density of Particles in Hydrometer Analysis

Temp (°C)	2.45	2.50	2.55	2.60	2.65*	2.70	2.75	2.80	2.85
16	.00484	.00476	.00468	.00461	.00454	.00447	.00441	.00434	.00429
17	.00478	.00470	.00462	.00455	.00448	.00441	.00435	.00429	.00423
18	.00472	.00464	.00456	.00449	.00442	.00436	.00430	.00423	.00418
19	.00466	.00458	.00451	.00444	.00437	.00430	.00424	.00418	.00413
20	.00460	.00453	.00445	.00438	.00432	.00425	.00419	.00413	.00408
21	.00455	.00447	.00440	.00433	.00426	.00420	.00414	.00408	.00403
22	.00449	.00442	.00434	.00428	.00421	.00415	.00409	.00404	.00398
23	.00444	.00437	.00429	.00423	.00416	.00410	.00404	.00399	.00393
24	.00439	.00432	.00424	.00418	.00411	.00405	.00400	.00394	.00389
25	.00434	.00427	.00420	.00413	.00407	.00401	.00395	.00390	.00384
26	.00429	.00422	.00415	.00408	.00402	.00396	.00391	.00385	.00380
27	.00424	.00417	.00410	.00404	.00398	.00392	.00386	.00381	.00376
28	.00420	.00412	.00406	.00400	.00393	.00387	.00382	.00377	.00372
29	.00415	.00408	.00401	.00395	.00389	.00383	.00378	.00373	.00367
30	.00410	.00404	.00397	.00391	.00385	.00379	.00374	.00368	.00363

Solid Density of Soil Particles of Silt and Clay Fraction (g/cm³)

*Column most frequently used.

Size Analysis of Indurated Sandstones

The texture of consolidated rocks for which disaggregation is undesirable must be studied in thin section under the polarizing microscope (using a micrometer eyepiece calibrated for each magnification), on photographic prints (e.g., enlargements directly from thin sections or peels—see Chapter 4; even magnified photocopies), by a variety of direct image analyzers (e.g., Mazzullo and Kennedy 1985), or by examination of rock surfaces under binocular microscopes, measuring projectors or the scanning electron microscope (e.g., Minnis 1984 describes a technique that automates SEM and X-ray spectroscopic mineral identification). Severe limitations are inherent in such textural studies because a thin section or rock surface provides an essentially two dimensional planar cross-section of the rock. Because they cut very few grains in the plane of their short and intermediate axes (the dimensions that determine their sieve size), quantitative grain size data cannot be directly compared with data compiled from sieve (or settling tube) analysis. Because SEM involves high magnifications (minimum c. × 50), it is suitable for the finest sands and the silts, but few grains of coarser sizes can be seen in a single field of view. Hand specimens or several surfaces at right angles should be examined for sphericity and packing properties, which can influence apparent size distributions. Generally it is the original, depositional texture that is desired, and diagenetic modifications to shape (dissolution, overgrowths) must be "subtracted"; in some cases the diagenetic modifications are too extensive or obscure for reasonable determinations to be made. Despite these problems, qualitative and semi-quantitative *intercomparisons* of thin-section grain size data are generally feasible if there is no prominent fabric of inequant grains, and particularly when data are compiled by the same operator (to avoid operator bias) and from rocks that do not differ greatly in packing or grain sphericity (however, precision is not very high, even with percentage estimation charts—Dennison and Shea 1966). Techniques for quantitative thin-section analysis, including modal analysis for compositional data, are provided in Krumbein (1935), Chayes (1956), and Carver (1971); see also Griffiths (1967).

Several attempts have been made to relate size distributions determined from thin sections to those from sieve analysis (see, e.g., Rosenfeld, Jacobsen, and Ferm 1953; Friedman 1958; Kellerhals, Shaw, and Arora 1975; Adams 1977). The most convincing relationship was determined by Harrell and Eriksson (1979), who analyzed 84 arenites both in thin section and by sieving. Their technique involves measuring the apparent long dimension of 200 to 500 grains, and recording the data in sets at $^1/_2\phi$ or $^1/_4\phi$ intervals. Plot cumulative curves on probability graph paper (arithmetic abscissa), with each point on the fine side of the class interval; connect points with straight lines and make no extrapolations. Apply conversion formula (Table 7-10). *Note:* The arenites were quartz-rich, and more testing is required to determine whether the conversion factors are valid for sandstones in which other components are common; almost certainly no conversion factor is possible for any sediments that contain common inequant grains (such as many rock and fossil fragments).

Table 7-10.   Conversion Factors for Thin-Section to Sieve Sizes via the Method of Harrell and Erikson (1979)

General formula:

$$(\text{sieve data}) = a + b \times (\text{thin-section data})$$

Graphic Parameters	a	b	
M_d (median)	0.121	1.030	accurate
M_z	0.227	0.973	accurate
σ_1	−0.029	1.015	reasonable
Sk_1	0.049	0.593	approximate
K_G	0.394	0.706	not significant at 95% confidence level

ϕ Cumulative Percentiles	a	b	ϕ Cumulative Percentiles	a	b
2	0.164	1.137	50	0.121	1.030
5	0.156	1.097	64	0.254	0.969
9	0.178	1.066	75	0.325	0.952
16	0.127	1.075	84	0.452	0.895
25	0.117	1.064	91	0.579	0.846
36	0.094	1.054	95	0.772	0.801
			98	0.989	0.770

Note: Not applicable to samples with common inequant grains and poor sorting.

TREATMENT OF SIZE DATA

Data obtained from size analyses of sediments can be utilized in various ways (e.g., Fig. 7-14). Graphs are useful to represent entire distributions pictorially (see Fig. 7-15). Note that histograms and frequency curves derived from histograms are misleading; cumulative frequency curves are generally used, plotted on probability paper when accurate interpolation from the "tails" is necessary to obtain the statistical parameters. Plotting of curves permits sample interrelationships—differences and similarities—to be seen throughout their range; modes can be found and polymodality seen; defective sieves may be detected ("kicks" at the same ϕ value in a bundle of curves); qualitative textural classification of the sample (**PS** Chapter 5) can be determined readily by visual inspection.

For statistical analysis, either the entire body of data is used (e.g., the *Method of Moments,* e.g., Wentworth 1936) or selected values from the cumulative curve are inserted into formulae. For analyses based on graphs, various formulae can be applied to obtain summary parameters that represent characteristics of the grain size distributions. The parameters are mathematical artifacts and represent only a few characteristics of each distribution. Folk (1966), Moiola and Weiser (1968), Jones (1970), and Tucker and Vasher (1980) review various methods of statistical analysis. Select the most suitable for the

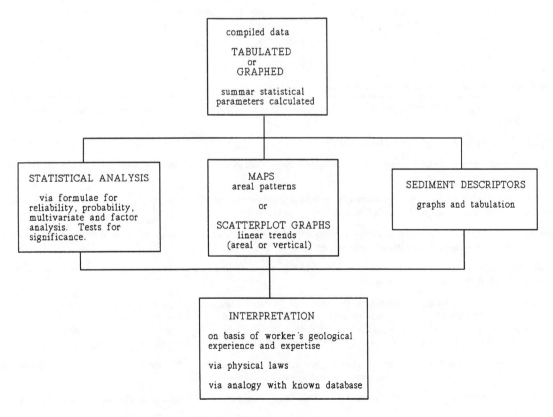

Figure 7-14.   Generalized options available for utilizing grain size data.

particular purpose (commonly the same as the data with which you wish to compare), and report the method selected when writing up the results of analysis. Note that in all calculations, weight % is used rather than frequency %. Statistical formulae are designed fundamentally for the latter, but counting of grains is impractical and the assumption is made that there is no significant difference in grain densities between samples—an assumption that is invalid if comparisons are attempted between quartz-rich sediments and blacksands, greensands, or some rock fragment, or shell-rich sands.

Method of Moments

The Method of Moments is widely used because it can be fully computerized (see broad discussion by Friedman 1979). Table 7-11 shows an example of moment calculations. It is claimed to have marked advantages over graphical analysis in speed and freedom from some bias inherent in the assumption of a fundamental lognormal size distribution (e.g., Leroy 1981). However, use of the method requires the complete size distribution of each sample—there can be no "open tail" (i.e., no unsized "pan fraction": data must be extrapolated to 100% at an arbitrarily selected size at the fine end if there is a fraction too fine for practical measurement). Also, size values used in the computations assume a "normal" (Gaussian) distribution within each class that has been measured (in the example, the value of d is midway in each size class); this assumption is generally false and can distort results. The mode (most common grain size) cannot be determined, nor can any polymodality in the distribution.

Graphical Analysis

Graphical analysis is relatively time-consuming because values must be interpolated from the curves (however, computers can be programmed to take the values). Numerical results of the graphical formulae are considered by many to be statistically less valid than those obtained by the Method of Moments, but the advantages of graphical representations are major, and if the graphs are available, why not use them. The graphical statistical parameters of Folk and Ward (1957; Table 7-12) are most widely used. Although results only approximate the moment values, because 5–8% of the distribution at each end is ignored, there are actual advantages in their use when samples have "open tails" or when there is likely to be experimental error in measuring the size of the coarsest or finest grains.

Statistical parameters can be easily and rapidly estimated after only slight experience with graphs. The mode(s) occurs at the inflection point of the steep part(s) of the curve (most easily found on graphs with arithmetic ordinates). The median of the sand fraction (versus the median of the whole sample, which is merely the ϕ value at 50 wt %) is obtained by taking the average of the percentage values at -1 and $+4\phi$. To obtain an estimate of the standard deviation consistent with the qualitative estimate of "sorting" made from looking at the actual sample, look at the central two-thirds of the curve (that part of the curve between 16% and 84%). If it spans less than 1ϕ, the sample is well sorted; if it spans $1–2\phi$, the sample is moderately sorted; if it spans more than 2ϕ, the sample is poorly sorted. For skewness (a measure of symmetry of the distribution), sketch a line between the points on the

Figure 7-15.   Graphic presentation of size data (on facing page).

1. The abscissa scale is arithmetic, and therefore bars of $1/2\phi$ intervals span different widths. Note that with sedimentary geological studies, the coarse size is to the left on the abscissa.

2. Diagram 1 plotted correctly for bar graphs, on the basis of unit areas.

3. The abscissa scale is geometric (although it is expressed in millimetre values). Data are the same as those in diagram 1 with one size class added. Note that the modal class appears different from that in 1.

4. 5. 6. Data are the same. Sketches 4 and 5 use an abscissa scale with 1ϕ intervals, starting at different points (-1ϕ for 4, -1.33ϕ for 5); 6 is subdivided at $1/2\phi$ intervals. Note the very different shapes that result from varying the abscissa scale.

7. This frequency polygon graph was constructed by joining the midpoints of the bars shown in 3. The graph is subject to all the same problems as demonstrated in the bar graphs—the shape will vary depending on the abscissa scale.

8. A cumulative curve can be constructed using bar graph data (of diagram 3). The shape of the curve will be independent of the abscissa scale. In practice, raw data are used directly to define points for the curve.

9. Cumulative curves may be drawn using probability ordinates. Because of the standard usage of the Udden-Wentworth grade scale, either an arithmetic abscissa (with ϕ values) or a $\log_2$ abscissa scale (i.e., semi-log paper when millimetre values are plotted, as shown) can be used. If the limits of size measurement are attained before the data line reaches 100% (as in the case of the till figured), it may be necessary to make an arbitrary assumption that the fine tail has a normal distribution and a minimum size of 14ϕ permitting a straight-line extrapolation from the last data point to 100% at 14ϕ.

10. A frequency curve derived from a histogram. Such graphs have the same problems as histograms, varying shape depending on the abscissa scale used. They are more misleading than histograms because one thinks of selecting a unique modal size.

11. Construction of the unique frequency curve. A cumulative curve is necessary, then a complex graphical technique must be employed (see Krumbein and Pettijohn 1938).

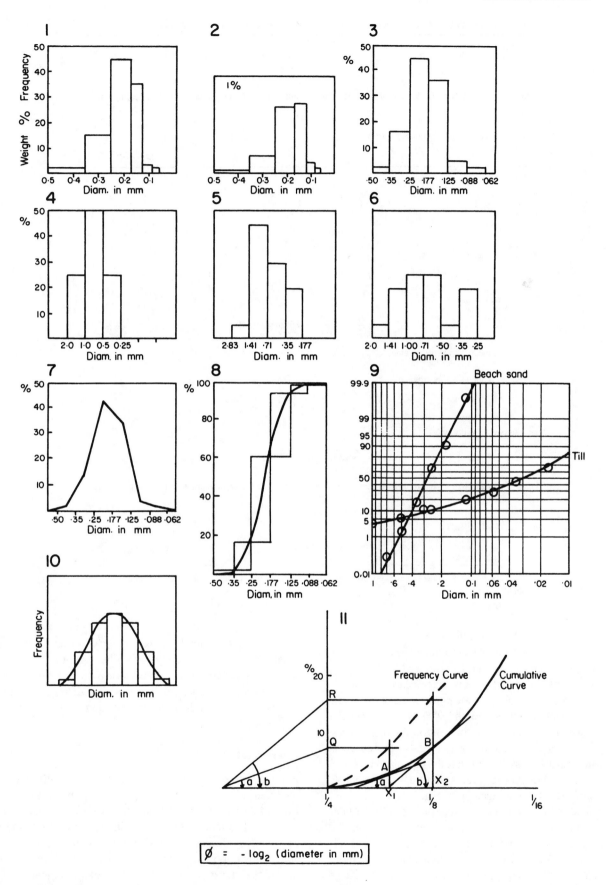

Figure 7-15.

Table 7-11.   Example of the Method of Moments Calculation for Grain Size Statistics

The general formula for the nth moment is

$$\log n = \frac{\Sigma(fd^n)}{N}$$

where f = frequency (weight %), d = log diameter, and N = number of measurements (100 when dealing with percents).

Given: an ideal "normal" frequency distribution:

ϕ	d	f	fd	d^2	fd^2	d^3	fd^3	d^4	fd^4
−3 to −2	−3.5	3	−10.5	12.25	36.75				
−2 to −1	−2.5	10	−25.0	6.26	62.5				
−2 to −1	−1.5	22	−33.0	2.25	49.5				
−1 to 0	−0.5	30	−15.0	0.25	7.5				
0 to +1	+0.5	22	+11.0	0.25	2.75				
+1 to +2	+1.5	10	+15.0	2.25	22.5				
+2 to +3	+2.5	3	+0.75	6.25	18.75				
		$N=100$	$\Sigma fd = 50$		$\Sigma fd^2 = 200$				

This first moment is the mean:

$$\frac{\Sigma fd}{N} = \frac{-50}{100} = -0.50\phi = X$$

This is a log mean, hence the ϕ value; the antilog$_2$ gives the millimeter value.

The second moment is the variance:

$$\Sigma fd^2 - \frac{(\Sigma fd)^2}{N-1} = 200.15 - 25 = 1.77\phi$$

The standard deviation is

$$\sigma = \sqrt{\text{variance}} = \sqrt{1.77} = 1.33\phi$$

The third moment will give a value interpretable as skewness, and the fourth moment a measure of kurtosis:

$$\text{3rd moment} - \text{skewness} = \frac{\Sigma f(d) - X^3}{100\sigma^3}$$

$$\text{4th moment} - \text{kurtosis} = \frac{\Sigma f(d-X)^4}{100\sigma^4}$$

Note: No one has yet managed to use kurtosis values to good effect; hence these values are commonly not computed.

curve at 16% and 84%. If the line lies very close to the (overall) median, the distribution is near symmetrical; if the line lies to the right of the median, it is fine-skewed (+ skewness); if it lies to the left of the median, it is coarse-skewed (− skewness). For kurtosis (a measure of the peakedness of the distribution), the spread between 5% and 95% is 2.44 times the spread between 25% and 75% for a normal curve (mesokurtic); if the ratio is much less than 2.44, the curve is platykurtic with a K_G less than 1.00; if the ratio is greater than 2.44, the curve is leptokurtic (excessively peaked) with a K_G greater than 1.00. Try your hand with Fig. 7-16.

Note: The sedimentologist uses measures of standard deviation as indicative of *sorting* processes—the lower the value, the better sorted and thus the more selective the deposition or winnowing. However, the civil engineer is more concerned with the concept of *grading* (in the context of overall grain size rather than relative distribution of sizes in a bed)—a well-graded aggregate for construction purposes is a distribution sufficiently spread (poorly sorted) so that the small particles will occupy spaces between the larger ones and thus provide greater strength or reduce the quantity of cement necessary. Specific (optimal) size-distribution ranges are listed as standards in the construction industry.

Interpretation

For any *interpretation* of the results of quantitative size analyses, a suite of samples must generally be examined: little of value will result from studying merely one or two samples. The major use is for sediment description and for discovering similarities, differences, or trends in sample suites. It is important to decide initially on the purpose of the investigation; assessing the suitability of a sediment for road or concrete aggregate requires a different procedure than assessing down-river textural trends. Much effort has been ex-

Table 7-12.   Formulae and Verbal Scales for the Folk and Ward (1957) Grain Size Parameters Determined from Graphs (cumulative curves)

Mode: Most frequently occurring particle size. Inflection point(s) on the steep part(s) of the cumulative curve. Found precisely only by trial and error method: discover the point with the maximum wt% within a $1/2\phi$ interval centered on it.

Graphic Mean: $M_z = \dfrac{\phi 16 + \phi 50 + \phi 84}{3}$

Inclusive Graphic Standard Deviation: $\sigma_1 = \dfrac{\phi 84 - \phi 16}{4} + \dfrac{\phi 95 - \phi 5}{6.6}$

<0.25φ	very well sorted
0.35 to 0.50φ	well sorted
0.50 to 0.71φ	moderately well sorted
0.71 to 1.0φ	moderately sorted
1.0 to 2.0φ	poorly sorted
2.0 to 4.0φ	very poorly sorted
>4.0φ	extremely poorly sorted

Inclusive Graphic Skewness:

$$Sk_1 = \frac{\phi 16 + \phi 84 - 2\phi 50}{2(\phi 84 - \phi 16)} + \frac{\phi 5 + \phi 95 - 2\phi 50}{2(\phi 95 - \phi 5)}$$

$$= \frac{\phi 84 - \phi 50}{\phi 84 - \phi 16} - \frac{\phi 50 - \phi 5}{\phi 95 - \phi 5} \quad \text{(see Warren 1974)}$$

+1.0 to +0.3	very fine-skewed
+0.3 to +0.1	fine-skewed
+0.1 to −0.1	near-symmetrical
−0.1 to −0.3	coarse-skewed
−0.3 to −1.0	very coarse-skewed

Graphic Kurtosis: $K_G = \dfrac{\phi 95 - \phi 5}{2.44(\phi 75 - \phi 25)}$

<0.67	very platykurtic
0.67 to 0.90	platykurtic
0.90 to 1.11	mesokurtic
1.11 to 1.50	leptokurtic
1.50 to 3.00	very leptokurtic
>3.00	extremely leptokurtic

pended for the last 50 years in discovering depositional environments of "unknowns" by comparing their size distribution characteristics with a data base obtained from analyzing samples from "known" environments; success has been limited because sampling procedures have largely been inadequate and because workers have been expecting too much—it is processes and their energy levels that dictate size distribution characteristics.

Studies that carefully relate distributions to processes in known environments (e.g., by plotting parameters against geographic position) have had marked success (e.g., Allen 1971; Erlich et al. 1980), but they are not easily applied to complex ancient environments and some workers evaluate applications on a general basis as reasonably hopeless (e.g., Erlich 1983). A few generalizations can be made about distributions that are characteristic of some environments (such as beaches, which are generally well sorted), but because a variety of processes act in most environments at varied energy levels, application of such studies to samples from unknown environments is of limited value. An important and commonly overlooked requirement for any textural analysis is to ensure that conclusions from the study take the sampling procedure into account—e.g., do the interpretations apply to a single process that affected a single sedimentation unit, or to the aggregate of processes that dictated the textures of a multilayer assemblage? If all these considerations are taken into account, quantitative size analysis can be a powerful tool in acquiring information on the various populations that can be recognized in many sediment suites.

Fundamental studies of the relationship between grain size (and shape) distributions and processes are surprisingly few; they have shown that there are great complexities (see, e.g., Davis and Erlich 1970; Moss 1972; Middleton 1976; Middleton and Southard 1978; Erlich et al. 1980). Grain size depends on the character of the source rocks, weathering processes, abrasion, and selective sorting during transportation.

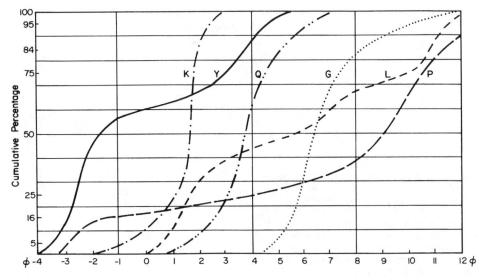

Figure 7-16.   Hypothetical cumulative curves for practicing estimations of statistical parameters.

During transport, the proportion of grains traveling by traction (continuously in contact with the substrate), intermittent suspension (bouncing along the substrate), and suspension (in the transporting medium above the substrate) depends on: (1) the competency and capacity of the transporting agent, which reflect factors such as volume, velocity, bed roughness, and turbulence; (2) the range of grain sizes available, which reflects both ultimate and local source factors (e.g., no gravels will be transported if none are locally available, however strong the current), and (3) complex boundary condi-

tions between the different transporting mechanisms (particularly the transition from traction to intermittent suspension).

One method of interpreting process from grain size distributions has been a simple analysis of the complete cumulative curve, separating populations that appear to comprise straightline segments of the curve (e.g., Visher 1969, and Fig. 7-17a). Insofar as these "populations" are artifacts of the method of analysis, interpretations still depend on comparing characteristics of "unknowns" with a data base of "knowns." Another

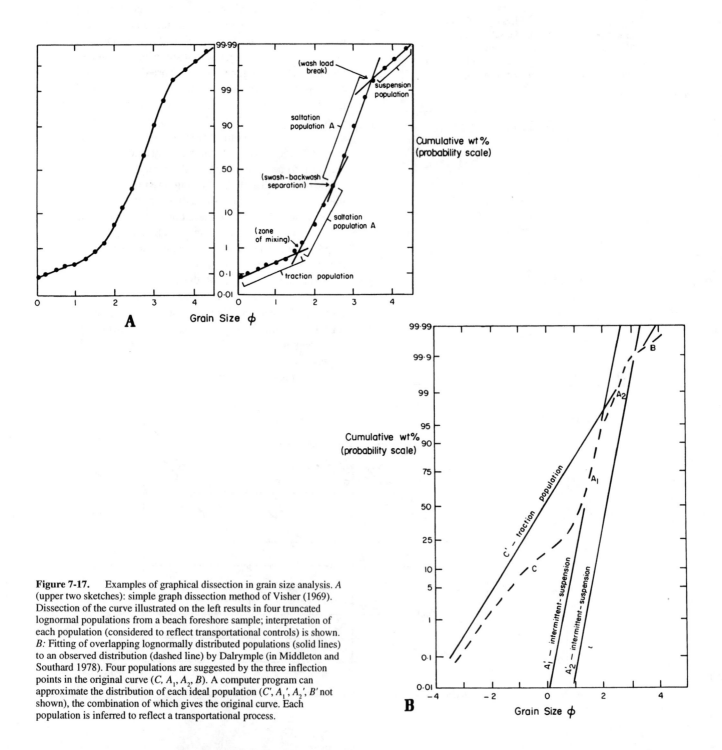

Figure 7-17. Examples of graphical dissection in grain size analysis. *A* (upper two sketches): simple graph dissection method of Visher (1969). Dissection of the curve illustrated on the left results in four truncated lognormal populations from a beach foreshore sample; interpretation of each population (considered to reflect transportational controls) is shown. *B:* Fitting of overlapping lognormally distributed populations (solid lines) to an observed distribution (dashed line) by Dalrymple (in Middleton and Southard 1978). Four populations are suggested by the three inflection points in the original curve (C, A_1, A_2, B). A computer program can approximate the distribution of each ideal population (C', A_1', A_2', B' not shown), the combination of which gives the original curve. Each population is inferred to reflect a transportational process.

approach is to dissect the frequency curves into assumed log-normal populations and to analyze each of these separately (see Middleton and Southard 1978, and Fig. 7-17b). However, original sedimentary grain populations are unlikely to have lognormal distributions, and the false assumptions made about the ideal distributions may invalidate these approaches. (They may not—if the results do prove useful, there is every reason to use them!)

Any trends that can be determined in any of the relevant statistical parameters can help distinguish source materials from deposits, determine net transport paths, indicate acting processes, and identify depositional environments. Hence, in all cases where the data come from an approximately contemporaneous unit, geographic plots of parameters can be very informative. For example, with selective (incomplete) erosion from any source material, the transported sediment will be finer, better sorted, and more negatively skewed than the source, and the residual (lag) deposit coarser, better sorted, and more positively skewed (see McLaren and Bowles 1985). With selective deposition from any sediment distribution, the deposit may be either finer or coarser than the source material, but sorting will be better and the skewness more positive.

Plots of various statistical parameters against each other may also depict trends or clustering of samples that can be used in other ways (e.g., Cronan 1972; Folk and Ward 1957; McCammon 1976; Passega 1964, 1977; McManus in Tucker 1988; Fig. 7-18). Unfortunately, in most studies

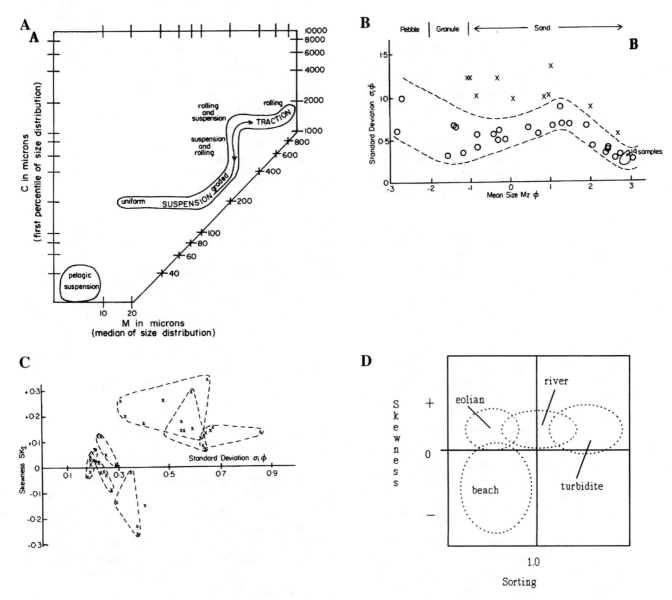

Figure 7-18. Examples of scatterplots using statistical grain size data. *A:* CM diagram—a guide to depositional process attempted by some workers. *B* and *C:* Examples of selected Folk and Ward (1957) statistical parameters. (After Andrews and van der Lingen 1969). Part *B* plots standard deviation against mean size (unimodal samples = *o*'s; polymodal samples = *x*'s). A sinusoidal field for unimodal samples appears to be a widespread phenomenon. Part *C* plots skewness versus standard deviation using only unimodal samples (fields enclose samples from the same locality). *D:* Approximate fields for sands from different environments, in terms of skewness and standard deviation.   (*A:* after Passega 1972; *D:* after Bjorlykke 1984.)

there is appreciable overlap between fields plotted on the scattergrams. It is particularly rare to find significant differences with the higher statistical parameters (skewness and kurtosis).

In many cases, maps or plots of the gross ratios between gravel, sand, silt, and/or clay contents may provide sufficient evidence for process or energy-level discrimination (e.g., Sly, Thomas, and Pelletier 1983). Maps of areal distributions of percentage coarser or finer than a stated size may prove particularly useful for trend identification. A relatively new field for grain size interpretation has been for interpreting process in volcanic deposits (e.g., attempts to discriminate pyroclastic surge, flow, airfall), as well as distance from the vent (e.g., Lirer and Vinci 1991).

TEXTURE OF CARBONATE SEDIMENTS

Loose carbonate grains and slightly cemented carbonate sediments are most commonly studied under the binocular microscope or with scanning electron microscopy. Quantitative size analyses are rarely attempted using the methods described above for detrital sediments because carbonate grains tend to break easily (many are commonly partially weathered in the outcrop), their hydrodynamic response depends on density (which varies due to common differences in internal texture and shape), and weathering/diagenetic effects commonly modify the grains even in modern environments.

The texture of fully indurated limestone hand specimens is clarified by moistening the surface with a clear oil (lasts longer than water, and will coat the grains even if the rock is permeable). Etching with dilute HCl results in differential relief of components because solubility varies with density and internal structure of the particles, and with the crystallographic orientation of the carbonate minerals; dolomite and other insoluble residues will stand out particularly well.

Stains for different carbonate minerals (Chapter 8) can also assist textural study because, for example, different organic hardparts may be composed of different minerals. A fluorescent dye penetrant applied to etched surfaces (washed in acetone, then the dye allowed to penetrate for c. 10 min before applying the emulsifier) may provide enhanced resolution of obscure fabrics as well as sedimentary structures when viewed in ultraviolet light (Ali and Weiss 1968). Acetate peels (see Chapters 4 and 5) are easily prepared if smooth surfaces are available or can be made (e.g., with sandpaper); detail is as good as in thin sections, and broader areas can be examined.

Nonetheless, perhaps by tradition, thin section studies (see Chapter 8) remain the most common method for studying carbonate sediments in the laboratory. A discussion of general textural attributes of carbonate sediments and the common classification schemes used to characterize them appears in **PS** Chapter 8.

GRAIN SURFACE TEXTURES

Grain surface textures can be determined with hand lens and binocular microscope on gravels and some of the coarsest sands to a useful level (e.g., striations), but for detailed work and all fine grains, high magnifications are necessary.

Scanning Electron Microscopy (SEM)

Scanning electron microscopes became available in 1965; because SEM analysis permits nondestructive magnification of surfaces, normally from c. 50× up to 100,000× or more, it has been applied to a wide variety of studies since that time. For sediment studies, the SEM has largely replaced the larger transmission electron microscope (TEM, with magnifications up to 300,000× or more) since the late 1960s because of its greater depth of field, wide and rapid range of magnification, and ease of preparing and scanning samples from individual grains to relatively large surfaces or rock. Nonetheless, the TEM remains a useful device for studies of very fine particles, such as colloids, because it produces images of very high resolution. (The complex and fallible procedure for replicating and shadowing grains on acetate film for electron microscopy is described by Krinsley and Margolis, in Carver 1971). SEM principles are well presented in Nixon (1969) and Smart and Tovey (1982). Essentially a metal specimen stub is coated with an adhesive (e.g., Table 7-13), the sample mounted on the stub and thinly coated with a conducting metal under vacuum, and the stub plus sample inserted in a vacuum chamber of the SEM and bombarded with electrons; in modern "environmental" SEMs that operate with less vacuum in the sample chamber, no sample coating is necessary. Three-dimensional effects are created by the luminescing sample surface. An excellent review of potential applications to, and previous work on, sedimentological problems is provided by Trewin in Tucker (1988).

Probably the best known geological application has been in the area of quartz-grain surface textures (e.g., Krinsley and Doornkamp 1973), but there are a multitude of studies ranging from soil microstructures (e.g., Jones and Squair 1989) to acetate peels of limestones (e.g., Brown 1986). SEM studies can also be made of pores and microfractures in rocks, or of borings in carbonate grains; in these cases, the specimens are impregnated, then the sediment dissolved to leave a molding of the original voids.

The coatings applied to grains and rock surfaces for SEM analysis are generally too thin to interfere with optical microscope analyses; hence the same specimen may be used even with transmitted light petrographic microscopes (Chapter 8) if the initial mounting procedures are carefully planned (e.g., Wilding and Geissinger 1973). Such combined studies permit the maximum amount of information to be obtained from the specimen. Specimen preparation for SEM analysis is listed in Table 7-14.

Consult the books supplied with the specific SEM machine for details of scanning procedures and the pros and

Table 7-13.   Advantages and Disadvantages of Common Adhesives for Mounting Samples for SEM Analysis

Adhesive	Advantages	Disadvantages
Araldite	Effective. Poor earthing easily overcome by lacing adhesive with copper filings.	Specimens mounted permanently. May introduce vapors into coating chamber; ensure glue is well cured before coating.
Colloidal silver/ graphite	Excellent earth, specimens recoverable.	Absorption may be a problem with more porous specimens. Amyl acetate dispersant toxic, often draws silver into viewing area. Expensive
Double-sided cellotape	Quick and cheap. Specimens sometimes recoverable.	Often partially lifts under vacuum. Earthing qualities poor. Mounted material has a short shelf life.
Copper print	Excellent.	Toxic vapors, mount in fume cupboard.
Elmer's glue and PVA	Some reorientation of specimens possible. Specimens recoverable.	Glue may be absorbed through more porous specimens. Poor earth, negatives often contrasty.
Mikrostik	Quick.	Glue may be absorbed by more porous specimens. More effective earth than cellotape.

Source: Prepared by K. M. Swanson, University of Canterbury.

cons of the variations possible. In general, for detrital sediments a beam current of 10^{-11} amp and accelerating voltage of 10–25 kV appears to be optimal. Different details can be seen at different angles relative to the primary beam; hence, e.g., scans are commonly at 0° (thickness differences, cleavage traces, scratches and depressions in the surface) and 45° (for grain-edge phenomena) angles. In many cases, the specimen/detector angle will be determined by signal intensity, particularly when secondary electron generation is low; higher tilts generate better signals. For photography, scanning speeds of 50 seconds are satisfactory (see Wilding and Geissenger 1973).

Phase and Differential Interference Contrast Microscopy

Differential contrast light microscopy can provide excellent resolution of surface detail in specimens under the polarizing microscope (e.g., see Hoffman and Gross 1970; Warnke and Gram 1969). Light waves are artificially separated then rejoined with special guidance devices. Synchronized pairs of objective lens systems are inserted into otherwise standard polarizing microscopes. Differences in refractive indices can be easily recognized by strong color contrasts and can be measured with standard compensators. Unlike the case with SEM, images commonly reflect internal anisotropism (e.g., different refractive indices of components, cleavages, and fractures) and not entirely geometry. Thus this technique can provide extra information with anisotropic grains, but adds little with isotropic grains. A problem is the depth-reversal illusion created by the way the brain interprets images.

Phase contrast microscopy can also effectively distinguish mineral and crystal edges, cleavage and fracture surfaces that do not show up well in other microscopy techniques. Light waves are separated into two superimposed images by refraction in the particles themselves. Exchangeable ring diaphragms are attached to standard polarizing microscopes, and a substantially greater contrast is attained than with brightfield microscopy (e.g., between slightly different grain refractive indices). Positive phase contrast is normal, where transparent grains with higher refractive indices than the embedding medium are darker than the background, and vice versa (hence differences in brightness can be used to estimate refractive indices and help to identify the mineral components). Multiple crystal growth outlines may be detected.

FABRIC STUDIES

Grain-packing studies involve such practical difficulties in collection of loose sediment (repacking generally occurs during sampling) and in attempting three-dimensional analysis that the most common methods bypass the problem and merely examine porosity and permeability (see below). Fundamental concepts with respect to packing arrangements (but of spheres only) are discussed by Graton and Fraser (1953).

Preferred grain orientations are most easily studied with gravel sizes and macrofossils in the field (e.g., Rust 1972). As long as individual grains can be removed, their three-dimensional shape can be observed and the *L* and *S* axial orientations recorded and treated as discussed under sedimentary structure analysis (Chapter 6). Grains are removed, *L, I,* and *S* axes are marked and the grains are re-

Table 7-14.   Examples of Procedures for the Preparation of Sediment Samples for SEM Study

For rock surfaces:

1. Prepare a sample 3–10 mm thick, and chip the surface to be examined; surfaces must be clean and dry; hence any initial cutting of the specimen must not be made with saws using lubricating oil or *thorough* removal of that oil must be achieved. Surface may be polished and etched (see preceding discussions). Beware of contamination and preparation-caused development of abnormal relief, tension, or compressional fractures.

2. If necessary, grind the undersurface sufficiently flat so that there will be no large air spaces when it is glued to the mounting stub. (If not possible, araldite or super glue may be used at single spots only instead of step 5 below.)

For individual grains (e.g., surface texture studies):

1. Remove any organic matter, carbonates, and grain coatings (see Chapter 5), wash in distilled water, and air-dry. Inspect remaining sample under the binocular microscope and select grains for study.

For all specimens: From this stage onward, use tweezers or tongs to avoid transferring skin oils to the sample and stub.

3. To dry the specimen thoroughly, soak in absolute alcohol and expose to a short period of ultrasonic radiation. (Test subsample initially in ultrasound—another method of drying, such as freeze-drying, must be found if the cavitation disrupts it!) Remove and air-dry at 35–40°C for c. 2 hours.

4. Clean mounting stub in alcohol (or the more toxic agents such as acetone) and ultrasonic device. Remove and place in perspex stub holder.

5. Apply glue to stub surface, leaving a small area clear for labeling. Label both this space (scratch with steel scribe) and the undersurface of the stub (india ink); having a label on the top of the stub is invaluable when working under the SEM, particularly when using a multistub holder. The best glues are composed of a good electrical earthing compound. See Table 7-13 for advantages and disadvantages of various common adhesives. Place specimen on surface, orienting as required if it is irregularly shaped. (If multiple small specimens are to be viewed, ensure that the adhesive does not cover the upper surfaces, e.g., by allowing the adhesive to dry, sprinkling specimens on the surface, and warming until they *begin* to sink, or gently blow them onto tacky glue at a low angle of incidence, or *gently* push each one into tacky adhesive with a soft brush.) Air-dry (warm oven).

6. Place stub under binocular microscope and *carefully* paint colloidal silver, copper print, or graphite around the base of the specimen to ensure a conductive bridge between the sample coating (step 8) and the stub base. (This step is not recommended for porous specimens—the colloid must not penetrate to the viewing surface!)

Caution: Copper print and colloidal silver produce toxic fumes; use a fume cupboard.

7. Place stub in desiccator with fresh silica gel until ready to view.

8. Coat specimen with conducting medium (generally gold or carbon, 100–500 angstroms thick) and examine. The coating provides a conductive path to the stage and thereby to earth, preventing excessive electron charge buildup; it assists secondary electron generation for improved resolution, and it provides greater thermal conductivity, preventing uneven heat-generated expansion of the specimen. The coating device (most commonly a diode sputterer) is a vacuum chamber attached to a nitrogen source, in which high-voltage generates a "cloud" of the coating substance. Details of the device used and thickness of the coating vary depending on the equipment available, the nature of the specimen, and the method of SEM examination (e.g., if EDAX—Energy Dispersive X-ray Analysis—is to be performed, see Chapter 8). (If the specimen is highly porous, two stages of coating are required on a special stub holder, with the specimen oriented normally then at 45° to the cathode).

If single specimens are to be viewed at different orientations, mount bent strands of copper wire onto stub with one segment vertical, place a drop of glue on the top of the vertical segments, and mount specimens on them. Brush colloidal silver between the horizontal segment of wire and the stub to ensure a good earthing contact. Bend wire to suit. Stub "maps" should be made of the position of individuals when multiple specimens are mounted—or each specimen provided with a scratched label.

Source: Method of K. M. Swanson, University of Canterbury.

placed, then the axial directions are recorded with a compass. Results of these measurements (as for all orientation data) are generally presented and analyzed on stereograms and current roses (see discussion in Chapter 6). Major problems arise with sand sizes and any lithified sediments, when the three-dimensional shapes cannot be determined. Various techniques have been developed to deal with these, as well as with removable gravels, in the field and laboratory (see especially Bonham and Spotts, in Carver 1971). Careful sample collection is particularly crucial for laboratory analysis, because the precise spatial orientation of the sample itself (and any subsamples) needs to be indelibly recorded (see Chapter 4). Grains must be inequant to have a preferred orientation. In general, analysis of indurated sediments must rely on two-dimensional analysis of two surfaces exposed or cut at right angles so that the relation to the fabric can be analyzed and preferred orientations on these resolved statistically. Measurements are most commonly made from photographic enlargements or with the Universal Stage (see Chapter 8), but there are a few instruments that utilize electronic or photometric techniques (e.g., see discussion in Mueller 1967; Taira and Lienert 1979), and the X-ray texture goniometer is finding increasing use in many laboratories. Fabric characteristics can be initially evaluated by binocular microscope examination of rock slices, several cm thick, with surfaces ground and polished and coated with glycerol or oil; these slices can also be subjected to X-ray radiography (see discussion in Chapter 6). A few studies of fabric in muds have been undertaken (e.g., Kuehl, Nittrouer, and DeMaster 1988).

POROSITY AND PERMEABILITY

Porosity and permeability, secondary properties of sediments, are determined by the distribution of interparticle spaces and are of vital importance to the retention and movement of fluids and gasses. Porosity is the percentage of voids in a volume of rock; pores may be primary or secondary (diagenetic, generally created by solution). It is determined by measuring the volume of gas displaced from a known bulk volume of rock (determined by displacement with total immersion in mercury) or sediment (which has generally been disturbed in the sampling procedure); the increase in volume of gas with reduction of pressure is measured in a sealed vessel. Permeability is a measure of the ability of a fluid to pass through the sediment—generally the rate at which a fluid passes through a core of rock or a bed of sediment. It depends on the geometry and interconnectedness of the pores, reflecting both depositional and diagenetic controls, grain size, shape, and fabric. The permeability is calculated by Darcy's Law and expressed as millidarcy units.

An overview of principles and methods is provided by Curtis (in Carver 1971). Estimates of porosity can be made from thin section using stained impregnating materials and other techniques (e.g., Halley 1978; Yanguas and Dravis 1985; Ruzyla and Jezek 1987; Lundegard 1992). Davis (1954) described a method for determining porosity from bulk density;

Manger, Cadigan, and Gates (1969) a method for determining immobile water; and Fatt, Maleki, and Upadhyay 1966 for dead-end pore space. Pittman and Duschatko 1970 discussed a method of making pore casts for SEM analysis. Eijpe and Weber (1971) discussed a technique for measuring permeability of consolidated and unconsolidated sediments; also see Netto (1993) and other papers in the Selected Bibliography.

SELECTED BIBLIOGRAPHY

General

Carver, R. E., 1971, *Procedures in Sedimentary Petrology.* Wiley-Interscience, New York, 653p.
Griffiths, J. C., 1967, *Scientific Method in Analysis of Sediments.* McGraw-Hill, New York, 508p.
Hoffman, R., and L. Gross, 1970, Reflected light differential-interference microscopy: Principles, use and image interpretation. *Journal of Microscopy* 91:149–72.
Krumbein, W. C., and F. J. Pettijohn, 1938, *Manual of Sedimentary Petrography.* Appleton-Century-Crofts, New York, 549p.
Mueller, G., 1967, *Methods in Sedimentary Petrology.* H.-U. Schmincke (trans.), Hafner, New York, pp. 38–9.
Pettijohn, F. J., 1975, *Sedimentary Rocks.* Harper & Row, New York, 628p.
Tucker, M. (ed.), 1988, *Techniques in Sedimentology.* Blackwell Scientific Publications, Oxford, 394p.

Fabric

Ali, S. A., and M. P. Weiss, 1968, Fluorescent dye penetrant technique for displaying obscure structures in limestone. *Journal of Sedimentary Petrology* 38:681–2.
Graton, L. C., and H. J. Fraser, 1953, Systematic packing of spheres—with particular relation to porosity and permeability. *Journal of Geology* 43:785–909.
Kuehl, S. A., C. A. Nittrouer, and D. J. DeMaster, 1988, Microfabric study of fine-grained sediments: Observations from the Amazon subaqueous delta. *Journal of Sedimentary Petrology* 58:12–23.
Rust, B. R., 1972, Pebble orientation in fluvial sediments. *Journal of Sedimentary Petrology* 42:384–8.
Taira, A., and B. R. Lienert, 1979, The comparative reliability of magnetic, photometric, and microscopic methods of determining the orientations of sedimentary grains. *Journal of Sedimentary Petrology* 49:759–72.

Shape

Barock, E. J., 1974, Coarse sediment morphometry: A comparative study. *Journal of Sedimentary Petrology* 44:663–72.
Barrett, P. J., 1980, The shape of rock particles, a critical review. *Sedimentology* 27:291–303.
Cailleux, A., 1945, Distinction des galets marins et fluviatiles. *Societe Geologique de France Bulletin* 15:375–404.
Catacosinos, P. A., 1965, Tables for the determination of sphericity and shape of rock particles. *Journal of Sedimentary Petrology* 35:354–65.
Crofts, R. S., 1974, A visual measure of shingle particle form for use in the field. *Journal of Sedimentary Petrology* 44:931–4.
Dobkins, J. E., Jr., and R. L. Folk, 1970, Shape development on Tahiti-Nui. *Journal of Sedimentary Petrology* 40:1167–1203.

Ehrlich, R., and B. Weinberg, 1970, An exact method for characterization of grain shape. *Journal of Sedimentary Petrology* 40:205–12.

Flemming, N. C., 1965, Form and function of sedimentary particles. *Journal of Sedimentary Petrology* 35:381–90.

Folk, R. L., 1972, Experimental error in pebble roundness determination by the modified Wentworth method. *Journal of Sedimentary Petrology* 42:973–4.

Gale, S. J., 1990, The shape of beach gravels. *Journal of Sedimentary Petrology* 60:787–9.

Harrell, J., 1984, Roller micrometer analysis of grain shape. *Journal of Sedimentary Petrology* 54:643–5.

Illenberger, W. K., 1991, Pebble shape (and size). *Journal of Sedimentary Petrology* 61:756–67.

King, C. A. M., and J. T. Buckley, 1968, The analysis of stone size and shape in Arctic environments. *Journal of Sedimentary Petrology* 38:200–14.

Krumbein, W. C., 1941, Measurement and geological significance of shape and roundness of sedimentary particles. *Journal of Sedimentary Petrology* 11:64–72.

Li, Z., and P. D. Komar, 1986, Laboratory measurements of pivoting angles for applications to selective entrainment of gravel in a current. *Sedimentology* 33:413–23.

Moss, A. J., 1962, The physical nature of common sandy and pebbly deposits, Parts 1 & 2. *American Journal of Science* 260:337–73 and 261:297–343.

Moss, A. J., P. H. Walker, and J. Hutka, 1973, Fragmentation of granitic quartz in water. *Sedimentology* 20:489–511.

Patro, B. C., and B. K. Sahu, 1974, Factor analysis of sphericity and roundness data of clastic quartz grains: Environmental significance. *Sedimentary Geology* 11:59–78.

Perez, F. L., 1987, A method for contouring triangular particle shape diagrams. *Journal of Sedimentary Petrology* 57:763–5.

Rittenhouse, G., 1943, A visual method for estimating two-dimensional sphericity. *Journal of Sedimentary Petrology* 13:79–81.

Rosenfeld, M. H., and J. C. Griffiths, 1953, An experimental test of visual comparison techniques in estimating two-dimensional sphericity and roundness of quartz grains. *American Journal of Science* 251:553–85.

Sames, C. W., 1966, Morphometric data of some recent pebble associations and their application to ancient deposits. *Journal of Sedimentary Petrology* 36:126–42.

Shepard, F. P., and R. Young, 1961, Distinguishing between beach and dune sands. *Journal of Sedimentary Petrology* 31:196–214.

Sneed, E. D., and R. L. Folk, 1958, Pebbles in the lower Colorado River, Texas, a study in particle morphogenesis. *Journal of Geology* 66:114–50.

Spalletti, L. A., 1976, The axial ratio C/B as an indicator of shape selective transportation. *Journal of Sedimentary Petrology* 46:243–8.

Swan, B., 1974, Measures of particle roundness. *Journal of Sedimentary Petrology* 44:572–7.

van Genderen, J.L., 1977, Nomograms for morphometric gravel analysis. *Sedimentary Geology* 17:285–94.

Wadell, H., 1932, Volume, shape, and roundness of rock particles. *Journal of Geology* 40:443–51.

Winkelmolen, A. M., 1971, Rollability, a functional shape property of sand grains. *Journal of Sedimentary Petrology* 41:703–14.

Methods of Size Analysis

Behrens, E. W., 1978, Further comparisons of grain size distribu-
tions determined by electronic particle counter and pipette techniques. *Journal of Sedimentary Petrology* 48:1213–8.

Billi, P., 1984, Quick field measurement of gravel particle size. *Journal of Sedimentary Petrology* 54:658–60.

Boggs, S. J., 1969, Relationship of size and composition in pebble counts. *Journal of Sedimentary Petrology* 39:1243–6.

Burke, K., and S. J. Freeth, 1969, A rapid method for the determination of shape, sphericity, and size of gravel fragments. *Journal of Sedimentary Petrology* 39:797–8.

Chapman, R. E., 1981, Calibration equation for settling tubes. *Journal of Sedimentary Petrology* 51:644–6.

Coates, G. F., and C. A. Hulse, 1985, A comparison of four methods of size analysis of fine-grained sediments. *New Zealand Journal of Geology and Geophysics* 28:369–80.

Colby, B. C., and R. P. Christensen, 1956, Visual accumulation tube for size analysis of sands. *Journal of the Hydraulics Division, Proceedings of the American Society of Civil Engineers,* Paper 1004, 17p.

Connor, C. W., and J. C. Ferm, 1966, Precision of linear and areal measurements in estimating grain sizes. *Journal of Sedimentary Petrology* 36:397–402.

Cook, D. O., 1969, Calibration of the University of Southern California automatically recording settling tube. *Journal of Sedimentary Petrology* 39:781–86.

Creager, J. S., and R. W. Steinberg, 1963, Comparative evaluation of three techniques of pipette analysis. *Journal of Sedimentary Petrology* 33:462–6.

Cui, B., and P. D. Komar, 1984, Size measures and the ellipsoidal form of clastic sediment particles. *Journal of Sedimentary Petrology* 54:783–97.

Davis, M. W., and R. Erlich, 1970, Relationship between measures of sediment-size-frequency distributions and the nature of sediments. *Geological Society of America Bulletin* 81:3537–48.

Dennison, J. M. V., and J. H. Shea, 1966, Reliability of visual estimates of grain abundances. *Journal of Sedimentary Petrology* 36:81–9.

Ellingboe, J., and J. Wilson, 1964, A quantitative separation of non-carbonate minerals from carbonate minerals. *Journal of Sedimentary Petrology* 34:412–8.

Erlich, R., 1983, Size analysis wears no clothes, or have moments come and gone! *Journal of Sedimentary Petrology* 53:1.

Folk, R. L., 1966, A review of grain-size parameters. *Sedimentology* 6:73–93.

Felix, D. W., 1969, An inexpensive recording settling tube for analysis of sands. *Journal of Sedimentary Petrology* 39:777–80.

Fritz, W.J., and J.N. Moore, 1988, *Basics of Physical Stratigraphy and Sedimentology.* Wiley and Sons, New York, 371p.

Gibbons, G. S., 1967, Shell content in quartzose beach and dune sands, Dee Why, New South Wales. *Journal of Sedimentary Petrology* 37:869–878.

Gibbs, R. J., 1972, The accuracy of particle-size analyses utilizing settling tubes. *Journal of Sedimentary Petrology* 42:141–5.

Harrell, J., 1984, A visual comparator for degree of sorting in thin and plane sections. *Journal of Sedimentary Petrology* 54:646–50.

Kaddah, M. T., 1974, The hydrometer method for detailed particle-size analysis. I. Graphical interpretation of hydrometer readings and test of methods. *Soil Science* 118:102–8.

Kennedy, S. K., T. P. Meloy, and T. E. Durney, 1985, Sieve data—size and shape information. *Journal of Sedimentary Petrology* 55:356–360.

Komar, P. D., and B. Cui, 1984, The analysis of grain-size measurements by sieving and settling-tube techniques. *Journal of Sedimentary Petrology* 54:603–14.

Kraus, N. C. and L. Nakashima, 1986, Field method for determining rapidly the dry weight of wet sand samples. *Journal of Sedimentary Petrology* 56:550–1.

Krumbein, W. C., 1932, The mechanical analysis of fine-grained sediments. *Journal of Sedimentary Petrology* 2:140–9.

Krumbein, W. C., 1968, Statistical models in sedimentology. *Sedimentology* 10:7–24.

Lovell, C. J., and C. W. Rose, 1991, Wake-capture effects observed in a comparison of methods to measure particle settling velocity beyond Stokes' range. *Journal of Sedimentary Petrology* 61:575–82.

Ludwick, J. C., and P. L. Henderson, 1968, Particle shape and inference of size from sieving. *Sedimentology* 11:197–235.

McCammon, R. B., 1976, A practical guide to the construction of bivariate scatter diagrams. *Journal of Sedimentary Petrology* 46:301–4.

McCave, I. N., and J. Jarvis, 1973, Use of the Model T Coulter Counter in size analysis of fine to coarse sand. *Sedimentology* 20:305–15.

McCave, I. N., R. J. Bryant, H. F. Cook, and C. A. Coughanowr, 1986, Evaluation of a laser-diffraction-size analyzer for use with natural sediments. *Journal of Sedimentary Petrology* 56:561–4.

Maiklem, W. R., 1968, Some hydraulic properties of bioclastic carbonate grains. *Sedimentology* 10:101–9.

Mazzullo, J., and S. K. Kennedy, 1985, Automated measurement of the nominal sectional diameters of individual sedimentary particles. *Journal of Sedimentary Petrology* 55:593–5.

Moussa, M. I., 1973, Measuring volumes of sedimentary grains. *Journal of Sedimentary Petrology* 43:1171–3.

Prokopovich, N. P., and C. K. Nishi, 1967, Methodology of mechanical analysis of subaqueous sediments. *Journal of Sedimentary Petrology* 37:96–101.

Rogers, J. J. W., 1965, Reproducibility and significance of measurements of sedimentary-size distributions. *Journal of Sedimentary Petrology* 35:722–32.

Singer, J. K., J. B. Anderson, M. T. Ledbetter, I. N. McCave, K. P. N. Jones, and R. Wright, 1988, An assessment of analytical techniques for the size analysis of fine-grained sediments. *Journal of Sedimentary Petrology* 58:534–43.

Slatt, R. M., and D. E. Press, 1976, Computer program for presentation of grain size data by the graphic method. *Sedimentology* 23:121–31.

Sly, P. G., R. L. Thomas, and B. R. Pelletier, 1983, Interpretation of moment measures derived from water-lain sediments. *Sedimentology* 30:219–33.

Solomon, M., and R. Green, 1966, A chart for designing modal analysis by point counting. *Geologische Rundshau* 55:844–8.

Stein, R., 1985, Rapid grain-size analyses of clay and silt fraction by SediGraph 5000D: Comparison with Coulter Counter and Atterberg methods. *Journal of Sedimentary Petrology* 55:590–3.

Steinberg, R. W., and J. S. Creager, 1961, Comparative efficiencies of size analysis by hydrometer and pipette methods. *Journal of Sedimentary Petrology* 31:96–100.

Swan, D., J. J. Clague, and J. L. Luternauer, 1978, Grain-size statistics I: Evaluation of the Folk and Ward graphic measures. *Journal of Sedimentary Petrology* 48:863–78.

Swift, D. J. P., J. R. Schubel, and R. W. Sheldon, 1972, Size analysis of fine-grained suspended sediment: A review. *Journal of Sedimentary Petrology* 42:122–34.

Taira, A., and P. A. Scholle, 1977, Design and calibration of a photo-extinction settling-tube for grain size analysis. *Journal of Sedimentary Petrology* 47:1347–60.

Taira, A., and P. A. Scholle, 1979, Discrimination of depositional environments using settling tube data. *Journal of Sedimentary Petrology* 49:787–800.

Tanner, W. F., 1969, The particle size scale. *Journal of Sedimentary Petrology* 39:809–12.

Wang, C., and P. D. Komar, 1985, The sieving of heavy mineral sands. *Journal of Sedimentary Petrology* 55:479–82.

Wentworth, C. K., 1922, A scale of grade and class terms for clastic sediments. *Journal of Geology* 30:377–92.

Wentworth, C. K., 1936, The method of moments. *Journal of Sedimentary Petrology* 6:158–9.

Examples of Textural Analyses

Allen, G. P., 1971, Relationship between grain size parameter distribution and current patterns in the Gironde Estuary (France). *Journal of Sedimentary Petrology* 41:74–88.

Andrews, P. B., and G. J. van der Lingen, 1969, Environmentally significant characteristics of beach sands. *New Zealand Journal of Geology and Geophysics* 12:119–37.

Bork, K. B., 1970, Use of textural parameters in evaluating the genesis of the Berne Conglomerate (Mississippian) in central Ohio. *Journal of Sedimentary Petrology* 40:1007–17.

Cronan, D. S., 1972, Skewness and kurtosis in polymodal sediments from the Irish Sea. *Journal of Sedimentary Petrology* 42:102–6.

Erlich, R., P. J. Brown, Y. M. Yarus, and R. S. Przygocki, 1980, The origin of shape frequency distributions and the relationship between size and shape. *Journal of Sedimentary Petrology* 50:475–84.

Folk, R. L., and W. C. Ward, 1957, Brazos River bar: A study in the significance of grain-size parameters. *Journal of Sedimentary Petrology* 27:3–26.

Friedman, G. M., 1979, Address of retiring president of the International Association of Sedimentologists: Differences in size distributions of populations of particles among sand grains of various origins. *Sedimentology* 26:3–32.

Jones, T. A., 1970, Comparison of descriptors of sediment grain size distributions. *Journal of Sedimentary Petrology* 40:1214–5.

Klovan, J. E., 1966, The use of factor analysis in determining depositional environments from grain-size distributions. *Journal of Sedimentary Petrology* 36:115–25.

Leroy, S. D., 1981, Grain-size and moment measures: A new look at Karl Pearson's ideas on distribution. *Journal of Sedimentary Petrology* 51:625–30.

Lirer, L., and A. Vinci, 1991, Grain-size distributions of pyroclastic deposits. *Sedimentology* 38:1075–83.

McLaren, P., and D. Bowles, 1985, The effects of sediment transport on grain-size distributions. *Journal of Sedimentary Petrology* 55:457–70.

Middleton, G. V., 1976, Hydraulic interpretation of sand size distributions. *Journal of Geology* 84:405–26.

Middleton, G. V., and J. B. Southard, 1978 and 1984, *Mechanics of Sediment Movement,* 1st and 2d eds. Society for Sedimentary Geology Short Course No. 3, Tulsa, Okla.

Moiola, R. J., and D. Weiser, 1968, Textural parameters: An evaluation. *Journal of Sedimentary Petrology* 38:45–53.

Moss, A. J., 1972, Bed-load sediments. *Sedimentology* 18:159–219.

Moussa, M. T., 1977, Phi mean and phi standard deviation of grain-size distribution in sediments: Method of moments. *Journal of Sedimentary Petrology* 47:1295–8.

Passega, R., 1964, Grain size representation by CM patterns as a geological tool. *Journal of Sedimentary Petrology* 34:830–47.

Passega, R., 1972, Sediment sorting related to basin mobility and environment. *American Association of Petroleum Geologists Bulletin* 56:2440–50.

Passega, R., 1977, Significance of CM diagrams of sediments deposited by suspension. *Sedimentology* 24:723–33.

Reed, W. E., R. leFever, and G. J. Moir, 1975, Depositional environment interpretation from settling velocity (psi) distributions. *Geological Society of America Bulletin* 86:1321–8.

Solohub, J. T., and J. E. Klovan, 1970, Evaluation of grain-size parameters in lacustrine environment. *Journal of Sedimentary Petrology* 40:81–101.

Taira, A., and P. A. Scholle, 1979, Discrimination of depositional environments using settling tube data. *Journal of Sedimentary Petrology* 49:787–800.

Tucker, R. W., and H. L. Vasher, 1980, Effectiveness of discriminating beach, dune, and river sands by moments and the cumulative weight percentage. *Journal of Sedimentary Petrology* 50:165–72.

Visher, G. S., 1969, Grain size distributions and depositional processes. *Journal of Sedimentary Petrology* 39:1074–106.

Warren, G., 1974, Simplified form of the Folk and Ward skewness parameter. *Journal of Sedimentary Petrology* 44:259.

Studies on Conversion to Sieve Sizes of Other Size Measures

Adams, J., 1977, Sieve size statistics from grain measurement. *Journal of Geology* 85:209–27.

Chayes, F., 1956, *Petrographic Modal Analysis.* Wiley, New York, 113p.

Friedman, G. M., 1958, Determination of sieve-size distribution from thin section data for sedimentary studies. *Journal of Geology* 66:394–416.

Galehouse, J. S., 1969, Counting grain mounts: Number percentage vs. number frequency. *Journal of Sedimentary Petrology* 39:812–5.

Harrell, J. A., and K. A. Eriksson, 1979, Empirical conversion equations for thin-section and sieve-derived size distribution parameters. *Journal of Sedimentary Petrology* 49:273–80.

Kellerhals, R., J. Shaw, and V. K. Arora, 1975, On grain size from thin section. *Journal of Geology* 83:79–96.

Krumbein, W. C., 1935, Thin section mechanical analysis of indurated sediments. *Journal of Geology* 43:482–96.

Rosenfeld, M. A., L. Jacobsen, and J. C. Ferm, 1953, A comparison of sieve and thin section techniques for size analysis. *Journal of Geology* 61:114–32.

Van Der Plas, L., and A. C. Tobi, 1965, A chart for judging the reliability of point counting results. *American Journal of Science* 263:87–90.

SEM References

Brown, R. J., 1986, SEM examination of carbonate microfacies using acetate peels. *Journal of Sedimentary Petrology* 56:538–9.

Culver, S. J., P. A. Bull, S. Campbell, R. A. Shanesby, and W. B. Whalley, 1983, Environmental discrimination based on quartz grain surface textures: A statistical investigation. *Sedimentology* 30:129–36.

Goldstein, J. I., D. E. Newbury, P. Echlin, D. C. Joy, C. Fiori, and E. Lifshin, 1981, *Scanning Electron Microscopy and X-ray Microanalysis.* Plenum Press, New York, 673p.

Jones, B., and C. A. Squair, 1989, Formation of peloids in plant rootlets, Grand Cayman, British West Indies. *Journal of Sedimentary Petrology* 59:1002–7.

Krinsley, D. H., and J. Donahue, 1968, Environmental interpretation of sand grain surface textures by electron microscopy. *Geological Society of America Bulletin* 79:743–8.

Krinsley, D. H., and J. C. Doornkamp, 1973, *Atlas of Quartz Grain Surface Textures.* Cambridge University Press, Cambridge, 91p.

Minnis, M. M., 1984, An automatic point-counting method for mineralogical assessment. *American Association of Petroleum Geologists Bulletin* 68:744–52.

Nixon, W. C., 1969, Scanning electron microscopy. *Contemporary Physics* 10:71–96.

O'Brien, N. R., 1981, SEM study of shale fabric—a review. *Scanning Electron Microscopy* 1:569–75.

Pittman, E. D., and R. W. Duschatko, 1970, Use of pore casts and scanning electron microscopy to study pore geometry. *Journal of Sedimentary Petrology* 40:1153–7.

Smart, P., and N. K. Tovey, 1982, *Electron Microscopy of Soils and Sediments.* Clarendon Press, Oxford.

Warnke, D. A., and R. Gram, 1969, The study of mineral grain surfaces by interference microscopy. *Journal of Sedimentary Petrology* 39:1599–1604.

Welton, J. E., 1984, *The SEM Petrology Atlas.* American Association of Petroleum Geologists, Tulsa, Okla., 237p.

Wilding, L. P., and H. D. Geissinger, 1973, Correlative light optical and scanning electron microscopy of minerals: A methodological study. *Journal of Sedimentary Petrology* 43:280–6.

Porosity and Permeability

Bradley, J. S., R. W. Duschatko, and H. H. Hinch, 1972, Pocket permeameter: Hand-held device for rapid measurement of permeability. *American Association of Petroleum Geologists Bulletin* 56:568–71.

Davis, D. H., 1954, Estimating porosity of sedimentary rocks from bulk density. *Journal of Geology* 62:102–7.

Dodge, C. F., D. P. Holler, and R. L. Meyer, 1971, Reservoir heterogeneities of some Cretaceous sandstones. *American Association of Petroleum Geologists Bulletin* 55:1814–28.

Eijpe, R., and K. J. Weber, 1971, Mini-parameters for consolidated rock and unconsolidated sand. *American Association of Petroleum Geologists Bulletin* 55:307–9.

Fatt, I., M. Maleki, and R. N. Upadhyay, 1966, Detection and estimation of dead-end pore volume in reservoir rock by conventional laboratory tests. *Society of Petroleum Engineers Journal* 6:206–12.

Halley, R. B., 1978, Estimating pore and cement volumes in thin section. *Journal of Sedimentary Petrology* 48:642–50.

Lundegard, P. D., 1992, Sandstone porosity loss—a "big picture" view of the importance of compaction. *Journal of Sedimentary Petrology* 62:250–60.

Manger, G.E., R.A. Cadigan, and G.L. Gates, 1969, Irmay's saturation factor as an indication of an immobile fraction of pore water in saturated permeable sandstone. *Journal of Sedimentary Petrology* 39:12–17.

Netto, A. S. T., 1993, Pore-size distribution in sandstones. *American Association of Petroleum Geologists Bulletin* 77:1101–4.

Rassineux, F., D. Beaufort, A. Meunier, and A. Bouchet, 1987, A method of coloration by fluoresein aqueous solution for thin-section microscopic observation. *Journal of Sedimentary Petrology* 57:782–3.

Ruzyla, K., and D. I. Jezek, 1987, Staining method for recognition of pore space in thin and polished sections. *Journal of Sedimentary Petrology* 57:777–8.

Villinger, R. M., and G. A. Kuhn, 1993, Nondestructive porosity determinations of Antarctic marine sediments derived from resistivity measurements with an inductive method. *Marine Geophysical Research* 15:201–18.

Yanguas, J. E., and J. J. Dravis, 1985, Blue fluorescent dye technique for recognition of microporosity in sedimentary rocks. *Journal of Sedimentary Petrology* 55:600–2.

The Udden-Wentworth grade scale for grain sizes, with σ/mm conversion chart

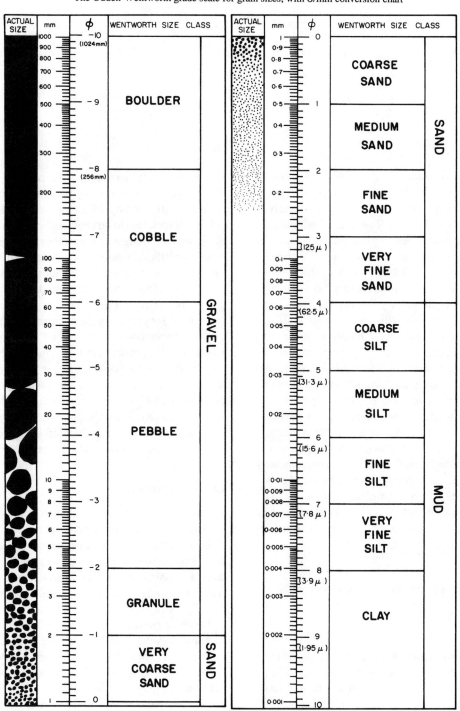

8

Mineralogy

Many methods exist for identifying mineral components from detrital sediments. The flowchart depicted in Fig. 8-1 illustrates routes to many of the commonly applied analytical instruments (also see Fig. 9-1). The method selected for mineral identification will depend on the nature of the sediment, the degree of detail required by the specific project, equipment and personnel available, and cost effectiveness (total cost in relation to time required for results). Clay mineral identification procedures are generally the most complex because of the extremely small size of the minerals and the difficulty in identifying species of highly variable mineral groups, and because it is difficult to prepare pure segregates from the common natural mixtures of clay minerals. Medium sand to coarse silt grains are mainly composed of individual mineral grains and are readily subjected to a wide range of identification procedures. Gravel particles are mainly composed of rock fragments, and most identifications are made by a geologist armed with a hand lens using a counting technique in the field as an approximation of volume percentage of constituents (see Chapter 4). Although no reliable quantitative instrumental methods exist to identify composition of bulk samples, individual gravel grains may be subjected to any of the procedures discussed below for sand particles.

COMMON ROCK-FORMING MINERALS OF SAND SIZE

Detrital Minerals

When dealing with loose sediments, preliminary steps are to separate the medium sand to coarse silt fractions (e.g., Chapter 7) and thoroughly cleanse the grains of any natural coatings (Chapter 5) prior to analysis; even then, the presence of inclusions and alteration products may result in some spuri-

ous data. Preliminary concentration may be achieved by electromagnetic means or by panning or flotation on the basis of specific gravity (discussed under "Heavy Minerals" below). If rocks are to be analyzed, either the sample must be disaggregated (Chapter 5) or subsamples must be cut from them. For bulk samples of both loose sediment and rocks, the most common mineral identification techniques are to identify components with transmitted light (for transparent grains) or reflected-light (for opaque grains) microscopy; procedures for preparing the necessary grain mounts for thin or polished sections are described in Chapter 5. Cathodoluminescence of rock slices is a refinement of optical microscopy that is useful particularly for diagenetic growths and replacements. For relatively pure mineral separates, X-ray diffraction (XRD) is the most common technique. X-ray fluorescence (XRF; see Chapter 9) techniques are commonly used for trace element geochemical studies that may serve to refine knowledge of mineral species.

Carbonate Minerals

Staining techniques (see this chapter) are probably the quickest and easiest ways of identifying the common varieties of carbonate minerals, and some can be applied in the field as well as in the laboratory. Experimentation and practice, with care in washing and use of fresh chemicals, are necessary with the staining techniques to obtain optimal results. Alternatively, X-ray diffraction (see below) or differential thermal analysis (see below) may be used. Element content in the carbonate crystal lattices (e.g., Mg vs. Ca) is revealed by stains, wet-chemical analysis, atomic absorption, emission spectroscopy, X-ray fluorescence, or electron probe analysis. Such studies may be informative as to the diagenetic history of the carbonates (e.g., Lindholm and Finkelman 1972), or there may be other ways of utilizing the information ob-

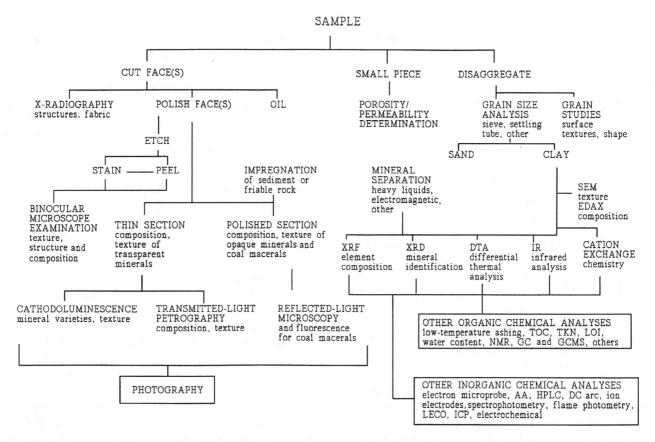

Figure 8-1.   Flowchart depicting common routes from a sediment sample to mineral identification of its components.

tained. Ratios of strontium to calcium in calcites may be measured for paleotemperatures because at higher temperatures crystal lattice disorder permits entry of greater proportions of the relatively large Sr^{2+} ion (e.g., **PS** Chapter 8). However, the Sr^{2+} ion enters the lattice of the less-ordered aragonite mineral at much lower temperatures, and because aragonite is unstable and converts to calcite, the resultant calcite can "inherit" an anomalously high Sr content; hence these elemental ratios are only useful when the original crystal chemistry is known to have been low-Mg calcite.

Complete digestion of carbonate-rich sediments or rocks in acetic acid or HCl (e.g., Sanford, Mosher, and Friend 1968) permits study of any insoluble components in the same way as for detrital minerals, but be aware that some noncarbonate minerals such as clays may be substantially affected by the acid (Chapter 5). Another problem is excessive effervescence and consequent destruction of delicate grains; ensure that acid is added slowly and/or is sufficiently dilute. See Ireland (in Carver 1971) for a thorough discussion of insoluble residues.

Dravis (1991) has reviewed the history and methodology of recently developed methods for studying depositional and diagenetic fabrics in dolostones and recrystallized limestones with diffused PPL and with blue-light fluorescence microscopy.

OPTICAL MICROSCOPY

Transmitted Light Microscopy

The petrographic microscope (Fig. 8-2) is a standard invaluable tool for both the geologist and the soil scientist (see Brewer 1964; Hartshorne and Stuart 1970; Phillips 1971; Bullock et al. 1984; Shelley 1985; Kemp 1985 for good descriptions of principles and applications). Its use permits easy and rapid identification of any minerals through which light can pass, in addition to the textural characteristics and relationships of the minerals. Light from a substage light source is polarized before it reaches the grain mount or thin section, and another polarizer can be inserted between the grains and the eyepiece so that the operator can view in either plane-polarized (PPL) or crossed-polarized (CPL) transmitted light. Another option is to insert a substage light condenser and a Bertrand lens (in the microscope tube) to view the interference figure produced by that optical arrangement. Because light waves are variously rotated and refracted as they pass through the atomic structure of grains, different minerals have different birefringence (coloring CPL) and interference figures; thus, they can be identified from their various properties (e.g., Winchell 1965; Phillips and Griffen 1981; Shelley 1985). Adams, MacKenzie, and Guilford (1984) depict a wide array of sedimentary rocks as seen in

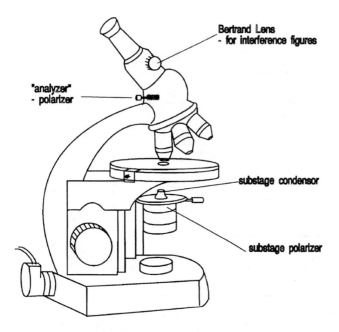

Figure 8-2.   Sketch of a petrographic microscope, designed for the study of thin sections of rock in transmitted light.

thin section (see also references in **PS** Chapter 6). Although mostly used with sand-size sediments, where optical properties can be fully determined and minerals identified, optical microscopy is also used with some gravels, where the main problem is representative sampling of the grain population, and with mudstones, where textural relationships of components can be examined even if the clay size minerals cannot be identified.

Although loose grains of a variety of sizes can be identified via transmitted light, practical considerations of light transmissibility (interference colors of the same mineral vary with thickness) and the focal range of the microscope dictate that grain thicknesses in general should fall within the medium silt to very fine sand range (0.015–0.125 mm); however, coarser particles can be studied by refractive index (RI) techniques alone. To simplify operator procedures, rock and soil samples are prepared in thin sections (see Chapter 5) that are a standard 0.03 mm thick. If the goal is to identify selected loose minerals from a sediment (e.g., segregated heavy minerals, see below), grain mounts (see Chapter 5) are prepared and immersed in various oils of known refractive index (RI) (commonly several indices must be determined, since the RI may vary with crystallographic orientation; see Phillips 1971; Shelley 1985); for precise RI determinations, the spindle stage (see below) should be used. Standard sets of oils are normally prepared and these can be mixed for precise matches; special refractometer microscopes are used to determine the RI of the oils.

Spindle Stage Technique

The spindle stage is an inexpensive, easily constructed device that permits utilization of a hypodermic needle to mount a single mineral grain (in the range 0.074–0.149 mm), which

can then be immersed in a RI oil (e.g., held in an attached glass plate) and rotated 180° about a horizontal axis as well as 360° about the vertical axis of the microscope. See Bloss and Light 1973 for construction details, Shelley 1985 for a simple explanation of operations, and Bloss 1981 for a comprehensive discussion of principles and procedures. All of the mineral RIs can be measured precisely, as well as its important optical properties. If using this method, ensure that the glue used for mounting does not coat the grain!

Oil Immersion Microscope Study

Finer resolution and increased brightness can be achieved in high-magnification studies by replacing air with oil in the gap between the objective lens of the microscope and the slide (in special cases, oil may be used between the light source condenser below the microscope stage and the slide). The principle involved is equalization of the refractive indices of the various media, which results in less disturbance of the light path and a greater numerical aperture (NA). The total magnification should not exceed 1000 × NA, the slides should be 0.8–1.0 mm thick, and cover slips on the slide must be particularly thin (No. 1 type); otherwise focusing problems may occur. When applying the oil, touch a drop to the slide and allow it to flow from the applicator to avoid air bubbles; gradually lower the microscope tube until the lens contacts the oil. Only lenses labeled OEL or OIL should be used (most are 63 or 100 power), and the oil must be wiped off (carefully on the lens—not rubbed!) immediately after use with a clean tissue with a few drops of xylol or special mixtures of solvents. *Caution:* When using xylol, ensure adequate ventilation and do not permit excessive contact with skin.

Universal Stage

Identification in thin section of uncommon minerals or of mineral species (e.g., the specific variety of plagioclase feldspar) is greatly facilitated by mounting the thin section into a universal stage attachment for the petrographic microscope (Fig. 8-3). The slide is mounted between two hemispheres of precision-ground glass that maintain a uniform light path for the microscope tube as the section is rotated about multiple axes (the number varies with the type of U-stage). Because the slide can be placed in almost any orientation (limits occur to the angle of rotation about some axes), optical and crystallographic characteristics can be precisely determined (e.g., crystal faces can be detected and rotated into perpendicular position); thus, minerals such as varieties of feldspars can be identified. The U-stage is also valuable for determining preferred orientations of crystals and for mineral fabric studies (particularly useful for diagenetic modifications involving compaction or pressure-solution effects, cement and overgrowth developments, and recrystallization). Good discussions of the procedures for U-stage study can be found in Wahlstrom (1951), Phillips (1971), and Shelley (1985); see Gilbert and Turner (1949) for a discussion of some applications in sediment studies.

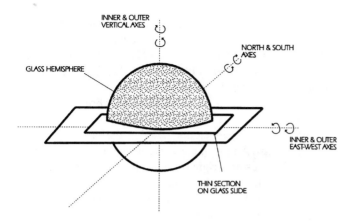

Figure 8-3. Sketch of the essential elements of a universal stage. The device is mounted on a special holder on the stage of a petrographic microscope (Fig. 8-2) and permits a thin-section sample to be rotated to almost any angle for viewing of the optical properties of minerals.

CATHODOLUMINESCENCE

Cathodoluminescence refers to light emission by compounds in response to their bombardment by electrons (e.g., the luminescence from chemical phosphors in a television tube). Thorough discussion of principles is provided in Long and Agrell (1965), and of applications to mineralogical studies in Walker (1985) and Miller (in Tucker 1988, chapter 6). Examples of applications in sediment studies are Sippel (1968) for arenites; Cercone and Pedone (1987) for sources of error in studying carbonates; Roberts and Graves (1965) for analysis of modern sediments in cores; Ryan and Szabo (1981) for rapid determination of light minerals in detrital sands. Although cathodoluminescence studies are expanding, fundamental understanding of the causes and implications of the phenomena in minerals is still limited and generally applicable relationships are few. Sites of crystal lattice imperfections or distortions preferentially luminesce; hence, e.g., portions of minerals with different ions in the lattice will show different colors. Overgrowths developed in diagenesis, under conditions that are commonly very different to the parent grain, will have a different lattice order and trace element content, and the difference will be very obvious in the light waves emitted under electron bombardment. Mechanically damaged crystals (e.g., where scratches are made by coarse abrasives during surface grinding) commonly luminesce brightly.

Basic equipment is a vacuum box (with lead glass window at the top) that can be mounted on the stage of a petrographic microscope. Samples are mounted on a movable glass tray in the box, and an attached cathode tube bombards the sample; in most commercial models the diameter of the electron beam can be varied, permitting varied levels of mineral excitation. Beam penetration is limited to a few tens of microns. During excitation, samples are heated and gasses are produced; the ions of these gasses increase the electron beam current and the vacuum must be adjusted to keep it constant; in addition, long-term viewing in one area may cause the

specimen to crack as it expands differentially or may cause minerals to undergo thermal changes. Because intensity of luminescence is generally low, the microscope used must have very high light transmissivity, all optics must be kept spotless, and the equipment should be operated in a darkened room. Because of the heating effect, the uncovered thin sections cannot be mounted with Canada Balsam or Lakeside Cement. Highly polished (and clean!) surfaces provide the best optical resolution.

Caution: Excitation from the electron beams creates X-rays, and although cathodoluminescence chambers are designed for the purpose, checks should be made regularly for radiation, particularly around the window, and operating voltages should be kept within the safety levels stated in the operating manual.

General applications in sediment studies include (see Miller in Tucker 1988 for a more thorough discussion):

1. Rapid differentiation of minerals and varieties of the same mineral, in cases where grains are similar under the ordinary petrographic microscope or where they are very fine; although one particular mineral type (e.g., feldspar) does not show a uniform color even in the same thin section, the range of colors shown differs from that of other minerals. Varieties of the same mineral derived from different sources (or subjected to different original crystallization or deformation histories) will show different colors.
2. Heightened visibility of textural characteristics, particularly in recrystallized carbonate (and evaporite?) rocks; what are vague ghostly outlines under PPL may appear as sharply outlined fossils or other allochems.
3. Diagenetic and geochemical studies being greatly aided, because the growth history of cements, authigenic minerals, recrystallized minerals, and stylolites can be determined in fine detail. Different trace element contents in the same mineral result in different colors.
4. Small-scale structural features, such as fractures, joints, veins, become readily apparent. The presence of such features is of considerable importance in studies of rock permeability and strength.

REFLECTED LIGHT MICROSCOPY

Reflected light microscopy is not commonly applied by sedimentologists because it is an analytical procedure for identification of opaque minerals that are generally rare in sediments. However, studies of the opaque fraction of heavy mineral suites and of the genesis of some sedimentary ore deposits make use of this technique. Gregg and Karakus (1991) describe a method of configuring cathodoluminescence and reflected light equipment that permits both techniques to be applied to the same sample. Because of the specialized nature of reflected light microscopy and the rarity of applications in sedimentology to date, we refer interested readers to Galopin and Henry (1972), Craig and Vaughan (1981), and Ineson (1989).

COAL MICROSCOPY

Petrology does not replace chemical analysis in the characterization of coal, but has the advantage that the necessary equipment is available in a typical earth science department. Because coal consists principally of organic material without precise form or composition, considerable variability is possible. Whereas bulk chemistry effectively homogenizes the sample, petrology allows the potentially diverse coal constituents to be evaluated in detail. Petrology has special value for differentiation of individual components of commercial coal blends, but this is an industrial rather than a geological aspect. Many petrological techniques are available; some are outlined briefly in the context of their geological applications.

Macerals are microscopic organic constituents, somewhat analogous to individual minerals in a detrital sediment. Macerals are classified according to their form and optical characteristics, which also have chemical and genetic significance. All standard petrological techniques are undertaken using polished specimens viewed in incident light. Conventional maceral analysis uses white light and oil immersion (oil immersion reduces the total amount of light reflected but improves resolution by enhancing the contrast between macerals). In this mode, three groups of macerals are recognized. *Inertinite* is relatively highly reflecting and originates principally as plant material that has been oxidized before burial. *Liptinite* is weakly reflecting waxy or resinous material, derived from specific hydrogen-rich materials such as cuticle. Most coals consist principally of *vitrinite*, which has intermediate reflectance and originates as plant tissues that have not been oxidized. (Most coals that have accumulated from in situ flora, with limited transport of vegetation within the immediate environment, contain abundant vitrinite, with or without inertinite, and are termed *humic*.) Many submacerals are defined within these groups (e.g., Stach et al. 1982). Procedures for maceral analysis are clearly described in ISO 7404/3, which is suitable for geological applications. The Australian standard method (AS2856, 1986) simplifies classification of the vitrinite submacerals and is convenient for industrial applications but places potentially undesirable limits on determination of details that may have geological significance.

Although white light and oil immersion allow adequate characterization for many purposes, more specialized techniques can provide valuable additional information. For example, liptinite generally occurs in relatively small amounts and can be difficult to identify due to small size, poor structural preservation, low reflectance, and confusion with mineral matter. Therefore, if equipment is available, many petrologists supplement their white light analyses with a count undertaken in short wavelength (blue) light, which induces strong fluorescence of liptinite macerals. Air, rather than oil, immersion is best used in this case. Also, whereas vitrinite as a group is distinct, individual submacerals can be difficult to distinguish from each other due to poor contrast. Although definition can be enhanced by special immersion media or appropriately filtered fluorescence systems, oxidative etching of the polished surface is the favored method for identification. Viewed in air with white light, the microstructure of both tissues and matrix is dramatically revealed by etching, and blue light analysis of liptinites can still be undertaken.

Vitrinite reflectance, measured in monochromatic green (546 nm) light and oil immersion, increases more regularly in relation to burial temperature than is the case for other maceral groups. Although considerable variability can result from differences in original vegetation and peatification, vitrinite reflectance has been an important method for rapid assessment of thermal history during petroleum exploration, and remains a standard procedure despite the advent of chemical methods such as rock-eval pyrolysis (Chapter 9). Despite considerable debate, there is general agreement that temperature, rather than time or pressure, is the principal influence on coal metamorphism, which is usually referred to as coalification or maturation. Because coals are composed of hydrocarbons, they respond chemically to temperatures lower than those normally associated with mineral metamorphism. The transition from brown or subbituminous coals to black coals with coking properties and the potential to generate petroleum occurs at temperatures less than 100°C.

The principal chemical reactions during coalification involve condensation of aromatic ring structures with consequent release of aliphatic hydrocarbons. Coalification is not reversible, and the rank of a sample represents the maximum temperature attained during burial. Coal rank therefore ideally complements some other thermal indicators, particularly fission tracks in apatite, which can provide an uplift history but do not record information about maximum burial if temperatures have exceeded approximately 100–120°C. Many coal properties are influenced by coalification and therefore provide information on coal rank. However, all such properties are a function of organic material that is partly dependent on original peat chemistry as well as coalification history. Therefore, no parameter precisely qualifies coal rank. Commonly used rank indicators include vitrinite reflectance, moisture, specific energy, and total carbon. Ideally more than one indicator is used, and interpreted in the context of all significant controls, of which thermal history is only one. In the absence of actual coal seams, dispersed organic matter can be concentrated to yield sufficient vitrinite for an analysis, even from a marine sediment, although in such cases results tend to be imprecise. When it is carefully evaluated in relation to chemical parameters, vitrinite reflectance can provide valuable data for reconstruction of basin development (e.g., related to hydrocarbon maturation or to structural history). Best results are usually attained above 0.5% reflectance, and some workers prefer to limit measurement to specific submaceral types. Vitrinite reflectance of industrial feedstock coal is also routinely determined and used as a predictor of coking behavior (all vitrinite submacerals are included in the analysis).

Esterle, Moore, and Hower (1991) describe the application to peat studies of reflected light petrography routinely utilized for coal examination. Samples in c. 5 × 7 cm blocks are prepared by quick freezing the peat (by immersion in liquid nitrogen), then freeze-drying in a vacuum to remove the water; an automated polisher is used in preference to hand polishing to achieve the consistently high-quality, mirror-smooth surface necessary for detailed examination of plant components with oil-immersion objectives.

HEAVY MINERALS

Heavy minerals are those minerals in sediments with a specific gravity (S.G.) greater than the medium used to separate them from the light minerals! In most texts, the arbitrary S.G. limit of 2.85 is set because that is the S.G. of the heavy liquid *bromoform*, which was the most commonly used medium in the first century of heavy mineral separations. However, the S.G. of bromoform varies with its temperature and the amount of acetone or other compounds that may not have been totally removed, and in recent decades a variety of media with other S.G.s have been used. Hence the initial sentence is not as daft as it reads. Be aware of the limits set by the nature of the separating procedure used for data with which your results are to be compared, and report the S.G. at which differentiation was carried out in any analysis.

Mention also is made in this section of other techniques that may be applied to the heavy or light fractions of samples (or even the initial bulk sample). For example, many minerals that separate into the heavy group are paramagnetic, that is, they become magnetized in, and proportionately to, the strength of a magnetic field (Table 8-1). Hence, after heavy liquid separation, they are commonly subjected to magnetic separation procedures.

No single technique results in a perfect separation: inclusions or bubbles within minerals, grain coatings, the effects of weathering, and differences inherent between species of the same pure mineral are the common causes for imperfect separations. Experimentation is necessary with any particular suite of minerals, and samples may require several passes through a single procedure before good separations are achieved. For some studies, analysis of only one or two minerals may be desired; appropriate modification of the following techniques may allow their simple and rapid separation. Once separated and segregated, the various minerals have to be identified; procedures for mounting grains for microscopy are discussed in Chapter 5, and the use of transmitted and reflected light microscopes has been outlined earlier in this chapter.

Initial Preparation

The following steps may be followed in order, or at some intermediate stage (depending on the sample) it may be appropriate to concentrate the heavy minerals by a panning procedure with a standard field gold pan (e.g., Theobald 1957) or superpanner (see below).

1. Disaggregate the sample, remove all soluble salts and organic matter, and remove by decantation or wet-sieving all of the clay size fraction (or mud fraction if silt particles are not to be studied); clean the remaining grains of any coatings that may be present (see Chapter 5). Removal of organic matter is particularly important because organic compounds can be very difficult to cleanse from the heavy liquid. (If chemical techniques are used in disaggregation, be aware of the destruction or modification of minerals that may have occurred, and ensure the grain assemblage is well washed and filtered.)

2. Dry the sample.

3. Sieve the sands into the grades chosen for analysis. In general, heavy minerals are of medium sand or finer size; most of those previously studied have been in the finer sand sizes, but those in the silt range may be the target of an individual study. Choose one size grade from the class just finer than the mode of the total sand frequency distribution (heavy minerals should be most abundant in this grade) and another grade at the same size for all samples (to facilitate the comparison of suites).

4. Microsplit a representative 1–3-g subsample (depending on abundance of the heavy minerals). Larger samples result in mass trapping effects in the heavy liquid or require wastefully large volumes of liquid.

5. Weigh the subsample to 0.01 g.

Heavy Liquid Separation

Many liquids are available for mineral separations (e.g., see Carver 1971); costs and availability of the liquids vary markedly between the liquids and for the same liquids in different countries, and procedures vary in detail depending on the liquid used. Particularly common since the 1970s is tetrabromoethane (TBE); either acetone or benzene is used as a solvent with it. Apart from cost (which should include an evaluation of the laboratory time and equipment needed for recovery of the liquids for reuse), important considerations are the miscibility of the liquid with common solvents (for recovery and for adjusting S.G. to desired values), the chemical reactivity (the liquid used should not react with or take into solution any sample constituent), and surface tension of the fluid (high surface tension can create major problems of balling and clotting of the finest sediments). Toxicity of many of the commonly used compounds is of concern to many workers; to avoid this problem, it is possible to utilize aqueous solutions of sodium polytungstate (see Gregory and Johnston 1987), although the cost of the compound is high and the viscosity of the solution is higher than desirable (filtration rates are much slower than with most other compounds). A sodium polytungstate solution with a density between 2.96 and 3.06 can be produced easily by dissolving 1.312 kg of the salt in 250 mL of distilled water; density can be raised to 3.1 by evaporation or lowered with the addition of water.

Table 8-1.   Specific Gravity and Magnetic Susceptibility Ranges of Some Minerals Commonly Separated by Heavy Liquid or Magnetic Methods

Heavy Liquid column (vertical labels with arrows, left to right): Bromoform, Tetrabromoethane, Methylene iodide, Thallous formate, Thallous formate-malonate

Group II S.G.	Group II Mineral	Group III S.G.	Group III Mineral	Group IV S.G.	Group IV Mineral
2.3	Glauconite			2.0–2.4	Zeolites
				2.3	Gypsum
				2.5	Leucite, Kaolin
				2.5–2.6	Alkali feldspars
				2.6–2.7	Na-Ca feldspars
				2.6–2.7	Scapolite group
				2.62	Chacedony
2.63	Cordierite	2.8		2.6	Nepheline
		2.9	Chlorites	2.65	Quartz
		2.8		2.72	Calcite
		3.0–3.5	Chamosite	2.85	Dolomite
				2.86	Phlogopite
			Biotites	2.8	Muscovite
3.1	Iron-rich biotite	3.4	Amphibole	3.0	Tremolite
	Actinolite		Pyroxene	3.3	Enstatite
3.2–3.6	Iron-rich Amphiboles	3.0–3.25	Tourmaline	3.1	Apatite
	and Pyroxenes	3.4	Epidote	3.2	Flourite, Andalusite
			Olivine (wide range	3.23	Sillmanite
		3.5	of S.G.) Chloritoid	3.5	Sphene
3.7	Melanite	3.65–3.75	Staurolite	3.5	Topaz
		3.8	Pleonaste*	3.6–3.7	Spinel
		4.0	Garnet (wide-range	3.6	Kyanite
3.8	Siderite		S.G.)	3.9	Anatase
	limonite				Brookite
				4.0	Pervoskite
	Garnet				Corrundum
4.0	Sphalerite				
4.3	Almandine	4.2	Ferriferous	4.2	Rutile
			Rutile	3.5–4.5	Leucoxene
4.4	Chromite*	4.4–5.1	Chrome spinel		
4.6	Xonotine	4.8	Pyrolusite	4.5	Barite
		4.9	Marcasite	4.7	Zircon
4.8	Ilmenite*	5.0–5.3	Monazite		
5.2	Hematite	5.3–7.3	Scheelite		
			Columbite*	5.0	Pyrite*
				7.0	Cassiterite
7.0	Cassiterite			7.2–7.6	Wolframite
				14.0–19.0	Platinum
				19.0–21.0	Gold

Note: Good conductors are noted by asterisks, and these can be separated by electrostatic methods used by industry. Note that the same mineral may vary significantly in magnetic properties and specific gravity because of compositional variation, coatings, alteration products, or the presence of inclusions. Group organization classes minerals that have relatively similar magnetic susceptibilities. Group I (not listed) comprises the highly magnetic minerals magnetite (S.G. 5.17), titanoferrite (S.G. 4.85), and pyrrhotite (S.G. 4.65); these are the minerals that must be separated by hand magnet and must not be fed into the Frantz separator. Group II comprises moderately magnetic minerals. Group III comprises weakly magnetic minerals (sometimes separable in the Frantz). Group IV comprises the nonmagnetic minerals.

Source: After Krumbein and Pettijohn 1938; Milner 1962.

Most heavy liquids can be diluted to any desired lesser S.G.—a quick method is to add a solvent to the fluid until a pure mineral grain of the desired specific gravity floats in the midst of the solution; thus, the following procedures can be modified to produce much more specific S.G. separations. The following equation can be used to produce a liquid of desired S.G.:

$$x = \frac{y(\text{S.G. required} - \text{S.G. liquid})}{\text{S.G. dilutant} - \text{S.G. required}}$$

where x = volume of dilutant

y = volume of heavy liquid.

(Note: The S.G. of common dilutants is acetone 0.79 g/cc; ethyl alcohol 0.79 g/cc; benzene 0.88g/cc.)

Caution: Many heavy liquids are volatile, particularly when mixed with a solvent such as acetone, and they produce toxic fumes; work with all of them in a well-ventilated fume cupboard, and use a respiratory mask if you still inhale vapor. Acetone fumes alone can damage cell membranes—do not inhale much vapor or permit excessive contact with skin (natural oils are rapidly dissolved and dermatitis may result).

Bromoform separation (other heavy liquid procedures vary only slightly): Bromoform ($CHBr_3$) has a specific gravity of 2.85 at 20°C, which changes 0.023/°C; common light rock-forming minerals (quartz, feldspar, calcite) float in it. Bromoform decomposes in light and should be stored in the dark. It is very expensive and steps should be taken to recover it in a storage bottle. Alcohol, acetone, and carbon tetrachloride are soluble in bromoform and should be used to rinse the final remains of bromoform into a separate "washings bottle" for later separation and recovery. Table 8-2 lists the proce-

---

Table 8-2.   Routine Procedure for Separating Heavy Minerals Using Bromoform, as an Example of the Heavy Liquid Method

1–5. Prepare the sample as discussed earlier in the text (page 135).

6. Check the specific gravity of bromoform before each use (with a specific gravity bottle, a special hydrometer, or a refractive index method).

7. Place the bromoform in a funnel with a pinch-clamped rubber tube or glass tap that feeds into another funnel in which filter paper can be placed (Fig. 8-4); label separate storage and washings beakers for bromoform recovery.

8. Add the subsample; stir occasionally to avoid mass-trapping effects (ensure grains are not removed on the stirring rod). Keep a watch glass over the funnel between stirrings. Allow 10–20 min for separation.

9. Drain liquid with heavy minerals (with accompanying bromoform) into No. 1 filter paper on the lower funnel; retain pure bromoform in storage bottle.

10. Drain the rest of the bromoform together with the light minerals into another filter paper; use the same storage bottle.

11. Replace the storage bottle with the bromoform washings bottle.

12. Using methyl alcohol or acetone, wash residual bromoform and remaining grains from the upper funnel onto the lower filter paper, then wash the bromoform from the filter paper.

13. Wash the bromoform from the filter paper containing the heavy minerals into the same washings bottle. Both grains and filter paper should be clean of bromoform.

14. Place the filter papers and grains into bowls; air-dry at a temperature no greater than 50°C for 1–2 hr.

15. Weigh both light and heavy fractions to 0.001 g.

16. Calculate the weight percentage of the "heavies" for the size grade used.

17. Reclaim the bromoform from the washings bottle (see Table 8-3 or place a beaker containing the washings in cold brine—bromoform freezes at 9°C and the alcohol or acetone can then be removed).

An alternative procedure for steps 7–12 is useful when few separations are necessary, grains are very small, or there are many grains with a specific gravity close to that of the heavy liquid.

7. Place 10 mL bromoform in a centrifuge tube that has a constricted end; add grains (a much smaller quantity than with the funnel method). Wash grains into the tube with another 10 mL bromoform. Cork the tubes.

8. Balance the centrifuge; ensure equivalent weights are in opposing tube holders.

9. Gradually build up the spin of the centrifuge (there is no need to exceed approximately half speed); hold for 5–15 min, then gradually reduce the spin to zero.

10. Extract the centrifuge tube(s) and follow step (a), step (b), or step (c).

(a) Insert a stopper to seal off the lower portion of the tube with its heavies. Pour off the bromoform and light minerals through a funnel with filter paper; rinse out the remaining grains (with bromoform or with alcohol). Remove and rinse the stopper. Remove the heavies and bromoform. (A rubber policeman may be used to extract recalcitrant grains.)

(b) Insert the constricted base of the centrifuge tube into cold brine. Bromoform freezes at 9°C , and the HMs will be trapped in the frozen fluid. Extract the light minerals and fluid bromoform as in step (a), then thaw and extract the heavies and bromoform. (If frozen bromoform begins to thaw too early, refreeze the lower portion of the tube.)

(c) Use a micropipette with a pipette bulb to withdraw the heavy minerals.

Note: Perform the separation in a fume cupboard.

Table 8-3.   Routine Method for Recovering the Expensive Heavy Liquid Bromoform After Mineral Separation

Equipment

    large (up to 5 L) reagent bottle
    1 stopper with 2 holes for glass tubing
    short length of rubber tubing
    pinchcock
    1 long and 1 short length of glass tubing
    ring stand and ring holder
    beaker
    heavy liquid washings
    distilled water

Procedure

1. Fit the bottle mouth with the stopper and tubing; seal carefully. The long tubing should almost reach the bottom of the bottle and protrude only 3–5 cm outside the stopper. The short tube should be flush with the inside of the stopper and protrude outside far enough to adequately receive the rubber tubing (a glass tube with a stopcock is even better). Place the pinchcock on the rubber tubing.

2. Place heavy liquid washings in the bottle (no more than one fourth of the bottle's volume), and fill the bottle with distilled water (to about three fourths to seven eighths of the bottle's volume).

3. Be sure the pinchcock is on the rubber tubing. Holding an index finger over the stopper end of the long glass tube (to prevent it from filling with liquid), insert the stopper firmly into the bottle top. Still holding a finger over the tubing, invert the bottle. If the bottle is properly filled, the level of the liquid in the inverted bottle will be below the open end of the long tube, allowing air to enter as the liquid drains.

4. Holding the stopper securely (and still with a finger over the open end of the long tube), shake the bottle vigorously. Place the bottle (still inverted) on the ring stand, and allow the heavy liquid to separate from the water for 1–2 hr.

5. Drain out the separated heavy liquid through the rubber tubing, rinse the bottle thoroughly with water, and repeat the process several times.

6. When reclamation is complete (heavy liquid should be clear in contrast to its earlier cloudy appearance), filter the heavy liquid through several thicknesses of filter paper. Reclamation will not normally bring the specific gravity of the heavy liquid back to its original value, but it should be within 0.05 of the original.

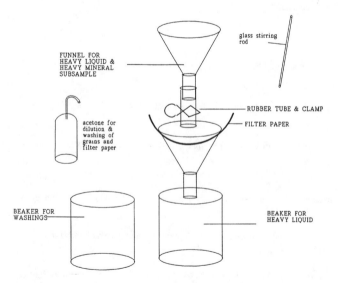

Figure 8-4. Sketch of equipment commonly used in heavy-liquid separation of heavy minerals from loose sediments.

dure for separation, and Table 8-3 a procedure for reclaiming the expensive heavy liquids.

ELECTROMAGNETIC SEPARATION

Electromagnetic separation is applicable either to the original dry bulk sample or to either light or heavy mineral separates; preliminary separation into subclasses will make mineral identification easier, but a similar argument applies in reverse—initial magnetic separation may simplify heavy liquid separation!

1–5. As for preparation procedure on page 135.

6. Remove ferromagnetic grains with a hand magnet (with thin paper between the magnet and grains). Weigh and store separately. *Do not allow these grains into the Frantz isodynamic magnetic separator!* Set the forward and side slopes as required (see Fig. 8-5—begin with 15° side, 25° forward tilt). Set the milliamperage at a

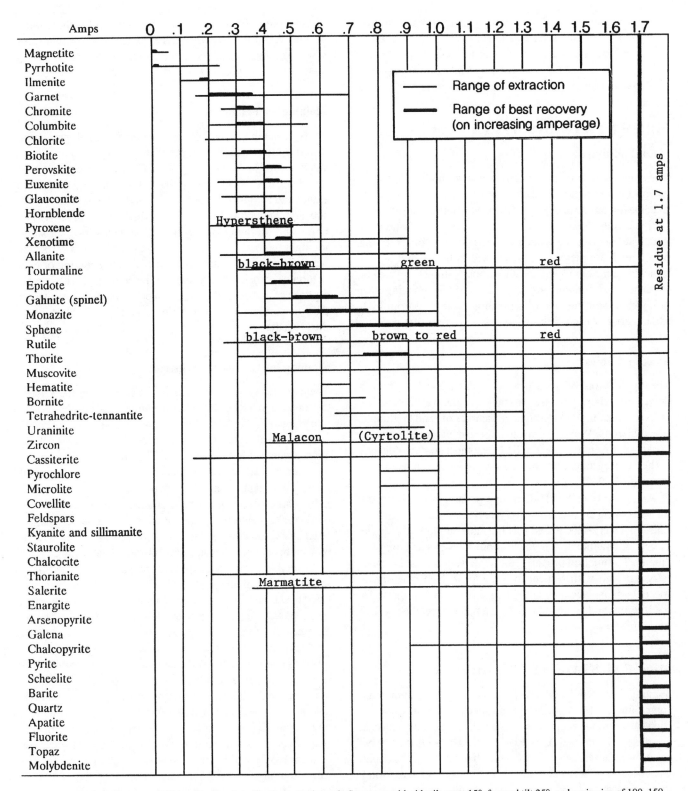

Figure 8-5. Magnetic susceptibilities of minerals in the Frantz Isodynamic Separator with side tilt set at 15°, forward tilt 25°, and grain size of 100–150 μm.   (After Rosenblum 1958.)

low value (e.g., 0.3 mA) first, then increase as required for subsequent separations. (For the cleanest separation, the grains have to be run through at one setting several times.) *Do not have a watch or other magnetizable device in the room while the Frantz is working.*

MINER'S PAN AND SUPERPANNER

Hand Panning

The simplest way of concentrating heavy minerals is by the use of a miner's pan or "dish." The method is not quantitative, inasmuch as some HMs may be lost unless considerable skill is acquired and great care is used. However, hand panning can provide an excellent method for concentrating bulky field samples and may be used in the laboratory for rapid concentration of materials without complex devices or expensive/toxic chemicals (e.g., Gunn 1968). Panning also provides a rapid method of separating clean coal from high ash material or rock cuttings. Access to water is necessary.

Shaking Tables

The application of industrial-type shaking tables scaled down for laboratory use provides a suitable method of preconcentrating HMs from large samples of sand to granule size materials. Properly adjusted, these devices can be used to extract different ranges of S.G. The most common design is the Wilfley table, but many other types are in use (e.g., Pryor 1965; Wills 1988).

An example is the Haultan superpanner, which is useful for separation of minerals over very small density ranges. This device is a specialized shaking table employing a single trough, or channel, that is shaken both sideways and longitudinally. To function properly, the superpanner must be carefully adjusted with respect to water flow, feed rate, shake speed, and displacement and tilt. Due to a very low throughput, the method is unsuitable for samples exceeding a few grams in size.

MINERAL STAINING METHODS

Staining is a rapid and practical means of distinguishing some minerals from others in hand specimens, thin sections, and grain mounts. Many minerals in sediments and rocks can be selectively stained (e.g., Reid 1969). Distinction may be between minerals of very different character (e.g., feldspar from quartz) or between minerals in the same mineral group (e.g., feldspar or carbonate varieties). In this chapter we treat stains for only feldspars and the common sedimentary carbonates; some of the stains can be applied in conjunction with the preparation of acetate peels of sample surfaces (Chapters 4 and 5). Details of the techniques may vary, depending on the character of the rock; experimentation is commonly necessary to obtain good results. Be aware of the fact that pores, fractures, some clays, and rarely other minerals may retain the stains. Stains are usually brittle and flaky when dry; thus, precautions must be taken to avoid breaking them off of the specimen.

Staining for Feldspars

Feldspar-staining principles depend on etching of grains followed by treatment with dyes that react with potassium or calcium ions in the etched grains. The method given here is largely that of Houghton (1980). For other methods, see Hayes and Klugman (1959), Bailey and Stevens (1960), Wilson and Sedora (1979), and Friedman (in Carver 1971). Gross and Moran (1970) suggested an innovation for staining light minerals by mounting the grains in a black roofing tar; the striking contrasts in color facilitate counting or estimation of components. Stains can be combined with acetate replicas of etched surfaces (Chapter 4) for rock areas larger than the normal thin section (Bjorlykke 1966). Table 8-4 lists the procedure.

Staining for Carbonate Minerals

There are many staining methods for identifying the various carbonate minerals or for specific cations that may be present in carbonates (e.g., Wolf, Easton, and Warne 1957 as cited in Chilingar et al. 1967; Friedman 1959; Warne 1962; Dickson 1966; Friedman in Carver 1971). The greatest variety appear to occur in coal-bearing sequences, where a great range of microchemical environmental conditions may be represented. Other carbonate minerals of economic importance occur in unusual sedimentary deposits or in rocks that have been altered by hydrothermal solutions. Only tests for the most common carbonate varieties in sediments are given below.

Several stain tests are applicable in the field (e.g., use of an acidic solution of alizarin red), but for detailed studies a polished face is necessary. Staining can be combined with acetate peel replicas of etched surfaces (described in Chapter 5; e.g., Katz and Friedman 1965). A stain solution may generally be used only once; thus, it pays to carefully judge quantities used for one sample. Use distilled water for rinsing. Note that some of the procedures involve boiling the sample; if thin sections are to be used, an epoxy mounting medium is necessary. (If dolomite is to be distinguished from calcite, etching a thin section may reveal the difference under the petrographic microscope because it will remain close to the original thickness whereas calcite will be much thinner.) Staining is commonly carried out in a series of sequential steps (e.g., Fig. 8-6); examples of the procedures are listed in Table 8-5.

Evaporite Minerals

A method for determining anhydrite and/or gypsum in the field as well as the laboratory was described by Hounslow (1979):

Table 8-4.   Procedure for Staining Rock and Mineral Samples for Feldspar Minerals

Necessary chemicals

(a) For K-feldspars: an oversaturated solution of sodium cobaltinitrite made with about 5 g/10 mL distilled water. (Shelf life of 6–8 months if the solution is kept in a sealed dark glass bottle, but a fresh solution is advisable.)

(b) If plagioclase feldspars are to be stained: both a 2–5% solution of barium chloride ($BaCl_2$—indefinite shelf life) and a potassium rhodizonate solution. The latter is prepared by dissolving 0.1 g rhodizonic acid dipotassium salt in 30 mL distilled water; remove coagulated clumps of powder by filtering to avoid blotchy stains (shelf life approximately 1 hour).

(c) Hydrofloric acid (HF) of 52–55% or greater concentration.

Caution: HF is dangerous, even in dilute solutions or as fumes; Work under a fume hood, wear full safety clothing, and ensure that you are familiar with all safety and first aid procedures. Neutralize the residue with washing soda before disposal.

Procedure

1. Polish the rock slab using 400–600-grade carborundum powder; for fine detail, apply a final polish on wet 600–800-grit silicon carbide paper. An uncovered thin section or loose grain mount (Chapter 5) can also be used, as long as the upper portions of the grains are not covered with epoxy. When using loose grains, ensure that any coatings such as iron oxides or salts have been removed first (e.g., in an ultrasonic device or see Chapter 5 for other methods).

2. Flush the surface of the rock slab or thin section with acetone; quickly rinse it in tap water. Immerse in detergent solution; rinse with distilled water and air-dry.

3. Etch the surface with fresh, strong hydrofluoric acid fumes. *Note: Evaporation causes the acid to weaken, so check the concentration or use new acid every 45 min.* The sample should be suspended 1–2 cm above the acid, preferably c. 5 mm deep in a closed container. Etching time can range from c. 25–35 sec to 30 min or more, depending on the strength of the acid and the character of the sample. Avoid acid condensation on the surface (uneven etching may result and the sample is then dangerous to touch!); brief immersion of the entire surface may be necessary if fumes do not etch adequately, in which case be particularly careful to dip the sample in an alkaline solution before rinsing it. If thin sections are being etched, paint the glass surfaces with paraffin wax to protect them from etching.

4. Remove the sample (polythene forceps), and immerse immediately into the supersaturated solution of sodium cobaltinitrite. Ensure saturation is maintained by periodically shaking the solution. Remove the sample after 30–45 sec (variable). K-feldspar (orthoclase, microcline, sanidine) should now show a distinctive yellow stain.

5. Rinse the sample twice in tap water, and gently shake off excess water (or blot at one end with a paper towel).

The remainder of the procedure is for staining plagioclase feldspars.

6. Dip the sample quickly into the $BaCl_2$ solution. Agitate once or twice, and remove before 2 sec have elapsed. (Barium will substitute for calcium in plagioclase.)

7. Dip the sample immediately into a beaker filled with tap or distilled water; agitate it briefly, then agitate in another beaker of distilled water for 10 sec.

8. While the surface is wet, place several drops of potassium rhodizonate solution on it with an eyedropper or micropipette; tilt or agitate the sample to distribute the stain evenly. Leave for a few seconds (or until plagioclase grains are pink). The degree of stain is proportional to calcium content in plagioclase (pure albite will not stain).

9. Rinse the sample in a beaker of water and air-dry (e.g., compressed air). Examine; if necessary repeat step 8.

Note: To stain loose sand grains, place c. 300 grains in a 50 mL beaker. Any coatings must first be cleansed by treating them with c. 10 mL of 1:1 nitric:hydrochloric acid. *Caution!* Heat gently in fume cupboard to boiling point; cool, then decant acid into large volume of water before disposal. Wash sample several times. Transfer the grains to an HF-resistant plastic beaker or platinum crucible. Add 3 mL HF in fume cupboard (*caution!*) and stand for c. 3 min. Decant and wash as before. Add 5 mL cobaltinitrate solution and swirl for 1–2 min; let stand another 2–3 min. Decant and wash; dry in oven. Store grains in sealed glass vial. Under the binocular microscope, quartz should be clear, glassy, K-feldspar yellow, Ca-feldspar white and chalky.

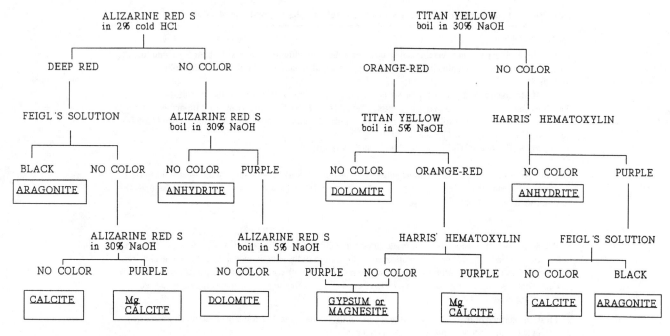

Figure 8-6.   Example of flowchart for staining identification of the most common carbonate and evaporite minerals.   (After Friedman 1959.)

1. Prepare a nitric acid solution by adding 10 mL concentrated acid *slowly* to 10 mL water. Prepare another solution of 10 g mercuric nitrate in 100 mL deionized water; stir until fine milky precipitate forms. *Slowly* add the nitric acid solution to the mercuric nitrate—while stirring—until the milky precipitate just dissolves (c. 8 drops of acid). Test resultant solution by adding a few drops to small sodium or potassium sulfate crystals; if a yellow precipitate does not form immediately, the solution is too acidic and more mercuric nitrate (c. 1 g) should be added. *Caution:* Nitric acid is corrosive and mercuric nitrate is extremely poisonous and must not touch the skin!
2. Filter the solution and store in dark bottle.
3. Immerse sample in solution for a few seconds or until a yellow precipitate forms. Gently rinse in distilled water and air-dry.

Both calcium sulfates are stained yellow.

Staining for Bioturbation
The carbohydrate polysaccharide is generally present in mucoid linings secreted by many burrowing organisms. In detrital sediments, the following stain procedure clearly picks out burrows with these linings (Risk and Szczuczko 1977):

1. Ensure uncontaminated sample with smoothed surface free of grease and oil (e.g., from finger contact).
2. Place surface face-down onto small supports in a solution of 1 wt.% periodic acid in distilled water; gently agitate solution and leave for c. 30 min, *or* paint solution onto surface.

3. Rinse for 30 sec in water.
4. Place face-down in Schiff's solution (commercially available) and leave in darkness for c. 30 min.
5. Rinse thoroughly for several minutes.

Dark stains against a light purple background will mark the burrows with mucoid linings (other burrows will not take the stain). If stored in darkness, the stains will hold for several years.

MODAL ANALYSIS OF THIN AND POLISHED SECTIONS AND GRAIN MOUNTS

Once the mineral components of a sediment have been identified, a general requirement is to determine the respective proportions of the various mineral species. It is vital that each class of component be mutually exclusive for the counting procedures, and it is in this respect that many inexperienced operators fail because of incorrect or inconsistent identification of matrix versus altered mineral versus rock fragment categories. Operators must be certain of consistent identifications at the magnification used during the counting procedures (usually medium-power objectives), both for qualitative and especially for quantitative analysis.

Qualitative estimations using percentage comparison charts have been mentioned in **PS** Chapter 6 and are worth making even when quantitative techniques are to be applied to ensure that the worker has an appreciation of the potential localization of components in the sample. Individual workers tend to be consistent after gaining experience making visual estimates, although reliability may not be high (e.g., Dennison and Shea 1966). These much more rapid studies can be

Table 8-5.   Procedures for Staining Carbonate Minerals

1. Polish a surface of the sample with 400–800 carborundum powder, depending on the rock texture and detail desired. Alternatively, uncover a thin section or mount loose grains on a glass slide.

2. Etch the polished surface with approximately 20 vol% acetic acid, or 1–10% HCl, avoiding brisk effervescence. Allow 30 sec or less if reaction is reasonably strong; if little or very little reaction occurs, etch for 45 sec in warm 10% HCl; experiment. (If using an uncovered thin section, etch in cold acid for 10–15 sec with approximately 1.5% HCl.)

3. Apply stains according to the flowsheet illustrated in Fig. 8-6.

Alizarin Red S

1. For hand specimens and loose grains, use 0.1 g sodium alizarin sulfonate per 100 mL 0.2% cold HCl (or 2 mL commercial HCl in 998 mL distilled water). For thin sections, use 0.2 g/100 mL.

2. Cover the etched surface with the solution for approximately 5 min (30–45 sec for thin sections).

3. Rinse gently but swiftly because the dye is soluble.

Deep red indicates calcite, witherite, high-magnesium calcite, or aragonite. Purple indicates ankerite, ferrodolomite, strontianite, or cerussite.

If the carbonate will not take the stain, it is siderite, dolomite, rhodochrosite, magnesite, or smithsonite (or it may be noncarbonate minerals such as anhydrite or gypsum).

Varieties of unstained or purple minerals may be distinguished by boiling them in a mixture of the stain and 30% NaOH for approximately 5 min (see Fig. 8-6). This test also distinguishes high-magnesium calcite from ankerite and calcite.

Titan Yellow: The Titan Yellow stain test can distinguish high-Mg calcite as well as dolomite and is selective on a very fine scale, so that details of its distribution can be seen under a microscope at high magnifications. The method of Choquett and Truswell (1978) is described below. If thin sections are to be stained, epoxy glue must be used because Lakeside Cement and Canada Balsam will dissolve in the treatment.

1. Dissolve 1.0 g Titan Yellow powder, 8.0 g NaOH, and 4.0 g of disodium EDTA in 1 L distilled water; store in dark bottle (shelf life 2 years).

2. Dissolve 200 g NaOH pellets *slowly* in 1 L distilled water; store in plastic bottle (shelf life indefinite). *Caution:* Heat and fumes are created; solution is corrosive to flesh and to glass. Use rubber gloves and fume cupboard.

3. Etch smooth face of specimen c. 30 sec in 5% acetic acid; air-dry.

4. Immerse specimen in stain solution for c. 20 min; air-dry.

5. Immerse specimen in NaOH solution (fixer) for c. 30 sec; air-dry.

Intensity of the pink to red color depends on Mg content (deeper colors with higher Mg) and on crystallographic orientation (deeper stain perpendicular to *c* axis). High-Mg calcite should be deep red.

Feigl's Solution

1. Add 1 g commercial Ag_2SO_4 to a boiling solution of 11.8 g $MnSO_4 \cdot 7H_2O$ in 100 mL distilled water.

2. After cooling, add 1 or 2 drops of dilute (10%) NaOH solution. Stand for 1–2 hr.

3. Filter out the precipitate and store the solution in a dark bottle.

4. After etching the sample, immerse it in room-temperature (or hot) solution for approximately 10 min.

Aragonite should turn black, and calcite will be unaffected but the staining is not always reliable.

Potassium Ferricyanide: This stain works only for carbonates that react with the acid medium.

1. Mix equal parts of 2% HCl and K-ferricyanide solution (2 g/100 mL distilled water). *Caution:* This solution is unstable and gives off poisonous HCN. Keep it under a hood and wear breathing apparatus.

2. Wash the sample gently but swiftly. Carbonates with Fe^{2+} are stained blue. Ankerite and ferrodolomite are dark blue in cold solution. Dolomites generally, and siderite always, are stained only with heating (up to 5 min for dolomite). Calcite and magnesite stain if they have any Fe^{2+}.

This solution may be mixed with alizarin red S (ARS) for a combined test. For thin sections, use three parts ARS to two parts potassium ferricyanide. Calcite should turn pale pink to red, depending on crystallographic orientation. Ferroan calcite should be the same or a pale to dark blue or mauve, purple, or royal blue. Dolomite should have no color and ferroan dolomite should be pale to deep turquoise. If another stain is then applied with 0.2 g ARS/100 mL of 1.5% HCl for 10–15 sec, calcite and ferroan calcite will turn very pale pink to red and ferroan dolomite will become colorless.

3. Rinse gently; do not touch the stained surface.

Meigen's Solution (modified)

1. Prepare a 5–10% solution of cobalt nitrate by mixing 15 g $Co(NO_3)_2$ in 100 mL distilled water.

2. Immerse the sample (not a thin section!) and boil 1–5 min (depending on the grain size).

Aragonite will turn dark violet. Dolomite, ankerite, magnesite, and siderite will remain unstained. Calcite will remain unstained if coarsely crystalline, and will turn lilac-rose if microcrystalline or light blue if boiled for a prolonged period. Boiling time may be critical.

---

given higher credibility by determining correction factors from quantitative analysis of a selection of samples after each individual worker has completed visual estimates of a suite of samples.

For quantitative results, most commonly linear traverses are made across the whole slide until on the order of 300–500 points on mineral grains have been counted to provide an estimate of the number frequency of the components. This number frequency is not the same as the number percentage (larger grains have a higher chance of being intercepted), area percent, or volume. For most characterization and inter-

pretation purposes, the number frequency is an adequate estimate of the relative abundance of components, but the values should not be treated statistically—if statistical treatment is necessary, alternate methods of counting must be employed (e.g., all grains must be counted, or all grains in "ribbons" across the slide; see discussion by Galehouse in Carver 1971). Traverse routes across thin sections must be selected carefully if there are any compositional segregations along bedding or other sedimentary structures. Full details of these procedures are discussed by Larsen and Miller (1935), Chayes (1956), and Galehouse (in Carver 1971); see also van der Plas and Tobi (1965). Despite the quantification that is possible with point counting, the problem of operator bias remains in the identification of minerals, rock fragments, and matrix (e.g., Welsh 1967).

To facilitate and automate counting procedures, a *mechanical stage* is attached to the petrographic microscope. This stage is commonly connected to a counting device that has multiple buttons and windows; each button is assigned a mineral composition (or null value for empty spaces) and the coupled window shows the total counts for that mineral. Each time any button is pressed, the stage automatically moves forward along the line of traverse, and the next count is made wherever the cross-hairs fall. The spacing between counts can be adjusted by changing gears in the mechanical stage, and should be adjusted to approximate the apparent diameter of the smallest of the common particles (greater spacing will bias against the number of times small grains are intercepted; note that the spacing when making point counts for *textural* analysis should be equal to or greater than the mean grain size).

This technique may be applied after mineral staining, although stains may partially obscure the particles (study the grain preparations carefully before staining or stain half the specimen or a separate representative subsample). It can also be applied to determine porosity of rock samples when a colored impregnating compound has been emplaced in the sample using a vacuum chamber prior to thin sectioning.

X-RAY DIFFRACTION

Apart from the naked eye and the microscope, X-ray diffraction (XRD) is the most widely used technique for mineral identification, particularly for minerals of very small size. In addition to mineral identification, XRD studies can also provide information on the degree of disorder in minerals, the nature and extent of isomorphous substitutions, crystallite size, and other characteristics of mineral crystals. XRD is also used in conjunction with other instruments for determination of multiple properties (e.g., Veblen, Guthrie, and Livi 1990). Although good modern instruments cost upward of U.S. $100,000, most have a long life expectancy and if they are well looked after they should rarely require major servicing. The Mineralogical Society of America volume by Bish and Post (1989) provides an excellent introduction to XRD procedures and applications and should be consulted if planning to apply X-ray powder diffraction methods.

Principles of Operation

The XRD method works because X-rays are diffracted by the atomic layers in crystals. In mineralogical studies, the diffracted X-rays are used to measure the dimensions of the various atomic layers in the crystals, and because each different mineral has a distinct set of atomic layer spacings (called *d* spacings), the suite of measurements can be used to identify the mineral. All crystalline minerals in a sample can be iden-

tified from one XRD scan, provided that they are present in sufficient abundance; however, XRD will not detect non-crystalline components (e.g., the allophane clays) in a sample because they have no regular atomic planes. The relationship between the wavelength of the X-rays used, the angle between the incident and the diffracted X-rays, and the distance between the atomic layers causing the diffraction is given by the Bragg's law equation:

$$n\lambda = 2d\sin\theta$$

where $n =$ an integer ($n = 1$ for first-order diffraction peaks, 2 for second-order peaks, etc.)

$\lambda =$ the wavelength of the radiation used, for $Cu_{k\alpha}$ radiation

$\lambda = 1.5418\text{Å}$

$d =$ the atomic layer spacing (in angstrom units) between the diffracting planes

$\theta =$ half the angle between the incident X-rays and the diffracted X-rays

For most geological applications of powder diffraction analyses, θ is the angle between the incident X-ray beam and the surface of the powder sample. Because the angle of incidence must equal the angle of reflection, as the sample is rotated in the goniometer relative to the incident beam, the detector must be rotated twice as far as the sample. Hence, the angle between the incident beam and the detector equals 2θ; this is the angle shown by most goniometers, and it must be halved to get the values of θ required for the Bragg's law calculation. For further details on the principles of powder diffraction, see Jenkins and DeVries (1967), Cullity (1978), and Reynolds (1989).

Instrumentation and Operation

The XRD system most widely used by geologists consists of an X-ray source, a goniometer to hold and rotate the sample, and an X-ray detection and data processing system (Fig 8-7). Most systems use a copper or a cobalt anode X-ray tube driven by a generator at up to 2.5 kW. The X-rays from the tube pass through a β filter (thin Ni foil for a Cu anode tube) and collimating slits onto the sample in the goniometer. From the goniometer, the diffracted X-rays pass through further collimating slits and a monochromator (fitted to most modern instruments to improve diffraction pattern resolution), then onto the detector (usually a proportional counter or a scintillation counter). The X-ray counts recorded by the detector are amplified, processed with a pulse height analyzer (to exclude counts due to photons with energies other than that of the X-rays used in the analysis, such as fluorescent X-rays), then passed to a chart recorder or a computer for further processing. Many modern instruments are also fitted with an autosampler (to automatically load samples in a batch), an autodivergence slit (as part of the collimation system), and a sample spinner to spin the sample in the goniometer about an axis normal to the sample surface. Because

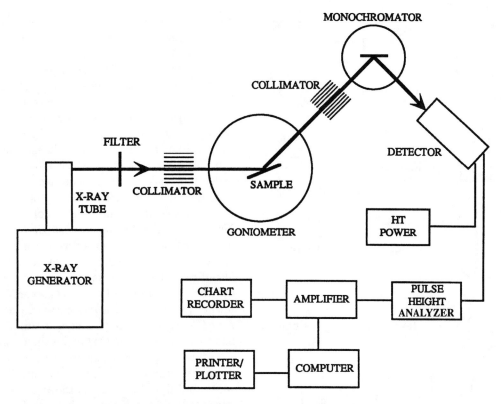

Figure 8-7.   Schematic diagram of the essential components of XRD equipment.

most modern instruments are computer driven and incorporate software for additional data processing (e.g., overlaying diffraction patterns, mineral peak searches, subtraction of particular mineral patterns from a mixed mineral diffraction pattern, etc.), both goniometer angles and count data are recorded digitally.

Some analysts use an alternative instrument set up for single crystal and mineral powder studies, where a camera (the common Debye-Scherrer camera, the Gandolfi camera, or the focusing Guinier camera) is used in place of the goniometer and electronic count-processing equipment (e.g., Jenkins 1989). Cameras are used much less commonly than goniometer-based systems, particularly in sedimentology. For further information on camera and goniometer-based diffraction procedures, see Cullity (1978) and Jenkins (1989).

Sample Preparation

Samples for powder diffraction studies must be finely ground (<5 μm) and completely homogeneous; coarse particles in the sample will substantially reduce peak-to-background ratios and will produce diffraction patterns with abnormal peak-intensity ratios. The fine powder is usually prepared for analysis either as a thin smear (by mixing the sample with a small volume of alcohol) on a glass slide (e.g., half a microscope slide), or by packing the powder into the well (about 2 mm deep) of a sample holder designed to fit into the goniometer. When an

autosample changer is used, sample holders compatible with the sample changer must be used.

When packing samples that may contain platy minerals (e.g, clays) into a sample holder, mineral grains may become preferentially oriented relative to the top of the holder. Whether some grains have a preferred orientation in the sample or the assemblage is randomly oriented is not usually a problem in mineral identification, but oriented grains will enhance peak-intensity ratios (e.g., the 001, 002, 003, etc. peaks for phyllosilicate minerals will be selectively enhanced if the grains are oriented with their *c* axis normal to the surface of the sample.) For some clay mineral studies, there are advantages in deliberately orienting the clay particles (e.g., by settling samples from a fluid onto the glass slide). Additional information can be obtained by pretreating samples by exposure to an atmosphere of ethylene glycol and/or firing them to 550°C (e.g., Fig. 8-8). A very readable discussion of sample preparation procedures and problems can be found in Bish and Reynolds (1989).

Data Interpretation

The first step in analyzing a diffraction pattern is to determine the *d* spacing corresponding to each peak. The *d* spacings can be calculated manually using Bragg's law, but all modern instruments with capability of transferring the diffraction data to a computer not only carry out this calculation

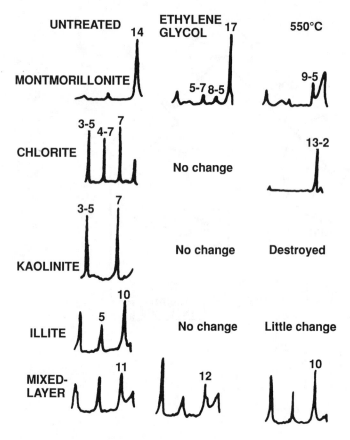

	UNTREATED	ETHYLENE GLYCOL	550°C
MONTMORILLONITE	14	5-7 8-5 17	9-5
CHLORITE	3-5 4-7 7	No change	13-2
KAOLINITE	3-5 7	No change	Destroyed
ILLITE	10 5	No change	Little change
MIXED-LAYER	11	12	10

Figure 8-8. X-ray diffraction patterns representative of the major clay mineral groups.

but also find the peaks on the diffraction pattern and tabulate the *d* spacing and peak intensity for each.

The next step is to determine which minerals are identified by which peaks. This operation can be performed by search-match routines on many computerized systems; however, search-match routines can suggest some very strange minerals, particularly if a mineral in the sample is part of a solid solution series (where the *d* spacing changes systematically with composition) or where the mineral is only a minor constituent of the sample. There is no doubt that search-match programs are very useful and save much time, but there is no solution to the problem of misidentification other than the application of liberal doses of common sense combined with any additional mineralogical or chemical information that may be available. Where search-match programs are not available, or where they provide obviously erroneous identifications, it is necessary to resort to the *ASTM* (American Society of Testing Materials) *Index* of X-ray powder diffraction patterns, and probably to a search manual such as the *Fink Mineral Index* (Bayliss et al. 1986). An example of a mineral record in the *ASTM Index* is shown in Fig. 8-9. These records provide some general information about the mineral, the operating conditions under which the diffraction pattern was recorded, and a table that shows the *d* spacings for the mineral, the *Miller Index* for the crystal plane with each *d* spacing and the relative intensity (*I***/I*) for each peak. Note that the relative intensities given are for randomly oriented crystallites of each mineral, and they apply only to pure min-

6 - 710

FeS_2					dA	Int	hkl	dA	Int	hkl
Iron Sulfide		Pyrite, syn			3.128	35	111			
					2.7088	85	200			
					2.4281	65	210			
Rad. $CuK_{\alpha 1}$	$\lambda = 1.5405$	**Filter** Ni		**d-sp**	2.2118	50	211			
					1.9155	40	220			
Cut off	**Int.** Diffractometer		**I/Icor.**		1.6332	100	311			
Ref. Swanson et al., Nat. Bur. Standards (U.S.), Circ. 539, 5 29 (1955)					1.5640	14	222			
					1.5025	20	230			
					1.4479	25	321			
Sys. Cubic		**S.G.** Pa3 (205)			1.2427	12	331			
a 5.417	**b**	**c**	**A**	**C**						
α	β	γ	**Z** 4		1.2113	14	420			
Ref. Ibid.					1.1823	8	421			
					1.1548	8	332			
Dx 5.01	**Dm** 5.02		**mp** 642°C		1.1057	6	422			
					1.0427	25	511			
					1.0060	8	250			
					0.9892	6	521			
					0.9577	12	440			
					0.9030	16	600			
Assorted notes on other physical and crystallographic properties included here					0.8788	8	611			
					0.8565	8	620			
					0.8261	4	533			
					0.8166	4	622			
					0.7981	6	631			

Figure 8-9. Example of a mineral reference card from the *ASTM Index*.

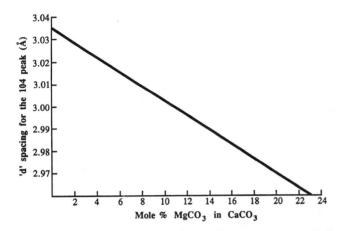

Figure 8-10.   Plot showing the relationship between the *d* spacing of the 104 peak to the mole % magnesium carbonate in calcite.

erals; the relative intensities for individual minerals in a mixed-mineral assemblage will vary according to the proportions of each mineral and peak overlaps (where two minerals have peaks at the same angle or *d* spacing) may produce cumulative peak intensities.

Once the minerals in a sample have been identified, it is possible to estimate their modal proportions based on the relative intensities of the 100% peak of each mineral (e.g., Snyder and Bish 1989). However, the process is not simple, because even with the use of several standards, it is necessary to correct for mass absorption effects (partly compensated for by spiking the sample with a known amount of standard), variation in X-ray reflectivity between different minerals, preferred orientations, crystallite size, and other factors. Semi-quantitative estimates of mineral proportions can be achieved relatively easily, but accurate determinations of mineral proportions require a lot of effort. X-ray diffraction studies also are not able to detect trace minerals in a sample. Some minerals with good X-ray reflectivity can be detected at concentrations down to about 0.1%, but for most minerals in a mixed-mineral sample the lower limit of detection is about an order of magnitude higher.

Accurately measured *d* spacings can be used to evaluate element ratios in minerals that form part of a solid-solution series (e.g., determination of the Fe/Mg ratio in tremolite-actinolite series minerals), or in minerals where isomorphous substitutions are common (e.g., the replacement of Ca by Mg in calcites; e.g., Fig. 8-10). Peak shape and relative peak intensities provide additional information on mineral properties, such as mean crystallite size in fine materials (e.g., Fig. 8-11), crystallographic perfection, and subvarieties of minerals (e.g., Fig. 8-12; see also Thorez 1975).

Advantages and Limitations

The principal use of X-ray diffraction is for mineral identification, particularly where clay size minerals are present, and in many laboratories it is extensively used for this purpose

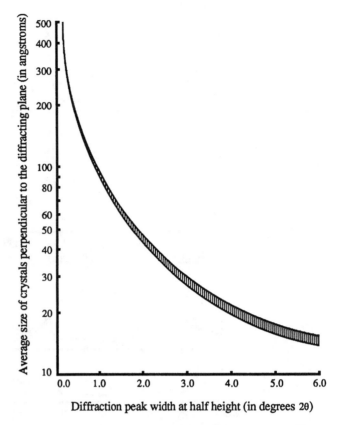

Figure 8-11.   Plot showing the relationship between mean crystallite grain size and the width of the diffraction peak at half height; for clays, the 001 peak is normally used for this purpose.

because sample preparation is reasonably straightforward and modern instruments are easy to operate and maintain. In addition, for many minerals it is possible to obtain additional mineralogical data on crystallite ordering, size, and isomorphous substitutions.

The major limitations of X-ray diffraction are that it will only record crystalline components (e.g., opal or allophanes are not detected) and the equipment is expensive. Beginners may see the steps in interpretion of the data as a further limitation, but this stage in the operation becomes much easier with experience.

DIFFERENTIAL THERMAL ANALYSIS AND THERMOGRAVIMETRIC ANALYSIS

Principle

Differential thermal analysis (DTA) is generally utilized for the identification of clays, organic matter-rich sediments, sulfidic sediments, or sediments containing other very fine-grained minerals that are amorphous to X-rays. Application of differential thermal analysis depends on two conditions:

1. That chemical or phase-change reactions (including dehydration) in the sample involve a heat change either in entropy, enthalpy, or both (e.g., Table 8-6). These heat

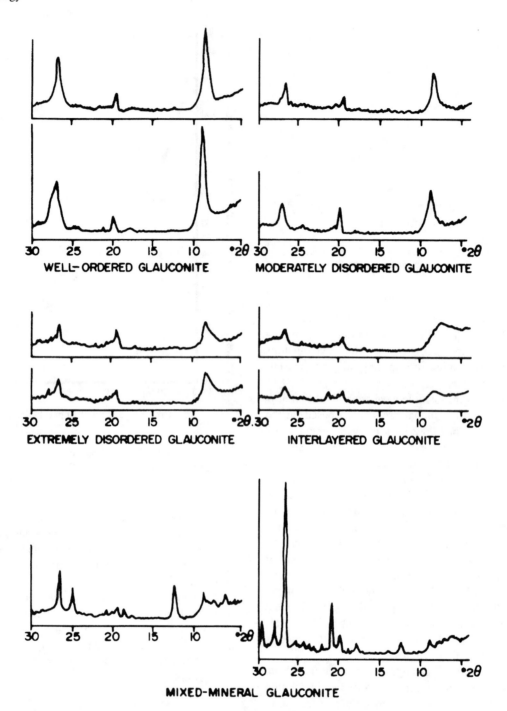

Figure 8-12.   Examples of XRD patterns for the crystallographic varieties of glauconite. The varieties are defined by McConchie and Lewis (1979, 1980) using the % expandable layers present and disorder coefficient ($DC =\tau b/(h - b)$) where τ (measured in degrees 2θ) = half-height line width of the 10Å XRD peak for an oriented sample previously heated to 400°C for 1 hr; h = intensity of the 10Å peak; b = background intensity (both h and b have the same arbitrary-length units). *Well-ordered glauconite*: $DC \leq 0.25$; sharp, symmetrical XRD peaks; comprising most glauconites with <10% expandable layers. *Disordered glauconite*: (1) Moderately disordered, with $0.25 < DC \leq 0.5$ and $\leq 40\%$ expandables; XRD pattern may show peak asymmetry. (2) Extremely disordered, with $DC > 0.5$ and $\leq 40\%$ expandables; XRD pattern will generally show pronounced peak asymmetry. *Interlayered glauconite*: >40% expandable layers; only one distinct mineral type apparent. *Mixed-mineral glauconite*: two or more distinct mineral varieties present, only one of which is glauconite (one of the previous classes).

Table 8-6.   DTA Reaction Temperature Ranges for Common Sedimentary Minerals

Mineral	Major Endothermic Peaks Temperature °C	Major Exothermic Peaks Temperature °C
Calcite	820–910	
Chlorite	100–150; 600–725; 800–900	
Dolomite	750–840; 850–920	
γ-Fe$_2$O$_3$		700–800
Gibbsite	250–380	
Goethite	325–425	750–950
Illite	150–200; 500–650; 850–920	920–980
Kaolinite	100–120; 500–650	920–1050
Lepidocrocite	250–325	310–400; 700–850
Magnesite	550–650	
Marcasite		350–550; 550–650
Montmorillonite	100–240; 600–700; 780–860	860–950
Muscovite	800–950	
Palygorskite	125–225; 400–600	850–950
Pyrite		400–500
Pyrophyllite	80–180; 550–750	
Pyrrhotite	350–380	500–650
Saponite	125–250; 725–925	900–980
Sepiolite	150–350; 500–750; 820–900	980–1040
Talc	850–1000	

changes can be either endothermic (heat consuming) or exothermic (heat releasing). Endothermic reactions include dehydration and dehydroxylation; structural decomposition and some structural transformations; melting, evaporating, and sublimation; demagnetization of ferromagnetic substances; some reduction reactions. Exothermic reactions include oxidation reactions; recrystallization; some structural transformations.

2. That these changes occur over a limited temperature range.

In DTA studies, two parameters are recorded: the temperature at which the reaction occurs, and the heat change involved. These measurements are made using a standard thermocouple and a differential thermocouple (Fig. 8-13),

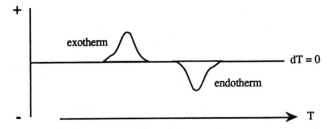

Figure 8-14.   Sketch of the relative positions of exothermic and endothermic "kicks" on the graph produced from differential thermal analysis of a sediment.

linked to a two-pen chart recorder. The position of each peak on the resulting graph shows the temperature range over which the reaction occurs, and the peak area shows the amount of heat involved in the reaction; the shape of each peak may be of further interpretative value, particularly when dealing with mixtures. Whether each reaction is endothermic or exothermic is indicated by whether the peak lies above (exothermic) or below (endothermic) the $dT = 0$ base line on the graph (Fig. 8-14). Hence, any reaction that involves a measurable heat change can in principle be studied by DTA, and its applicability does not depend on sample crystallinity as does the use of XRD methods.

Instrumentation

Heat Source

Most systems use a wire-wound furnace (Kanthel or Nichrome wire is good to 1300°C; Pt to 1750°C or Mo in an H$_2$ atmosphere to 2000°C); some fast-heating devices use microwave-induction furnaces. The furnaces may be wound in the common-coil mode or with the coil loops parallel to the

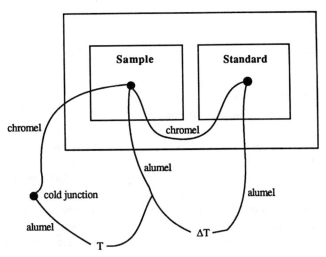

Figure 8-13.   Sketch of the essential elements of a sample holder for differential thermal analysis.

axis of the furnace core; the latter method is better because it avoids stray currents being transferred to the thermocouple by induction. If coils are wound perpendicular to the axis of the furnace core, the sample block must be wrapped in earthed metal foil to eliminate possible induction currents. The furnace must be capable of providing a uniform heating rate over the desired temperature range, or heating in a series of preset (0.5–10°C) steps; hot spots or nonuniform heating rates will adversely affect analyses and reproducibility. The furnace should be designed such that samples can be positioned in the same place for each analysis.

Temperature Controller

The controller should not be of the on/off type but should be able to vary the power supplied to the furnace; some modern systems control the furnace so that heating takes place in a series of preset incremental steps. It is also necessary to have a feedback system from an independent thermocouple to an inert material in the furnace to ensure that the heating rate is constant over the required temperature range.

Sample Holder

Most systems use a ceramic block (usually sintered alumina) with two circular or rectangular wells for the sample; this ceramic block is held in a larger sample block made of the same material. Other systems have used metal holders made of Ni, Pt, Cu, or an alloy such as Pt/Rh. The metal holders are simpler to make, are nonporous, and give little baseline drift, but their use results in smaller peaks because they act as a heat sink. Ceramic holders cause some baseline drift but they give better peaks and eliminate the need for additional insulation around the thermocouple wires. The size and shape of the sample holders will influence peak size and shape as will whether a lid is used or not. Whatever design is chosen, the same one must be used consistently for all comparative work. In general, the holder will contain c. 0.5 g of sample in one well and c. 0.5 g of standard in the other, sufficiently separated so that there is no heat conduction between them.

Thermocouples

The most common types of thermocouple used are Pt/PtRh alloy, chromel/alumel, or Pt/PtIr alloy; selection will depend on temperature and sensitivity requirements. Irrespective of the metal combination used, the wires should be < 0.4 mm in diameter (usually 0.2 mm); if they are thicker, they may conduct heat between standard and sample, and to the outside atmosphere. The wiring is sketched in Fig. 8-13 (although the cold junction shown there is desirable, it is not essential and most operators are happy with an air junction). The hot junctions of the thermocouples should be positioned as near to the center of the sample and standard as possible.

Temperature Recorder

The choice of recorder will depend on available finance; a two-pen chart recorder is preferable, with one pen for recording T and the other for recording ΔT.

Conditions, Sample Preparation, and Standards

Although not always necessary, most modern DTA studies are carried out in a controlled atmosphere; a controlled atmosphere should be used to prevent oxidation of organic matter, which produces a large exotherm at 250°C and above. Various standards can be used in DTA analyses, but calcined alumina is most common. Some analysts prepare standards from prefired specimens of their own sample, but they then run the risk of interference if any of the reactions that take place during prefiring are reversible.

The samples for study may be amorphous or crystalline and are prepared by grinding the sample material to 2–5 microns. The exact size used is not critical, but it must be the same for all samples in a study because particle size can affect the size and shape of DTA peaks. A small amount of the ground sample (commonly 0.1–0.5 g; Broersma et al. 1978) is packed into the sample well, with care to ensure that packing is uniform and to a standard compaction (packing can affect DTA peak shape). Uniform preparation is important at all stages, and it is advisable that a single operator prepare all the analyses in a sample suite. It is also important to ensure that samples are not heated above their melting or vitrification point, because any melt is likely to destroy the sample holder and the thermocouple.

Some analysts pretreat their samples (e.g., with H_2O_2 if organics are present), but care must be taken to ensure that the pretreatment does not produce new problems (e.g., see Chapter 5). A more common pretreatment is to store the sample in an atmosphere with about 50% humidity for 3–4 days before testing to ensure a standard initial moisture content. If only a small amount of the sample is available, it is possible to dilute it up to three times its initial volume using some of the standard material. During the analysis, heating rate is an important consideration. A fast rate will increase the peak area, but a slow heating rate gives the best formed peaks. A heating rate of about 10°C per minute is a widely used compromise.

Interpretations

From the DTA curve the first things to note are:

1. Whether the peaks are endothermic or exothermic
2. The size and shape of each peak
3. The temperature at which each peak occurs ·

These data can then be compared with published data (e.g., Fig. 8-15) as the first step in a mineralogical determination, or they can be used to provide a fingerprint for particular samples in a suite. In all samples, check whether there is more than one mineral type present. The presence of more than one mineral type may be suggested by asymmetric peaks, by the presence of peaks that cannot be accounted for by any monomineralic system, or by data from other analytical (e.g., XRD) work. DTA is seldom used alone in mineral

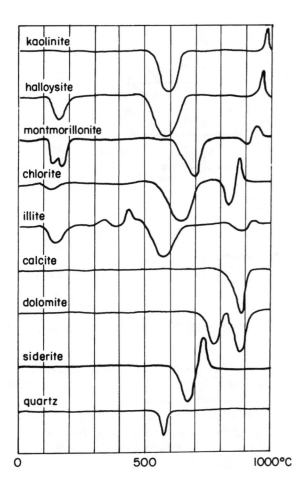

Figure 8-15.   Sketch examples of some DTA curves from common sedimentary minerals.

5. DTA can indicate the extent of some elemental substitutions in particular minerals.

Limitations

1. The furnace takes a long time to cool between runs.
2. There is some difficulty in distinguishing between components in many polymineralic samples, particularly where peak asymmetry occurs in a peak-overlap zone.
3. The presence of organic matter and some minerals can cause interference problems.
4. Some minerals that give good curves when pure do not show up well in mixtures.
5. There may be spurious effects in which one component in a mixture reacts with others.
6. The DTA curves can be strongly affected by small variations in sample preparation and analysis procedures.

Thermogravimetric Analysis

Many reactions (e.g., dehydration or dehydroxylation) involve not only a heat change but also a mass change. The mass change is measured in thermogravimetric analysis (TGA) in a similar way to that in DTA insofar as the change is recorded as a function of temperature. In contrast to determinations of loss on ignition (LOI) at 1000°C (see Chapter 9), commonly used in association with XRF analyses, in TGA the mass gained or lost is used to calculate the amount of each component gained or lost, or the amount of iron oxidized at each relevant temperature. Mass changes at known temperatures are usually attributed to specific reactions such as dehydration (loss of H_2O^+), dehydroxylation (loss of H_2O^-), oxidation of organic matter, oxidation of sulfides, loss of CO_2 from carbonates, or oxidation of Fe^{2+} to Fe^{3+}. Because TGA and other thermoanalytical techniques are less commonly used by sedimentologists than DTA methods, no further discussion of them is included here; additional information on these techniques can be found in MacKenzie (1982).

INFRARED SPECTROPHOTOMETRY

Principles

Infrared spectrophotometry is finding increasing use in sedimentology, particularly for examining the organic and fine grained inorganic fractions of sediments; it can be used to examine mineral components that are not sufficiently crystalline to be studied by X-ray diffraction. Infrared studies also can be used to obtain detailed mineralogical information that is not readily obtainable by other methods; e.g., determination of the Si/Al ratio in phyllosilicates (Lyon and Tuddenham 1960) or to study exchangeable cations in clays and zeolites (e.g., Brodskii and Zhdanov 1980). *The Infrared Spectra of Minerals* by Farmer (1974) provides a good intro-

analysis; it is usually used in combination with other techniques such as XRD or IR (infrared analysis).

In addition to using the DTA curve for mineral identification, the peaks on the curve can be related to specific reactions in the sample. Thus, if a series of known standards is also analyzed, the area under each peak can be used to indicate the amount of a reactant involved in a particular reaction (e.g., the amount of H_2O lost). In principle, it is possible to relate any variation in peak characteristics to specific differences in the chemistry of the starting minerals. Knowledge of the reactions taking place during DTA runs often can be improved by XRD examination of the mineralogy of the end products of firing.

Advantages Over Other Methods of Analysis for Fine Grained Samples

1. The equipment is inexpensive relative to that necessary for many other procedures.
2. The samples need not be crystalline.
3. Sample preparation and analysis are relatively simple.
4. DTA can detect lower concentrations of some minerals (e.g., calcite and dolomite) than XRD.

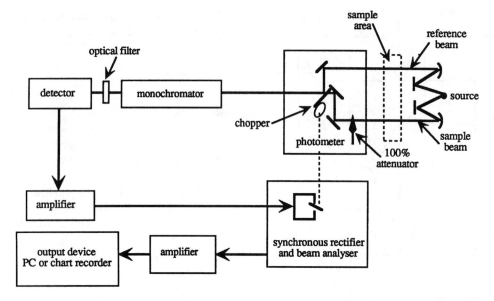

Figure 8-16.   Schematic diagram showing the principal components of an infrared specrophotometer.

duction to the principles of infrared spectrophotometry and their application in mineralogy. Figure 8-16 illustrates the mechanical principles.

The application of infrared spectrophotometry depends on the fact that chemical bonds can be stretched, bent, or rotated by photons of infrared radiation (the infrared region of the electromagnetic spectrum is from about 0.7 μm to >200 μm, but usually only the region between 2.5 μm and 25 μm is used in examining the infrared spectra of minerals); electronic transitional states are not affected. The infrared photon energy is related to wavelength by $E = hc/\lambda$ (where E is the photon energy, h is Planck's constant, c is the speed of light in a vacuum, and λ is the wavelength). The energy to which a specific bond reacts depends on the bond type and its atomic environment; in spectrophotometry, the wave number (i.e., the reciprocal of the wavelength) is often referred to instead of the wavelength.

If photons with a particular energy (i.e., wavelength) strike a molecule containing a bond that has the same natural vibration frequency, the photon energy is absorbed by the bond and increases its natural vibration. If radiation with a wavelength that does not match an interatomic bond (or a harmonic of the interatomic bond) strikes the molecule, it passes through without change. The characteristic vibration frequencies for individual bonds are determined by the masses of their atoms, their spatial geometry, and the bond strength. Thus, because infrared photons will be absorbed by a molecule only if their energy matches that of a bond in the molecule, a scan over a range of infrared wavelengths will produce an absorption spectrum (e.g., Figs. 8-17 and 8-18) that is characteristic of the molecules present in the sample. Application of infrared spectrophotometry in geology depends on being able to (a) relate the infrared frequencies absorbed to particular bonds; (b) relate these bonds to particular minerals; and (c) relate progressive changes in absorbed fre-

quencies to similar changes in mineralogy. Mineral identifications and the assignment of individual peaks to specific bond types is normally achieved by reference to published data for standards and the concentration of the bond type (or mineral where there is a direct relationship between a specific bond type and mineral concentration) is proportional to its absorbance (see Fig. 8-17).

Equipment

A wide range of modern and reasonably inexpensive wavelength-dispersive and more expensive Fourier-transform infrared spectrophotometers are available. Most modern spectrophotometers have built in microprocessors or are connected to a PC to enable plots to be labeled and "tidied up" (i.e., spikes are removed and a smoothing function may be applied), as well as to allow calculation of absorbance and integration of the area under peaks.

The basic wavelength-dispersive infrared spectrophotometer (e.g., Fig. 8-16) uses a source of infrared energy (heater), a monochromator to separate energy into its component frequencies, a detector (thermocouple) to measure the intensity of radiation in the sample and reference beams, and a means of recording the wavelengths absorbed as the range is scanned. Radiant energy from the source is divided into two beams: the sample beam (which passes through the sample) and the reference beam. The reference beam can be attenuated to set the 100% transmission baseline. Different instruments treat beam processing and detection in different ways but all use optical focusing components, a monochromator, and a detector to produce a scan of the wavelengths in each beam. When the energies of the sample and reference beams are equal, the thermocouple detectors produce a steady output voltage (the baseline voltage), but if both beams are not of equal intensity, the detectors will produce an output volt-

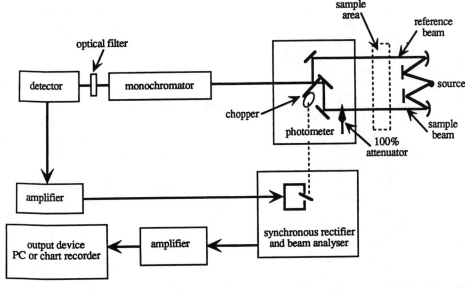

Figure 8-16.   Schematic diagram showing the principal components of an infrared specrophotometer.

duction to the principles of infrared spectrophotometry and their application in mineralogy. Figure 8-16 illustrates the mechanical principles.

The application of infrared spectrophotometry depends on the fact that chemical bonds can be stretched, bent, or rotated by photons of infrared radiation (the infrared region of the electromagnetic spectrum is from about 0.7 μm to >200 μm, but usually only the region between 2.5 μm and 25 μm is used in examining the infrared spectra of minerals); electronic transitional states are not affected. The infrared photon energy is related to wavelength by $E = hc/\lambda$ (where E is the photon energy, h is Planck's constant, c is the speed of light in a vacuum, and λ is the wavelength). The energy to which a specific bond reacts depends on the bond type and its atomic environment; in spectrophotometry, the wave number (i.e., the reciprocal of the wavelength) is often referred to instead of the wavelength.

If photons with a particular energy (i.e., wavelength) strike a molecule containing a bond that has the same natural vibration frequency, the photon energy is absorbed by the bond and increases its natural vibration. If radiation with a wavelength that does not match an interatomic bond (or a harmonic of the interatomic bond) strikes the molecule, it passes through without change. The characteristic vibration frequencies for individual bonds are determined by the masses of their atoms, their spatial geometry, and the bond strength. Thus, because infrared photons will be absorbed by a molecule only if their energy matches that of a bond in the molecule, a scan over a range of infrared wavelengths will produce an absorption spectrum (e.g., Figs. 8-17 and 8-18) that is characteristic of the molecules present in the sample. Application of infrared spectrophotometry in geology depends on being able to (a) relate the infrared frequencies absorbed to particular bonds; (b) relate these bonds to particular minerals; and (c) relate progressive changes in absorbed fre-

quencies to similar changes in mineralogy. Mineral identifications and the assignment of individual peaks to specific bond types is normally achieved by reference to published data for standards and the concentration of the bond type (or mineral where there is a direct relationship between a specific bond type and mineral concentration) is proportional to its absorbance (see Fig. 8-17).

Equipment

A wide range of modern and reasonably inexpensive wavelength-dispersive and more expensive Fourier-transform infrared spectrophotometers are available. Most modern spectrophotometers have built in microprocessors or are connected to a PC to enable plots to be labeled and "tidied up" (i.e., spikes are removed and a smoothing function may be applied), as well as to allow calculation of absorbance and integration of the area under peaks.

The basic wavelength-dispersive infrared spectrophotometer (e.g., Fig. 8-16) uses a source of infrared energy (heater), a monochromator to separate energy into its component frequencies, a detector (thermocouple) to measure the intensity of radiation in the sample and reference beams, and a means of recording the wavelengths absorbed as the range is scanned. Radiant energy from the source is divided into two beams: the sample beam (which passes through the sample) and the reference beam. The reference beam can be attenuated to set the 100% transmission baseline. Different instruments treat beam processing and detection in different ways but all use optical focusing components, a monochromator, and a detector to produce a scan of the wavelengths in each beam. When the energies of the sample and reference beams are equal, the thermocouple detectors produce a steady output voltage (the baseline voltage), but if both beams are not of equal intensity, the detectors will produce an output volt-

Table 8-6.   DTA Reaction Temperature Ranges for Common Sedimentary Minerals

Mineral	Major Endothermic Peaks Temperature °C	Major Exothermic Peaks Temperature °C
Calcite	820–910	
Chlorite	100–150; 600–725; 800–900	
Dolomite	750–840; 850–920	
γ-Fe₂O₃		700–800
Gibbsite	250–380	
Goethite	325–425	750–950
Illite	150–200; 500–650; 850–920	920–980
Kaolinite	100–120; 500–650	920–1050
Lepidocrocite	250–325	310–400; 700–850
Magnesite	550–650	
Marcasite		350–550; 550–650
Montmorillonite	100–240; 600–700; 780–860	860–950
Muscovite	800–950	
Palygorskite	125–225; 400–600	850–950
Pyrite		400–500
Pyrophyllite	80–180; 550–750	
Pyrrhotite	350–380	500–650
Saponite	125–250; 725–925	900–980
Sepiolite	150–350; 500–750; 820–900	980–1040
Talc	850–1000	

changes can be either endothermic (heat consuming) or exothermic (heat releasing). Endothermic reactions include dehydration and dehydroxylation; structural decomposition and some structural transformations; melting, evaporating, and sublimation; demagnetization of ferromagnetic substances; some reduction reactions. Exothermic reactions include oxidation reactions; recrystallization; some structural transformations.

2. That these changes occur over a limited temperature range.

In DTA studies, two parameters are recorded: the temperature at which the reaction occurs, and the heat change involved. These measurements are made using a standard thermocouple and a differential thermocouple (Fig. 8-13),

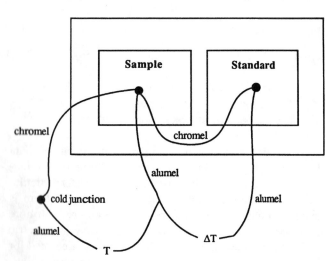

Figure 8-13.   Sketch of the essential elements of a sample holder for differential thermal analysis.

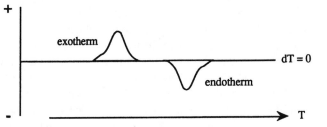

Figure 8-14.   Sketch of the relative positions of exothermic and endothermic "kicks" on the graph produced from differential thermal analysis of a sediment.

linked to a two-pen chart recorder. The position of each peak on the resulting graph shows the temperature range over which the reaction occurs, and the peak area shows the amount of heat involved in the reaction; the shape of each peak may be of further interpretative value, particularly when dealing with mixtures. Whether each reaction is endothermic or exothermic is indicated by whether the peak lies above (exothermic) or below (endothermic) the $dT = 0$ base line on the graph (Fig. 8-14). Hence, any reaction that involves a measurable heat change can in principle be studied by DTA, and its applicability does not depend on sample crystallinity as does the use of XRD methods.

Instrumentation

Heat Source

Most systems use a wire-wound furnace (Kanthel or Nichrome wire is good to 1300°C; Pt to 1750°C or Mo in an H₂ atmosphere to 2000°C); some fast-heating devices use microwave-induction furnaces. The furnaces may be wound in the common-coil mode or with the coil loops parallel to the axis of the furnace core; the latter method is better because it avoids stray currents being transferred to the thermocouple by induction. If coils are wound perpendicular to the axis of the furnace core, the sample block must be wrapped in earthed metal foil to eliminate possible induction currents. The furnace must be capable of providing a uniform heating rate over the desired temperature range, or heating in a series of preset (0.5–10°C) steps; hot spots or nonuniform heating rates will adversely affect analyses and reproducibility. The furnace should be designed such that samples can be positioned in the same place for each analysis.

Temperature Controller

The controller should not be of the on/off type but should be able to vary the power supplied to the furnace; some modern systems control the furnace so that heating takes place in a series of preset incremental steps. It is also necessary to have a feedback system from an independent thermocouple to an inert material in the furnace to ensure that the heating rate is constant over the required temperature range.

Sample Holder

Most systems use a ceramic block (usually sintered alumina) with two circular or rectangular wells for the sample; this ceramic block is held in a larger sample block made of the same material. Other systems have used metal holders made of Ni, Pt, Cu, or an alloy such as Pt/Rh. The metal holders are simpler to make, are nonporous, and give little baseline drift, but their use results in smaller peaks because they act as a heat sink. Ceramic holders cause some baseline drift but they give better peaks and eliminate the need for additional insulation around the thermocouple wires. The size and shape of the sample holders will influence peak size and shape as will whether a lid is used or not. Whatever design is chosen, the same one must be used consistently for all comparative work. In general, the holder will contain c. 0.5 g of sample in one well and c. 0.5 g of standard in the other, sufficiently separated so that there is no heat conduction between them.

Thermocouples

The most common types of thermocouple used are Pt/PtRh alloy, chromel/alumel, or Pt/PtIr alloy; selection will depend on temperature and sensitivity requirements. Irrespective of the metal combination used, the wires should be < 0.4 mm in diameter (usually 0.2 mm); if they are thicker, they may conduct heat between standard and sample, and to the outside atmosphere. The wiring is sketched in Fig. 8-13 (although the cold junction shown there is desirable, it is not essential and most operators are happy with an air junction). The hot junctions of the thermocouples should be positioned as near to the center of the sample and standard as possible.

Temperature Recorder

The choice of recorder will depend on available finance; a two-pen chart recorder is preferable, with one pen for recording T and the other for recording ΔT.

Conditions, Sample Preparation, and Standards

Although not always necessary, most modern DTA studies are carried out in a controlled atmosphere; a controlled atmosphere should be used to prevent oxidation of organic matter, which produces a large exotherm at 250°C and above. Various standards can be used in DTA analyses, but calcined alumina is most common. Some analysts prepare standards from prefired specimens of their own sample, but they then run the risk of interference if any of the reactions that take place during prefiring are reversible.

The samples for study may be amorphous or crystalline and are prepared by grinding the sample material to 2–5 microns. The exact size used is not critical, but it must be the same for all samples in a study because particle size can affect the size and shape of DTA peaks. A small amount of the ground sample (commonly 0.1–0.5 g; Broersma et al. 1978) is packed into the sample well, with care to ensure that packing is uniform and to a standard compaction (packing can affect DTA peak shape). Uniform preparation is important at all stages, and it is advisable that a single operator prepare all the analyses in a sample suite. It is also important to ensure that samples are not heated above their melting or vitrification point, because any melt is likely to destroy the sample holder and the thermocouple.

Some analysts pretreat their samples (e.g., with H_2O_2 if organics are present), but care must be taken to ensure that the pretreatment does not produce new problems (e.g., see Chapter 5). A more common pretreatment is to store the sample in an atmosphere with about 50% humidity for 3–4 days before testing to ensure a standard initial moisture content. If only a small amount of the sample is available, it is possible to dilute it up to three times its initial volume using some of the standard material. During the analysis, heating rate is an important consideration. A fast rate will increase the peak area, but a slow heating rate gives the best formed peaks. A heating rate of about 10°C per minute is a widely used compromise.

Interpretations

From the DTA curve the first things to note are:

1. Whether the peaks are endothermic or exothermic
2. The size and shape of each peak
3. The temperature at which each peak occurs ·

These data can then be compared with published data (e.g., Fig. 8-15) as the first step in a mineralogical determination, or they can be used to provide a fingerprint for particular samples in a suite. In all samples, check whether there is more than one mineral type present. The presence of more than one mineral type may be suggested by asymmetric peaks, by the presence of peaks that cannot be accounted for by any monomineralic system, or by data from other analytical (e.g., XRD) work. DTA is seldom used alone in mineral

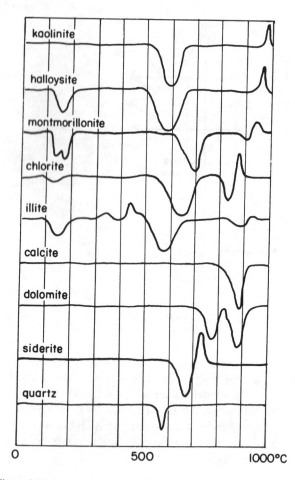

Figure 8-15.   Sketch examples of some DTA curves from common sedimentary minerals.

analysis; it is usually used in combination with other techniques such as XRD or IR (infrared analysis).

In addition to using the DTA curve for mineral identification, the peaks on the curve can be related to specific reactions in the sample. Thus, if a series of known standards is also analyzed, the area under each peak can be used to indicate the amount of a reactant involved in a particular reaction (e.g., the amount of H_2O lost). In principle, it is possible to relate any variation in peak characteristics to specific differences in the chemistry of the starting minerals. Knowledge of the reactions taking place during DTA runs often can be improved by XRD examination of the mineralogy of the end products of firing.

Advantages Over Other Methods of Analysis for Fine Grained Samples

1. The equipment is inexpensive relative to that necessary for many other procedures.
2. The samples need not be crystalline.
3. Sample preparation and analysis are relatively simple.
4. DTA can detect lower concentrations of some minerals (e.g., calcite and dolomite) than XRD.

5. DTA can indicate the extent of some elemental substitutions in particular minerals.

Limitations

1. The furnace takes a long time to cool between runs.
2. There is some difficulty in distinguishing between components in many polymineralic samples, particularly where peak asymmetry occurs in a peak-overlap zone.
3. The presence of organic matter and some minerals can cause interference problems.
4. Some minerals that give good curves when pure do not show up well in mixtures.
5. There may be spurious effects in which one component in a mixture reacts with others.
6. The DTA curves can be strongly affected by small variations in sample preparation and analysis procedures.

Thermogravimetric Analysis

Many reactions (e.g., dehydration or dehydroxylation) involve not only a heat change but also a mass change. mass change is measured in thermogravimetric analysis (TGA) in a similar way to that in DTA insofar as the change is recorded as a function of temperature. In contrast to terminations of loss on ignition (LOI) at 1000°C (see Chapter 9), commonly used in association with XRF analysis, TGA the mass gained or lost is used to calculate the amount of each component gained or lost, or the amount of iron oxidized at each relevant temperature. Mass changes at high temperatures are usually attributed to specific reactions such as dehydration (loss of H_2O^+), dehydroxylation (of H_2O^-), oxidation of organic matter, oxidation of sulfides, loss of CO_2 from carbonates, or oxidation of Fe^{2+}. Because TGA and other thermoanalytical techniques less commonly used by sedimentologists than DTA methods, no further discussion of them is included here. Additional information on these techniques can be found in MacKenzie (1982).

INFRARED SPECTROPHOTOMETRY

Principles

Infrared spectrophotometry is finding increasing use in sedimentology, particularly for examining the organic and grained inorganic fractions of sediments; it can also examine mineral components that are not sufficiently crystalline to be studied by X-ray diffraction. Infrared also can be used to obtain detailed mineralogical data that is not readily obtainable by other methods; estimation of the Si/Al ratio in phyllosilicates (e.g., Tuddenham 1960) or to study exchangeable cations and zeolites (e.g., Brodskii and Zhdanov 1980). *Spectra of Minerals* by Farmer (1974) provides

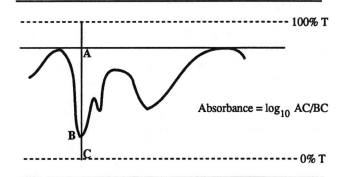

Figure 8-17. Section of an infrared spectrum showing the measurements used to calculate absorbance.

$$\text{Absorbance} = \log_{10} AC/BC$$

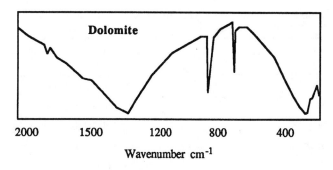

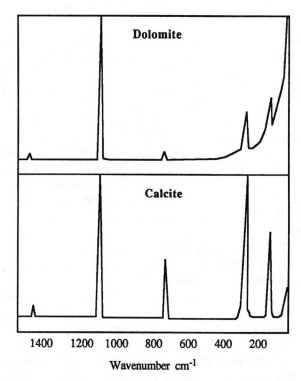

Figure 8-18. Sketch examples showing the appearence of infrared (upper) and Raman spectra (lower for two minerals) in common carbonates.

age different from the baseline voltage. The output voltage is amplified and passed to either a chart recorder or a computer to prepare a plot of infrared transmittance against wavelength; the plot can be used either quantitatively or qualitatively as desired.

An alternative to the wavelength-dispersion spectrophotometer is rapidly becoming the standard: the Fourier-transform infrared spectrophotometer (e.g., Bell, 1972). Fourier-transform infrared (FTIR) spectrophotometers use interferometry to examine the infrared spectra of the sample rather than the more traditional wavelength-dispersion procedure (described above). FTIR instruments have an advantage over dispersion instruments in that they observe the whole spectral range all the time, whereas dispersion instruments can record only part of the spectrum at any particular time; consequently, FTIR instruments provide a marked improvement in the signal-to-noise ratio. Offsetting this advantage is the fact that at wavelengths below about 3.5 μm, the large amount of spectral information makes deciphering the interferogram difficult, and in this part of the spectrum, wavelength-dispersive instruments still have the advantage.

Recently, some developments in reflectance infrared spectrometry have led to the production of field portable instruments and search-match software that make it possible to obtain almost instantaneous mineral identifications in the field for samples with no pretreatment. This technology is developing rapidly and will no doubt find increased use by sedimentologists working on fine-grained rocks.

Sample Preparation

Samples for use in geological studies are normally prepared as transparent potassium bromide (KBr) pellets. The KBr pellets are usually prepared by mixing 2 mg of dried sample (previously ground in an agate vibromill to ≤ 2 μm) with 500 mg of infrared-grade KBr. The mixture is blended in an agate vibromill then pressed in a stainless steel vacuum die (at about 10 tons per square inch for 2–5 min) to form a transparent disk. Any concentration of sample in KBr may be selected, but it is usual to choose one with absorption in the infrared beam of almost full-scale deflection in the region of

interest; for silicates, a dilution of 0.1–0.5% in KBr is suitable. It is important that the mixture is well blended or a poor-quality infrared spectrum of the sample will be obtained; it is also important that the die be scrupulously cleaned after use to remove all traces of KBr (which corrodes stainless steel) and to avoid any contamination from previous samples. Because so little is used, the samples are very sensitive to contamination, and considerable care is necessary at all stages of sample preparation and handling; care also must be taken to ensure that the subsample for analysis is representative of the original sample. It is possible to examine the infrared spectra of liquid samples, but this method is not common in geology.

Identification and Characterization of Minerals

The identification of unknown minerals by infrared spectroscopy usually requires comparison with the spectra of refer-

Table 8-7.   Infrared Absorption Bands for Some Common Anions in Sedimentary Minerals

Group	Absorption Peak Wave Number (cm^{-1})
CO_3^{2-}	800–880, 1410–1450
NO_3^-	800–860, 1340–1410
PO_4^{3-}	950–1100
SO_4^{2-}	610–680, 1080–1130
Silicates	900–1100

ence standards. A collection of reference spectra is obtained either through accumulation of mineral spectra from published literature, or through publishing houses specializing in large collections of infrared spectra of organic and inorganic compounds (e.g., Sadler Research Laboratories, Coblentz Society, American Society for Testing and Materials). Analysts working with sediments will also become familiar with the position of common peaks produced by various anions (e.g., Table 8-7) and will find useful quick reference tables of peak positions in books such as the *Handbook of Chemistry and Physics* (any recent edition of this standard reference published by the Chemical Rubber Co. Press, Cleveland, Ohio).

The procedure for quantitative determination of a mineral is straightforward, but in practice the results should be considered as semi-quantitative. The simple statement that the intensity of an absorption band varies with the amount of absorbing material present is not always valid, due to the influence of factors such as the effects of particle size, reproducibility of grinding, and disc preparation on the intensity of absorption bands. However, with careful sampling, grinding, accurate weighing and careful preparation of KBr discs, meaningful results are possible. To carry out a semi-quantitative determination, an absorption band that is distinctive for the mineral in the system being investigated is selected, and its absorbance is measured by drawing a straight line between the points of maximum transmission on each side of the band. A vertical line is then drawn on this baseline to pass through the point of minimum transmission of the band. The absorbance is then given by the logarithm of the ratio of the transmission at the point on the baseline vertically below the peak to the transmission at the absorption maximum (see Fig. 8-17). The mineral concentration in the sample is determined by reference to a calibration plot prepared by plotting absorbance and concentration for a series of standards.

Infrared analysis is rapid, requires very little sample, can be used with amorphous solids or liquids, gives some idea of crystal chemistry as well as mineralogy, and can give some clues to atomic geometry. However, the extreme care necessary to avoid contamination of the sample and to ensure a highly standardized method of sample preparation is a major disadvantage. The fact that mineralogical determinations are usually only semi-quantitative is also a limitation.

RAMAN SPECTROSCOPY

Raman spectroscopy (e.g., Farmer 1974) is used in a similar manner to infrared analysis and also uses very little sample (about 1–2 mg packed into a capillary tube). For some minerals, Raman spectroscopy provides much better peak resolution than infrared techniques (e.g., Fig. 8-18). As with infrared analysis, Raman scattering is related to bond type and atomic environment, but some bonds that produce no significant infrared peaks do produce good Raman lines (and vice versa).

Raman spectroscopy depends on the fact that when light hits a transparent substance some is scattered, and some of the scattered light photons undergo an energy change. Raman spectroscopy is the analysis of the fraction of the incident light that has undergone a change in wavelength. If the photon energy has been increased (i.e., the wavelength has decreased) *antistokes* lines are produced, and *stokes* lines are produced when photon energy is decreased (i.e., the wavelength has increased); the stokes lines are stronger than the antistokes lines because changes involving a decrease in photon energy are more common. Antistokes result when there is a gain in energy from an already excited bond, whereas stokes lines reflect a loss of energy to a bond in a lower vibration state.

In modern analytic instruments, a tunable dye laser is commonly used as a light source. The scattered light used in the analysis is intercepted perpendicular to the incident beam. Generally, a higher-frequency incident beam will produce stronger Raman scattered lines, but usually it is best to use an incident beam that has a similar color to that of the sample.

MÖSSBAUER SPECTROSCOPY

Principles

Application of the Mössbauer effect to mineralogical studies began in 1965 and has grown steadily ever since. Geological applications received a major boost from the publication of the 1973 book by Bancroft and are expanding rapidly. Mössbauer studies of geological materials offer "fingerprint" data on many minerals, indicate important differences between subvarieties of minerals, provide information on the distribution of ferric and ferrous iron between lattice sites in minerals, and give reliable data on the ferric/ferrous ratio in minerals. Mössbauer spectra, Mössbauer peak assignments, and discussion of the spectra for a very large range of minerals have been published since 1970; reference to information on particular minerals can be found quickly by consulting the Mössbauer Data Index, which records all published references on Mössbauer spectroscopy and on the Mössbauer spectra of minerals.

Over fifty isotopes are known to exhibit Mössbauer resonance, but almost all geological uses of the Mössbauer effect involve either iron (one of the best Mössbauer absorbers) or tin. Because iron is one of the geologically most widespread and important elements, the following discussion focuses on Mössbauer studies that use a ^{57}Co source and an iron absorber. Only the ^{57}Fe isotope of iron (natural abundance, 2.19%) in the absorber minerals is involved in the Mössbauer transitions. The ^{57}Co decays to ^{57}Fe by electron capture (half-life = 270 days) and the resulting ^{57}Fe is in a nuclear-excited state that decays to the ^{57}Fe ground state with the emission of three γ-rays (14.4, 123, 137 keV); X-rays are also emitted as a result of the electron vacancy created after electron capture. Of these photons, only the 14.4-keV γ-ray is used in Mössbauer studies. (Interaction between the excited nucleus and the extranuclear electrons may result in the release of conversion electrons with an energy equal to the nuclear-transition energy less the electron-binding energy; conversion electrons are used in some Mössbauer studies; e.g., Cohen 1976.)

In Mössbauer experiments, the 14.4-keV γ-ray photons need to be modulated to produce a known energy range using the Doppler effect; i.e., the ^{57}Co source is moved backward and forward (using a transducer driven at between 5 and 50 Hz by a very stable waveform) so that the γ-ray energy "seen" by the absorber is $(E_S) = (1 + V/C)E_\gamma$ where C is the speed of light and V is the source movement velocity (e.g., if $V = 1$ cm/sec, then the energy change imparted to the γ-ray $\Delta E = 4.8 \times 10^{-7}$ eV for ^{57}Fe). Hence, a plot of source velocity against γ-ray absorption is equivalent to a plot of ΔE against absorption. When a γ-ray from the moving source collides with an ^{57}Fe atom in the absorber, it may undergo recoil-free absorption if its energy is exactly that required by the absorbing atom plus or minus an amount that can be totally taken up by the lattice holding the absorber atom. Hence, any variation in lattice structure, temperature, or ligand type will alter the E_S required for recoil-free absorption and a Mössbauer absorption spectrum will be obtained.

Equipment

The equipment required for Mössbauer studies is simple and inexpensive. Basically it consists of a stable waveform generator (usually sine, triangular or saw-tooth waveforms) that supplies a signal to the drive amplifier, in turn driving a vibrator on which the ^{57}Co source is mounted. Radiation passing through the absorber is picked up by a detector, which has high sensitivity around 14.4 keV, and the signal is boosted in a preamplifier. The detected signal passes to the main amplifier and pulse height analyzer, in which the window is set to exclude all pulses not produced by γ-rays with an energy of about 14.4 keV. The "cleaned-up" signal finally enters a multichannel analyzer (usually 512 or 1024 channels), where pulses are allocated to sequential channels in a cycle matching that of the drive waveform exactly. A pulse generator and integrator link the multichannel amplifier and the drive waveform generator, to ensure that the sweep rate of the analyzer remains synchronized with the drive waveform. Modern instruments have all these components combined in a

single unit not much larger than a personal computer, and the output from the multichannel analyzer is fed directly to a small computer. Further details of experimental equipment can be found in Cohen (1976).

Calibration of the system for each drive velocity setting is achieved using a natural iron absorber in which the velocity of each of the six peaks is accurately known; calibration absorbers most widely used are certified by the U.S. National Bureau of Standards. Velocities for the six natural iron peaks are $V_1 = -5.3123$ mm/sec, $V_2 = -3.0760$ mm/sec, $V_3 = -0.8397$ mm/sec, $V_4 = +0.8397$ mm/sec, $V_5 = +3.0760$ mm/sec and $V_6 = +5.3123$ mm/sec.

Sample Preparation

Samples are normally prepared in one of three forms:

1. A randomly oriented powder packed into a hole cut in thin cardboard and sealed with transparent tape
2. A powder mixed with a boric acid binder and pressed into a steel ring (necessary for any low temperature studies)
3. A thin slice of the rock or mineral of interest

Where equipment is set up for operation in reflection mode, polished sections or slabs can also be used as absorbers. When working with a reflection arrangement or using the third sample preparation method, a lead mask can be used to restrict the area of the sample exposed to the source γ-rays (e.g., for single grain analysis).

With all methods, the optimum iron content of the sample is about 10 mg Fe/cm^2 of cross-sectional area of the sample, but satisfactory results can be obtained with iron contents between 0.5 mg Fe/cm^2 and 50 mg Fe/cm^2. If the iron content is too low, absorption is low and peak-to-background ratios are poor even when long counting times are used. If, on the other hand, the iron content is too high, mass absorption effects reduce peak-to-background ratios and resolution decreases.

Data Processing

It is possible to carry out routine work (e.g., Fe^{2+}/Fe^{3+} ratio determinations for silicates) using a direct analogue printout of the spectra, but it does not permit the accurate evaluation of the peak parameters (see next section). Most laboratories prefer to transfer the data (counts in each channel) directly to a computer for both calibration and parameter determination. During computer analysis of the raw data, a series of ideal Lorentzian lines are fitted to the spectrum by iterative adjustment to minimize the chi-square value for goodness of fit between the experimental data and the fitted curves. When a satisfactory fit is obtained, the parameters for the fitted curves are tabulated, and both raw data and fitted curves are plotted. Constraints can be applied to these Lorentzian curves (e.g., equal-area or particular-intensity ratios) where justified by other data or theoretical expectations.

During computer processing, it is also possible to de-blacken the data using the method of Dibar-Ure and Flinn (1971) to remove the source line width and distortions in peak amplitude caused by saturation. The resulting lines are narrower by about 0.1 mm/sec, and resolution is improved; however, the chi-square value is no longer a satisfactory indicator of goodness of fit.

The Measured Parameters

Isomer Shift

The isomer shift (IS), which is always quoted relative to the calibration standard used, results from the electrostatic interaction between the charge distribution of the nucleus and electrons that have a finite probability of being near the nucleus. On a Mössbauer spectrum it is normally the center shift that is measured and not the IS. The center shift represents the combined influence of the IS and second-order Doppler (SOD) shifts, due to thermal vibrations of the Mössbauer atoms (the second-order Doppler shifts are cumulative on the Doppler shift applied to the source), but because the SOD effects are small, the measured center shift is equated with the IS in most studies. For single peaks (i.e., no quadruple splitting), the IS is the displacement of the peak from the point of zero velocity (Fig. 8-19) for the standard; for pairs of peaks, it is displacement of their center of symmetry from the point of zero velocity for the standard (for

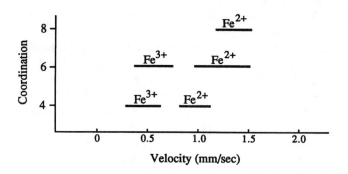

ISOMER SHIFT

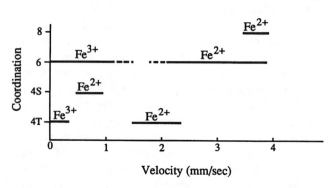

QUADRUPOLE SPLITTING

Figure 8-20. Velocity ranges for the isomer shift (IS) and quadrupole splitting (QS) in silicate minerals as a function of coodination number.

magnetically split spectra, see calculation in section on magnetic splitting).

The mathematical expression of factors producing the IS is complex and is not included here except to note that the IS is sensitive to changes in oxidation state, crystallographic coordination, and ligand type. For silicate minerals, the IS is a useful mineralogical parameter that reflects the Fe oxidation state and the Fe coordination (e.g., Fig. 8-20).

Quadrupole Splitting

Quadrupole splitting (QS) is caused by interactions between noncubic extranuclear electric fields and the nuclear charge that causes splitting of the nuclear energy levels (Fig. 8-21); for pairs of peaks it is measured as their separation (in mm/sec); for relationships involving magnetically split spectra, see the section on magnetic splitting. For silicates, QS primarily reflects changes in the oxidation state and coordination of iron and changes in ligand type; a general summary of the relationship between QS and crystal chemistry is shown in Fig. 8-20. Further details on the origin of quadrupole splitting in minerals can be found in Ingalls (1964), Bancroft (1973), and Cohen (1976) and a geological example is given in McConchie et al. (1979).

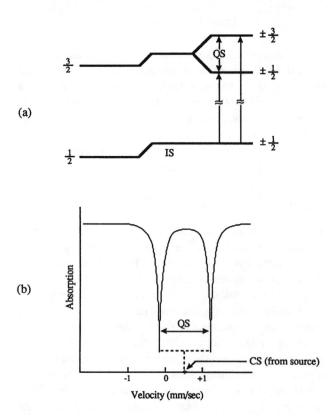

Figure 8-19. Nuclear energy levels split by quadrupole interactions. a: absorber energy levels; b: the resulting Mössbauer spectrum. QS is the quadrupole splitting and CS is the center shift, which is approximately equal to isomer shift. (IS).

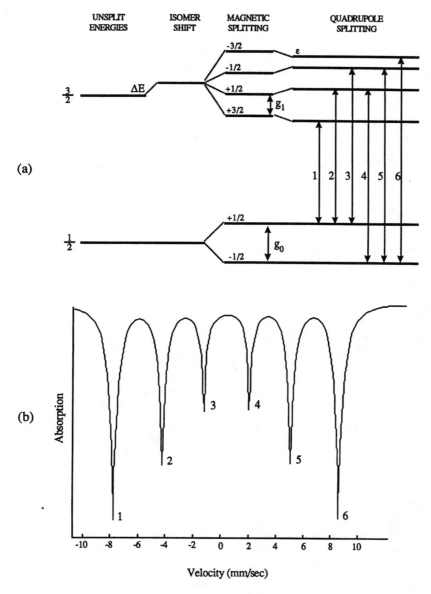

Figure 8-21. Nuclear energy levels split by quadrupole and magnetic interactions. *a:* absorber energy levels; *b:* the resulting Mössbauer spectrum (peak numbers are the same in both).

Half-Height Line Width

The half-height line width (T) of Mössbauer peaks is due to the natural line widths of both source and absorber; it also reflects variations in absorber site conditions. Minor variations in the lattice, particularly in the nature and abundance of localized defects and distortions, result in variations in T; in some studies, T can be used as an indicator of crystallographic homogeneity and lattice perfection. T is measured directly from each peak and is recorded in mm/sec.

Peak Intensity

Peak intensity (I) is normally recorded as a % absorption or an absorption probability, and has no absolute numerical value because it is influenced by many factors, including the iron density in the sample, the mass absorption coefficient of

the sample, and the number of different mineralogical varieties of iron present. Within a single spectrum, however, I is a function of the number of iron atoms in each oxidation state and lattice position in the sample. The product of I and T is used to indicate the proportion of iron atoms in the sample in each oxidation state and each lattice position.

Magnetic Splitting

A nucleus with spin I has a magnetic dipole that interacts with local or applied magnetic fields at the nucleus, and the degeneracy of nuclear energy levels can be completely removed. Where degeneracy is removed, a full six-line Zeeman split spectrum can be observed (e.g., for natural iron, hematite, goethite, magnetite, etc.). If the powder in the sample is randomly oriented, the area ratio of the six peaks is about 3:2:1:1:2:3.

The peak position for each of the six lines will be determined by the IS and QS, in addition to the magnetic splitting (Fig. 8-21). However, the ground state, excited state, and other magnetic splitting parameters can still be calculated as:

$$g_0 = 1/2(V_4 + V_5 - V_2 - V_3) \, kG$$
$$g_1 = 1/4(V_6 + V_3 - V_1 - V_4) \, kG$$
$$\varepsilon = 1/8[2(V_1 + V_6) - V_2 - V_3 - V_4 - V_5] \text{ mm/sec}$$
$$IS = 1/4(V_2 + V_3 + V_4 + V_5) + \varepsilon \text{ mm/sec}$$
$$\varepsilon = (QS/2)\cos^2\theta$$

where g_0, g_1 = the Zeeman splitting parameters for the ground and excited states respectively

V_1 to V_6 = the velocities for peaks 1 to 6 (Fig. 8-21) in mm/sec

ε = a function of QS and the angle (θ) between the magnetic field and the electric field gradient

The values of g_0 and g_1 can be used to calculate H_{eff} (the effective magnetic field "seen" by the nucleus and given in kilo Gauss) using the relation:

$$H_{eff} = H_{eff}{}^* \, (g_0/g_0{}^*) = H_{eff}{}^* \, (g_1/g_1{}^*)$$

where $H_{eff}{}^*$ = the effective field for natural iron (= 333 kG)

$g_0{}^*$ = the g_0 value for natural iron (= 3.9157)

$g_1{}^*$ = the g_1 value for natural iron (= 2.2363)

The Effect of Temperature

The temperature of the sample at the time the spectrum is taken will affect the magnitude of the measured parameters; correction factors can be applied to convert values obtained at one temperature to those expected at another. Thus, when examining data reported in the literature, it is important to note the temperature at which the spectra were taken. In the case of magnetically split spectra, it is even more important, because at particular temperatures major changes can take place. For example, at the Néel temperature, the six-line spectrum will disappear and only a two-line spectrum will be recorded. Further discussion of various temperature effects on Mössbauer spectra and their mathematical expression can be found in Kundig et al. (1966), Bancroft (1973), Cohen (1976), and McConchie et al. (1979).

Geological Applications

Fe²⁺/Fe³⁺ Determination

Fe^{2+}/Fe^{3+} ratio determinations are probably the most obvious application of Mössbauer spectroscopy for geologists and one of the easiest to carry out. Not only are standard wet chemical methods of Fe^{2+}/Fe^{3+} ratio determination difficult to carry out but they involve the destruction of the sample material. Mössbauer methods, on the other hand, are comparatively simple and leave the sample material available for

analysis by other methods. Furthermore, it has been shown that no matter how careful the analyst, wet chemical determinations of Fe^{2+}/Fe^{3+} ratio are frequently erroneous (e.g., Bancroft and Burns 1969; Hogg and Meads 1970; Bancroft et al. 1977). It has also been shown (e.g., Rozenson and Heller-Kallai 1977) that even mild reagents such as 1M ammonium acetate, used in exchangeable cation determinations, can change the Fe^{2+}/Fe^{3+} ratio in phyllosilicates. Hence, the process of wet chemical determination of the Fe^{2+}/Fe^{3+} ratio may change the very thing it is intended to measure. These problems do not apply in Mössbauer determinations of the Fe^{2+}/Fe^{3+} ratio, which are carried out by calculating the ratio of the sum of the areas of peaks due to Fe^{2+} to the sum of the areas of the peaks due to Fe^{3+} (e.g., Boulter et al. 1989). Which peaks are produced by Fe^{2+} and which by Fe^{3+} can also be determined without difficulty from published spectra, and in the case of silicates from data such as those shown in Fig. 8-20.

Crystallographic Site Determination

Mössbauer spectra can be used to determine the proportion of iron in each oxidation state that occupies each lattice site. In simple systems, this determination can be made using the data in Fig. 8-20, but in more complex systems additional mathematical analysis of the data may be necessary. With the additional mathematical treatment of the data, it is also possible to distinguish between cis and trans octahedral sites, and to determine site preferences. For some geological problems, this detailed data can reveal genetic and diagenetic information not revealed by other lines of investigation. The model for glauconite genesis proposed by McConchie et al. (1979) is an example of the application of this type of data.

Fingerprint Mineral Description

Because Mössbauer spectra can be used to define site occupancy and oxidation state so clearly, they can often be used to identify particular iron-bearing minerals by comparison with published spectra. An example of this application would be in distinguishing between ferri- and ferrostilpnomelane; this distinction is difficult to make with confidence by other means. Another example involves the identification of some iron oxides, iron hydroxides, or other minerals that may be present as particles too fine or too poorly crystalline to detect by X-ray diffraction. Many studies of the geochemistry of elements in soil profiles and gossans may benefit from a Mössbauer investigation of poorly crystalline mineral constituents (e.g., Childs and Baker-Sherman 1984).

Provenance Studies

In addition to using Mössbauer spectra to provide mineral fingerprints, it has been shown (e.g., Strens 1976) that if certain minerals are placed under pressure, both the IS and QS values on their Mössbauer spectra change. For most minerals, the increase in pressure causes an increase in s electron density resulting in a systematic decrease in the IS (see Strens 1976 for detailed explanation). Some of the minerals that show this trend are stable in the weathering environment (e.g., ilmenite, almandine, chromite) and are frequently found in rocks formed at depth. These minerals may have potential as geobarometers sensitive to increments of 10 Kb to about 150 Kb. In natural specimens, the magnitude of the change in IS would not be expected to be as great as that in the experimental studies of Strens (1976), due to elastic readjustment during the mineral's rise to the surface. Changes in Mössbauer spectral parameters due to the mode of origin of minerals such as ilmenite and chromite may provide useful information on provenance (e.g., Singhvi et al. 1974; Gendler et al. 1977; Lucas et al. 1989).

Laterite Dating

Most spectra of iron oxides and iron hydroxides show the full six-line Zeeman or magnetic splitting as in Figure 8-21, but some will show only a two-line superparamagnetic spectrum with IS and QS values less than 0.8 mm/sec. Whether the two-line superparamagnetic spectrum or the normal magnetically split six-line spectrum is observed will depend on the size of particles in the sample and the sample temperature (see Kündig et al. 1966); thus, for spectra taken at room temperature, whether a superparamagnetic or a ferromagnetic (six-line) spectrum is observed will depend on the size of the iron oxide or iron hydroxide crystallites (e.g., see figures in Kündig et al. 1966).

When iron oxides or iron hydroxides are first precipitated under natural conditions, they have a very fine crystallite size that increases slowly with aging. Hence, a very young sample will produce an entirely superparamagnetic spectrum whereas the spectrum for an old sample will show all six lines; spectra for samples with intermediate ages will contain a combination of the two end member spectral types (e.g., Hanstein et al. 1983); the ratio of the area of the superparamagnetic lines to that of the six ferromagnetic lines may often be a guide to the relative age of laterite samples. Further investigation is needed to determine whether the change in superparamagnetic crystallite abundance with aging is the same for different climatic areas, rock types, or sets of diagenetic conditions and to quantify the effect of sample heating (e.g., Johnston and Lewis 1983; McConchie and Smith 1989) on the loss of superparamagnetism.

SELECTED BIBLIOGRAPHY

General

Adams, A. E., W. S. MacKenzie, and C. Guilford, 1984, *Atlas of Sedimentary Rocks under the Microscope.* Longman Group, Ltd., London, 104p.

Bloss, F. D., 1981, *The Spindle Stage: Principles and Practice.* Cambridge University Press, Cambridge, 340p.

Bloss, F. D., and J. F. Light, 1973, The detente spindle stage. *American Journal of Science* 273-A:536–8.

Brewer, R., 1964, *Fabric and Mineral Analysis of Soils.* Wiley, New York.

Bullock, P., N. Federoff, A. Jongerius, G. Stoops, and T. Tursina, 1984, *The Handbook for Soil Thin Section Description.* Wayne Res., Wolverhampton.

Carver, R. E., 1971, *Procedures in Sedimentary Petrology.* Wiley-Interscience, New York, 653p.

Craig, J. R., and D. J. Vaughan, 1981, *Ore Microscopy and Ore Petrography.* Wiley, New York.

Davis, J. J., 1991 Carbonate petrography—update on new techniques and applications. *Journal of Sedementary Petrology* 61:626–8.

El-Hinnawi, E. E., 1966, *Methods in Chemical and Mineral Microscopy.* Elsevier, New York, 222p.

Esterle, J. S., T. A. Moore, and J. C. Hower, 1991, A reflected-light petrographic technique for peats. *Journal of Sedimentary Petrology* 61:614–16.

FitzPatrik, E. A., 1984, *Micromorphology of Soils.* Chapman and Hall, London.

Frondel, C., 1962, *Dana's System of Mineralogy*, 7th ed., vol. 3, *Silica Minerals.* Wiley, New York, 334p.

Galopin, R., and N. F. M. Henry, 1972, *Microscopic Study of Opaque Minerals.* McCrone Research Association, London.

Gilbert, G. M., and F. J. Turner, 1949, Use of the universal stage in sedimentary petrography. *American Journal of Science* 247:1–26.

Glover, J. E., 1964, The universal stage in studies of diagenetic textures. *Journal of Sedimentary Petrology* 34:851–4.

Hartshorne, V. F., and A. Stuart, 1970, *Crystals and the Polarising Microscope.* Arnold, London, 614p.

Hoffman, R., and L. Gross, 1970, Reflected light differential-interference microscopy: Principles, use and image interpretation. *Journal of Microscopy* 91:149–72.

Horowitz, A. S., and P. E. Potter, 1971, *Introductory Petrography of Fossils.* Springer-Verlag, New York, 302p.

Ineson, P. R., 1989, Introduction to Practical Ore Microscopy. Longman, United Kingdom.

Kemp, R. A., 1985, Soil micromorphology and the Quaternary. *Quaternary Research Association Technical Guide No. 2*, Cambridge, 80p.

Majewske, O. P., 1969, *Recognition of Invertebrate Fossil Fragments in Rocks and Thin Sections.* E. J. Brill, Leiden, The Netherlands, 101p (plus 106 plates).

Marshall, C. E., and C. D. Jeffries, 1945, Oil immersion studies. *Soil Science Society of America Proceedings* 10:397–405.

Milner, H. B., 1962, *Sedimentary Petrography*, vol. 1. Methods in Sedimentary Petrography. Allen & Unwin, London, 643p.

Moore, T. C., Jr., 1973, Method of randomly distributing grains for microscopic examination. *Journal of Sedimentary Petrology* 43:904–6.

Phillips, W. R., 1971, *Mineral Optics.* W. H. Freeman and Co., San Francisco, 249p.

Phillips, W. R., and D. T. Griffen, 1981, *Optical Mineralogy—the Nonopaque Minerals.* W. H. Freeman, San Francisco, 677p.

Risk, M. J., and R. B. Szczuczko, 1977, A method for staining trace fossils. *Journal of Sedimentary Petrology* 47:855–9.

Saggerson, E. P., 1975, *Identification Tables for Minerals in Thin Sections.* Longman, New York, 378p.

Sanford, J. T., R. E. Mosher, and J. E. Friend, 1968, A quantitative study of Silurian rocks from the Michigan Basin. *Geological Association of Canada Proceedings* 19:37–44.

Schneider, H. E., 1970, Problems of quartz grain morphoscopy. *Sedimentology* 14:325–35.

Scholle, P. A., 1978, A color illustrated guide to carbonate rock constituents, textures, cements and porosities. *American Association of Petroleum Geologists Memoir 27*, Tulsa Okla, 241p.

Scholle, P. A., 1979, A color illustrated guide to constituents, textures, cements and porosities of sandstones and associated rocks. *American Association of Petroleum Geologists Memoir 28*, 201p.

Shelley, D., 1985, *Optical Mineralogy.* Elsevier, New York, 321p.

Stach, E., M.-Th. Mackowsky, M. Teichmuller, R. Teichmuller, G. H. Taylor, and D. Chandra, 1982, *Coal Petrology*, 3d ed. Gebruder Borntraeger, Berlin.

Tickell, F. G., 1965, *The Techniques of Sedimentary Mineralogy.* Developments in Sedimentology 4, Elsevier, Amsterdam, 220p.

van der Plas, L., 1966, *The Identification of Detrital Feldspars.* Developments in Sedimentology 6, Elsevier, New York, 305p.

Wahlstrom, E. E., 1951, *Optical Crystallography.* Wiley, New York, 489p.

Warnke, D. A., and R. Gram, 1969, The study of mineral grain surfaces by interference microscopy. *Journal of Sedimentary Petrology* 39:1599–1604.

Warnke, D. A., and D. K. Stauble, 1971, An application of reflected-light differential-interference microscopy: Beach studies in eastern Long Island. *Sedimentology* 17:89–101.

Winchell, H., 1965, *Optical Properties of Minerals: A Determinative Table.* Academic Press, New York, 91p.

Zuffa, G. G., 1984, Optical analysis of arenites: Influence of methodology on compositional results. In G. G. Zuffa (ed.), *Provenance of Arenites.* Reidel, pp. 163–190.

Zussman, J. (ed.), 1967, *Physical Methods in Determinative Mineralogy.* Academic Press, London and New York, 514p.

Luminescence

Amieux, P., 1982, La cathodluminescence: Methode d'etude sedimentologique des carbonates. *Centres Recherches Exploratione-Elf-Aquitaine Bulletin* 6:437–83.

Barker, C. E., and O. C. Kopp, (eds.), 1991, *Luminescence Microscopy and Spectroscopy: Qualitative and Quantitative Applications.* Society for Sedimentary Geology Short Course 25, Tulsa, Okla., 194p.

Barker, C. E., and T. Wood, 1986, Notes on cathodoluminescence microscopy using the Technosyn stage, and a bibliography of applied cathodoluminescence. *U.S. Geological Survey Open-File Report 86–85*, 35p.

Calderon, T., M. Aguilar, F. Jaque, and R. Coy-Yll, 1984, Thermoluminescence from natural calcites. *Journal of Physics* 17:2027–38.

Cercone, K. R., and V. A. Pedone, 1987, Fluorescence (photoluminescence) of carbonate rocks: Instrumental and analytical sources of observational error. *Journal of Sedimentary Petrology* 57:780–2.

Dorobek, S. L., J. F. Read, J. M. Niemann, T. C. Pong, and R. M. Haralick, 1987, Image analysis of cathodoluminescent-zoned calcite cements. *Journal of Sedimentary Petrology* 57:766–70.

Dravis, J. J., and D. A. Yurewicz, 1985, Enhanced carbonate petrography using fluorescence microscopy. *Journal of Sedimentary Petrology* 55:795–804.

Fairchild, I. J., 1983, Chemical controls of cathodoluminescence of natural dolomites and calcites: New data and review. *Sedimentology* 30:579–83.

Gregg, J. M., and M. Karakus, 1991, A technique for successive cathodoluminescence and reflected light petrography. *Journal of Sedimentary Petrology* 61:613–4.

Hemming, N. G., W. J. Meyers, and J. C. Grams, 1989, Cathodoluminescence in diagenetic calcites: The roles of Fe and Mn as

deduced from electron probe and spectrophotometric measurements. *Journal of Sedimentary Petrology* 59:404–11.

Long, J. V. P., and S. O. Agrell, 1965, Cathodoluminescence of minerals in thin section. *Mineralogical Magazine* 34:318–26.

McDougall, D. J., 1968, *Thermoluminescence of Geological Minerals.* Academic Press, London, 675p.

Machel, H. G., 1985, Cathodoluminescence in calcite and dolomite and its chemical interpretation. *Geoscience Canada* 12:139–47.

Marshall, D. J., 1988, *Cathodoluminescence of Geological Materials.* Allen & Unwin, Winchester, Maine, 128p.

Mugridge, S. J., and H. R. Young, 1984, Rapid preparation of polished thin sections for cathodoluminescence study of carbonate rocks. *Canadian Mineralogist* 22:513–5.

Roberts, H. H., and W. E. Graves, 1972, Thermoluminescence: A tool for the environmental analysis of recent sediment cores. *Journal of Sedimentary Petrology* 42:146–9.

Rost, F. W. D., 1992, *Fluorescence Microscopy,* vol. 1. Cambridge University Press, New York, 253p.

Ryan, D. E., and J. P. Szabo, 1981, Cathodoluminescence of detrital sands: A technique for rapid determination of the light minerals of detrital sands. *Journal of Sedimentary Petrology* 51:669–70.

Sippel, R. F., 1968, Sandstone petrology, evidence from luminescence petrography. *Journal of Sedimentary Petrology* 38:530–54.

Soloman, S. T., and G. M. Walkden, 1985, The application of cathodoluminescence to interpreting the diagenesis of an ancient calcrete profile. *Sedimentology* 32:877–96.

Tucker, M., 1988, *Techniques in Sedimentology.* Blackwell Scientific Publications, Oxford, 394p.

Walker, G., 1985, Mineralogical applications in luminescence techniques. In *Chemical Bonding and Spectroscopy in Mineral Chemistry.* Chapman and Hall, London, pp. 103–40.

Yacobi, B. G., and D. B. Holt, 1990, *Cathodoluminescence Microscopy of Inorganic Solids.* Plenum Press, New York, 292p.

Heavy Mineral Techniques

Callahan, J., 1987, A nontoxic heavy liquid and inexpensive filters for separation of mineral grains. *Journal of Sedimentary Petrology* 57:765–6.

Greene, G. M., and L. E. Cornitus, 1971, A technique for magnetically separating minerals in a liquid mode. *Journal of Sedimentary Petrology* 41:310–2.

Gregory, M. R., and K. A. Johnston, 1987, A nontoxic substitute for hazardous heavy liquids—aqueous polytungstate ($3Na_2WO_4 \cdot 9WO_3 \cdot H_2O$) solution. *New Zealand Journal of Geology and Geophysics* 30:317–20.

Gunn, C. B., 1968, Field concentration of heavy minerals. *Journal of Sedimentary Petrology* 38:1362.

Henningsen, D., 1967, Crushing of sedimentary rock samples and its effect on shape and number of heavy minerals. *Sedimentology* 8:253–5.

Hunter, R. E., 1967, A rapid method for determining weight percentages of unsieved heavy minerals. *Journal of Sedimentary Petrology* 37:521–9.

Komar, P. D., K. E. Clemens, Z. Li, and S.-M. Shih, 1989, The effects of selective sorting on factor analysis of heavy-mineral assemblages. *Journal of Sedimentary Petrology* 59:590–6.

Lowright, R. H., 1973, Environmental determination using hydraulic equivalence studies. *Journal of Sedimentary Petrology* 43:1143–7.

McAndrew, J., 1957, Calibration of a Frantz isodynamic separator and its application to mineral separation. *Australian Institute of Mining and Metallurgy Proceedings* 181:59.

McLaughlin, W. A., and D. C. Berkshire, 1969, Nitrobenzene-tetrabromoethane solutions for the gravity separation of minerals. *Journal of Sedimentary Petrology* 39:1610–5.

Modarres, H. G., 1968, Simple and effective device for gravity separation of heavy mineral grains. *Journal of Sedimentary Petrology* 38:240–2.

Pryor, E. J., 1965, *Mineral Processing*, 3rd ed. Elsevier, Amsterdam.

Ramesam, V., 1966, Improved methods of heavy mineral separation and counting suitable for fine grained sandstone. *Journal of Sedimentary Petrology* 36:629–31.

Rittenhouse, G., 1943, Transportation and deposition of heavy minerals. *Geological Society of America Bulletin* 54:1725–80.

Rosenblum, S., 1958, Magnetic susceptibilities of minerals in the Frantz isodynamic separator. *American Mineralogist* 43:170–3.

Ross, C. S., 1926, Methods of preparation of sedimentary materials for study. *Economic Geology* 21:454–68.

Rubey, W. W., 1933, The size distribution of heavy minerals within a water-laid sandstone. *Journal of Sedimentary Petrology* 3:3–29.

Taggart, A. F., 1951, *Elements of Ore Dressing*. Wiley, New York.

Theobald, P. K., Jr., 1957, The gold pan as a quantitative geologic tool. *United States Geological Survey Bulletin 1071-A*, 54p.

Wang, C., and P. D. Komar, 1985, The sieving of heavy mineral sands. *Journal of Sedimentary Petrology* 55:479–82.

Wills, B. A., 1988, *Mineral Processing Technology: An Introduction to the Practical Aspects of Ore Treatment and Mineral Recovery*, 4th ed. Pergamon Press, New York.

Young, E. J., 1966, A critique of methods for comparing heavy mineral suites. *Journal of Sedimentary Petrology* 36:57–65.

Feldspar Stains

Bailey, E. H., and R. E. Stevens, 1960, Selective staining of K-feldspar and plagioclase on rock slabs and in thin sections. *American Mineralogist* 45:1020–6.

Bardsley, W. E., 1975, Modified technique for staining feldspar in grain mounts. *New Zealand Journal of Geology and Geophysics* 18:515–8.

Bjorlykke, K., 1966, The study of arenaceous sediments by means of acetate replicas. *Sedimentology* 6:343–5

Broch, O. A., 1961, Quick identification of potash feldspars, plagioclase and quartz for quantitative thin section analysis. *American Mineralogist* 46:752–3.

Bromley, R. G., 1981, Enhancement of visibility of structures in marly chalk: Modification of the Bushinsky oil technique. *Bulletin of the Geological Society of Denmark* 29:111–8.

Choquette, P. W., and F. C. Trusell, 1978, A procedure for making the titan-yellow stain for Mg-calcite permanent. *Journal of Sedimentary Petrology* 48:639–41.

Gross, D. L., and S. R. Moran, 1970, A technique for the rapid determination of the light minerals of detrital sands. *Journal of Sedimentary Petrology* 40:759–61.

Hayes, J. R., and M. A. Klugman, 1959, Feldspar staining methods. *Journal of Sedimentary Petrology* 29:227–32.

Houghton, H. F., 1980, Refined technique for staining plagioclase and alkali feldspars in thin section. *Journal of Sedimentary Petrology* 50:629–31.

Laniz, R. V., R. E. Stevens, and M. B. Norman, 1964, Staining of plagioclase feldspar and other minerals with F.D. and C. Red

No. 2. *United States Geological Survey Professional Paper 501-B*, pp. 152–3.

Morris, W. J., 1985, A convenient method of acid etching. *Journal of Sedimentary Petrology* 55:600.

Reid, W. P., 1969, Mineral staining tests. *Colorado School of Mines Mineral Industries Bulletin* 12:1–20.

Russell, R. D., 1935, Frequency percentage determination of detrital quartz and feldspar. *Journal of Sedimentary Petrology* 5:109–14.

Wilson, M. D., and S. S. Sedora, 1979, An improved thin section stain for potash feldspar. *Journal of Sedimentary Petrology* 49:637–8.

Carbonate Stains

Chilingar, G. V., H. J. Bissell, and R. W. Fairbirdge, 1967, *Carbonate Rocks, Developments in Sedimentology,* vols. 9A and 9B. Elsevier, Amsterdam, 471p., 413p.

Dickson, J. A. D., 1965, A modified staining technique for carbonates in thin section. *Nature* 205:587.

Dickson, J. A. D., 1966, Carbonate identification and genesis as revealed by staining. *Journal of Sedimentary Petrology* 36:491–505.

Evamy, B. D., 1963, The application of a chemical staining technique to a study of dedolomitization. *Sedimentology* 2:164–70.

Friedman, G. M., 1959, Identification of carbonate minerals by staining methods. *Journal of Sedimentary Petrology* 29:87–97.

Gabriel, A., and E. P. Cox, 1929, A staining method for the quantitative determination of certain rock minerals. *American Mineralogist* 14:290–2.

Katz, A., and G. M. Friedman, 1965, The preparation of stained acetate peels for the study of carbonate rocks. *Journal of Sedimentary Petrology* 35:248–9.

Lindholm, R. C., and R. B. Finkelman, 1972, Calcite staining: Semiquantitative determination of ferrous iron. *Journal of Sedimentary Petrology* 42:239–42.

Schneidermann, N., and P. A. Sandberg, 1971, Calcite-aragonite differentiation by selective staining and scanning electron microscopy. *Transactions of the Gulf Coast Association of Geological Societies* 21:349–52.

Steidtmann, E., 1917, Origin of dolomite as disclosed by stains and other methods. *Geological Association of America Bulletin* 28:431–50.

Warne, S. S. J., 1962, A quick field or laboratory staining scheme for the differentiation of the major carbonate minerals. *Journal of Sedimentary Petrology* 32:29–38.

Modal Analysis

Chayes, F., 1956, *Petrographic Modal Analysis*. Wiley, New York, 113p.

Dennison, J. M. V., and J. H. Shea, 1966, Reliability of visual estimates of grain abundances. *Journal of Sedimentary Petrology* 36:81–9

Galehouse, J. S., 1969, Counting grain mounts: Number percentage vs. number frequency. *Journal of Sedimentary Petrology* 39:812–5.

Galehouse, J. S., 1971, Point counting. In R. E. Carver (ed.), *Procedures in Sedimentary Petrology.* Wiley-Interscience, New York, pp. 385–408.

Larsen, E. S., and F. S. Miller, 1935, The Rosiwal method and the modal determination of rocks. *American Mineralogist* 20:260–73.

Mckinney, D. B., 1993, Computerized point-counting with the Swift mechanical point-counting stage and an easy-to-build serial interface. *Computers and Geosciences* 19:95–110.

van der Plas, L., and A. C. Tobi, 1965, A chart for judging the reliability of point counting results. *American Journal of Science* 263:87–90.

Welsh, W., 1967, The value of point-count modal analysis of greywackes. *Scottish Journal of Geology* 3:318–26.

X-ray Diffraction

Bayliss, P., D. C. Erd, M. E. Mrose, A. P. Sabina, and D. K. Smith, 1986, *Fink Mineral Index.* J.C.P.D.S. International Centre for Diffraction Data, Swarthmore, Pa., 467p.

Bish, D. L., and J. E. Post (eds.), 1989, *Modern Powder Diffraction.* Reviews in Mineralogy 20, Mineralogical Society of America, Washington, D.C., 369p.

Bish, D. L., and R. C. Reynolds, 1989, Sample preparation for X-ray diffraction. In D. L. Bish and J. E. Post (eds.), *Modern Powder Diffraction.* Reviews in Mineralogy 20, Mineralogical Society of America, Washington, D.C., pp. 73–99.

Brindley, G. W., and G. Brown (eds.), 1980, *Crystal Structures of Clay Minerals and Their X-ray Identification.* Mineralogical Society of London, Monograph 5.

Brown, G. (ed.), 1961, *The X-ray Identification and Crystal Structures of Clay Minerals.* Mineralogical Society of London, Jarrold and Sons Ltd.

Calvert, L. D., A. F. Sirianni, G. J. Gainsford and C. R. Hubbard, 1983, A comparison of methods for reducing preferred orientation. *Advances in X-ray Analysis* 26:105–10.

Cullity, B. D., 1978, *Elements of X-Ray Diffraction.* Addison Wesley Publishing Co., New York, 555p.

Hughes, R., and B. Bohor, 1970, Random clay powders prepared by spray drying. *American Mineralogist* 55:1780–6.

Hutchison, C. S., 1974, *Laboratory Handbook of Petrographic Techniques.* Wiley-Interscience, New York.

Jenkins, R., 1989, Instrumentation. In D. L. Bish and J. E. Post (eds.), *Modern Powder Diffraction.* Reviews in Mineralogy 20, Mineralogical Society of America, Washington, D.C., pp. 19–45.

Jenkins, R., and J. L. DeVries, 1967, *Practical X-ray Spectrometry.* Philips Technical Library, Springer-Verlag, New York, 182p.

Klug, H. P., and L. E. Alexander, 1974, *X-ray Diffraction Procedures.* Wiley, New York.

Reynolds, R. C., 1989, Principles of powder diffraction. In D. L. Bish and J. E. Post (eds.), *Modern Powder Diffraction.* Reviews in Mineralogy 20, Mineralogical Society of America, Washington, D.C., pp. 1–17.

Snyder, R. L., and D. L. Bish, 1989, *Quantitative Analysis.* In D. L. Bish and J. E. Post (eds.), *Modern Powder Diffraction.* Reviews in Mineralogy 20, Mineralogical Society of America, Washington, D.C., pp. 101–144.

Thorez, J., 1975, *Phyllosilicates and Clay Minerals: A Laboratory Handbook for Their X-Ray Diffraction Analysis.* Editions G. Lelotte, Dison, Belgium.

Tien, Pei-Lin, 1974, A simple device for smearing clay-on-glass slides for quantitative X-ray diffraction studies. *Clays and Clay Minerals* 22:367–8.

Veblen, D. R., G. D. Guthrie, Jr., and K. J. T. Livi, 1990, High-resolution transmission electron microscopy and electron dif-

fraction of mixed-layer illite/smectite: Experimental results. *Clays and Clay Minerals* 38:1–13.

Wilson, M. J., and D. R. Clark, 1978, X-ray identification of clay minerals in thin section. *Journal of Sedimentary Petrology* 48:656–60.

Differential Thermal Analysis

Broersma, A., P. L. de Bruyn, J. W. Jens, and R.J. Stol, 1978, Simultaneous DTA and DTG measurements on aluminum oxide monohydroxides. *Journal of Thermal Analysis* 13:341–55.

Coats, A. W., and J. P. Redfern, 1963, Thermogravimetric analysis. *Analyst* 88:906–24.

Daniels, T., 1973, *Thermal Analysis.* Kogan Page, London, 272p.

Hutchison, C. S., 1974, *Laboratory Handbook of Petrographic Techniques.* Wiley-Interscience, New York.

MacKenzie, R. C. (ed.), 1957, *The Differential Thermal Analysis of Clays.* Mineral Society of London, London, 456p.

MacKenzie, R. C. (ed.), 1970 and 1972, *Differential Thermal Analysis,* 2 vols. Academic Press, London.

MacKenzie, R. C., 1982, Thermoanalytical methods in clay studies. In J. J. Fripiat (ed.), *Advanced Techniques for Clay Mineral Analysis.* Developments in Sedimentology 34, Elsevier, Amsterdam, pp. 5–29.

Smykatz-Kloss, W., 1974, *Differential Thermal Analysis.* Springer-Verlag, Berlin, 185p.

Warne, S. St. J., and B. D. Mitchell, 1979, Variable atmosphere DTA in identification and determination of anhydrous carbonate minerals in soils. *Journal of Soil Science* 30:111–6.

Wendlandt, W. W., 1974, *Thermal Methods of Analysis,* 2d ed. Wiley, New York, 505p.

IR and Raman Spectroscopy

Bell, R. J., 1972, *Introductory Fourier Transform Spectroscopy.* Academic Press, New York.

Brodskii, I. A., and S. P. Zhdanov, 1980, Application of far infrared spectroscopy for a study of cation positions in zeolites. *Proceedings of the 5th International Conference on Zeolites,* pp. 234–41.

Farmer, V. C. (ed.), 1974, *The Infrared Specta of Minerals.* Mineralogical Society of London, London, 539p.

Fripiat, J. J., 1982, Application of far infrared spectroscopy to the study of clay minerals and zeolites. In J. J. Fripiat (ed.), *Advanced Techniques for Clay Mineral Analysis.* Developments in Sedimentology 34, Elsevier, Amsterdam, pp. 191–210.

Keller, W. D., 1950, The absorption of infrared radiation by clay minerals. *American Journal of Science* 248:264–73.

Launer, P. J., 1952, Regularities in the infrared spectra of silicate minerals. *American Mineralogist* 37:764–84.

Lyon, R. J. P., and W. M. Tuddenham, 1960, Determination of tetrahedra aluminium in mica by infrared absorption anlysis. *Nature* 185:374–5.

Nahin, P. G., 1955, Infrared analysis of clays and related minerals. *California Division of Mines Bulletin* 169:112–8.

Zussman, J., 1967, *Physical Methods in Determinative Mineralogy.* Academic Press, New York.

Mössbauer Spectroscopy

Bancroft, G. M., 1973, *Mössbauer Spectroscopy: An Introduction for Inorganic Chemists and Geochemists.* McGraw-Hill, New York, 252p.

Bancroft, G. M., and R. G. Burns, 1969, Mössbauer and absorption spectral study of alkali amphiboles. *Mineralogical Society of America Special Paper 2*, pp. 137–48.

Bancroft, G. M., T. K. Sham, C. Riddle, T. E., Smith, and A. Turek, 1977, Ferric/ferrous-iron ratios in bulk rock samples by Mössbauer spectroscopy—the determination of standard rock samples G-2, GA, W-1 and Mica-Fe. *Chemical Geology* 19:277–84.

Boulter, C. A., D. M. McConchie, and E. M. Clarke, 1989, Evidence from Mössbauer spectroscopy for intense localisation of metamorphic fluid flow during the formation of spaced cleavage. *Terra Nova* 1:365–9.

Childs, C. W., and J. G. Baker-Sherman, 1984, Mössbauer spectra and parameters of standard samples. *New Zealand Soil Bureau Scientific Report 66*, 50p.

Cohen, R. L. (ed.), 1976, *Applications of Mössbauer Spectroscopy*, vol. 1, Academic Press, New York, 349p.

Dibar-Ure, M. C., and P. A. Flinn, 1971, A technique for the removal of the "blackness" distortion of new geothermometer. *Earth and Planetary Science Letters* 16:346–54.

Hanstein, T., U. Hauser, F. Mbesherubusa, W. Neuwirth, and H. Spath, 1983, Dating of Western Australian laterites by means of Mössbauer spectroscopy. *Zeitschrift für Geomorphologie N.F.* 27:171–90.

Hogg, C. S., and R. E. Meads, 1970, The Mössbauer spectra of several micas and related minerals. *Mineralogical Magazine* 37:606–14.

Ingalls, R., 1964. The electric-field gradient tensor in ferrous compounds. *Physics Review* 113A:787–95.

Johnston, J. H., and D. G. Lewis, 1983, A detailed study of the trans-formation of ferrihydrite to hematite in an aqueous medium at 92°C. *Geochemica et Cosmochemica Acta* 47:1823–31.

Kündig, W., H. Bömmel, G. Constabaris, and R. H. Lindquist, 1966, Some properties of supported small α-Fe$_2$O$_3$ particles determined with the Mössbauer effect. *Physics Review* 142:327–33.

Lucas, H., M. Muggeridge, and D. M. McConchie, 1989, Iron in kimberlitic ilmenites and chromian spinels: A survey of analytical techniques. In J. Ross, A. L. Jacques, J. Ferguson, D. W. Green, S. Y. O'Reilly, R. V. Danchin, and A. J. A. Janse (eds.), *Kimberlites and Related Rocks: Proceedings of the 4th International Kimberlite Conference, Perth 1986*. Geological Society of Australia Special Publication 14, vol. 1, pp. 311–22.

McConchie, D. M., and C. B. Smith, 1989, Iron-oxides as palaeo-temperature indicators in Ellendale lamproite intrusions. In J. Ross, A. L. Jacques, J. Ferguson, D. W. Green, S. Y. O'Reilly, R. V. Danchin, and A. J. A. Janse (eds.), *Kimberlites and Related Rocks: Proceedings of the 4th International Kimberlite Conference, Perth 1986*. Geological Society of Australia Special Publication 14, vol. 1, pp. 520–7.

McConchie, D. M., J. B. Ward, V. H. McCann, and D. W. Lewis, 1979, A Mössbauer investigation of glauconite and its geological significance. *Clays and Clay Minerals* 27:339–48.

Rozenson, I., and L. Heller-Kallai, 1977, Mössbauer spectra of deoctahedral smectites. *Clays and Clay Minerals* 25:94–101.

Singhvi, A. K., D. K. Gupta, K. V. G. K. Gokhale, and G. N. Rao, 1974, Mössbauer spectra of ilmenites from primary and secondary sources. *Physica Status Solidi* 23:321–4.

Strens, R. G. J. (ed.), 1976, *The Physics and Chemistry of Minerals and Rocks*. Wiley, New York.

9
Chemical Composition

GENERAL CONSIDERATIONS

Choice of Analytical Technique

A wide variety of analytical techniques are used by sedimentologists to investigate the chemical composition of samples; here we only provide a basic introduction to the more common procedures. Good introductions to other of the less common, as well as the common, analytical techniques and instrumentation are provided in Olsen (1975); Bauer, Christian, and O'Reilly (1978); Fritz and Schenk (1979); Fripiat (1982); Albaiges (1982); Potts (1987); Clesceri, Greeberg, and Tressell (1989); see also the more specific references provided in the Selected Bibliography at the end of this chapter. Figures 9-1 and 9-2 provide examples of simple flowcharts showing how chemical analytical procedures can fit into sedimentological studies.

Ideally, the selection of a technique for a particular analysis should be based on the suitability of the method for the intended task, but in reality the selection is often based on whether or not the best equipment is available or can be either purchased or rented. Many sophisticated modern analytical instruments are expensive both to purchase and to operate. Consequently, when deciding which technique to use for a particular analysis, the best first step is to talk to other analysts doing similar work; out of necessity, many analysts have developed ingenious ways to get instruments to perform tasks beyond those for which they were designed. Additional considerations when selecting an instrument are whether the speed of operation, accuracy, precision, and sensitivity will be satisfactory for the task.

Precision refers to how repeatable the results are from a series of analyses of a particular sample; it is largely a function of instrumental stability and is usually expressed in terms of standard deviations (or as a standard error percent) from the mean. The concept of *accuracy* is different; accuracy refers to how closely the data obtained from the analysis approach the true (usually unknown) composition of the material analyzed. A set of analyses may have both high precision and high accuracy, whereas another set might have high precision and low accuracy. The accuracy of an analysis is not solely a function of the analytical equipment used; it will also depend on sampling procedure, preparation techniques, and the specific analytical methods employed. Accuracy is determined by the proximity of the results to the results of other precise methods applied to discover the same type of information. A guide to the likely accuracy of an analysis can be obtained by applying identical sample preparation and analytical procedures to a series of standards of known composition (e.g., the U.S. Geological Survey standard rock powders); the compositions of these and other geochemical standards have been determined from many analyses in many laboratories using multiple techniques (e.g., Flanagan 1973; Abbey 1983). Analysts are rightly concerned with achieving accurate results in the laboratory, but it must be remembered that in most geological work the greatest sources of error are associated with initial sampling, subsampling, sample preparation, and contamination related to these steps (see Chapters 3 and 5); there is no analytical instrument or procedure that can correct for poor sampling and sample preparation.

The *sensitivity* of an analytical technique refers to the smallest difference that can be quantitatively distinguished in the concentration of an element or compound between two samples. To determine sensitivity, a *blank* (i.e., a sample as similar as possible to that being studied, but containing none of the element or compound being analyzed) is compared with a made-up sample containing a very small amount of the element or compound to be analyzed; the results help to define the lower limit of detection (LLD) for the analytical procedure. There is considerable disagreement on how many standard deviations (determined from multiple analyses of

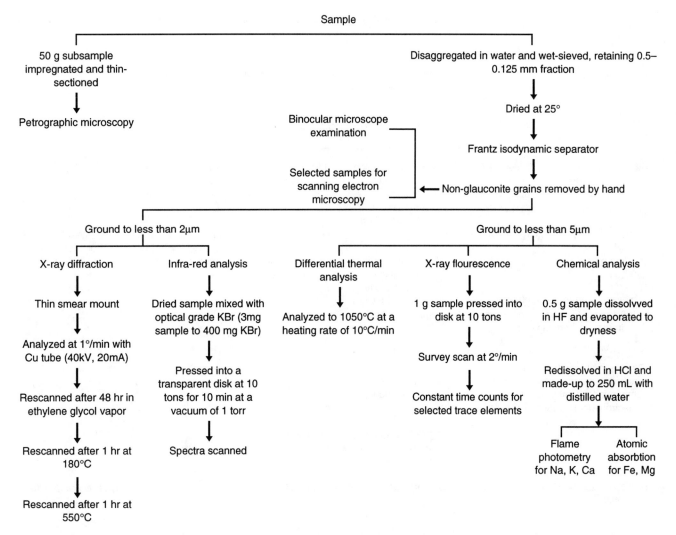

Figure 9-1.   Example of a flowchart showing the stages at which chemical compositional studies could be incorporated into a study of glauconite.

the blank) above the mean reading for the blank should be used to define the LLD (see discussion in Potts 1992). For most practical purposes, the LLD should be taken as the mean reading for the blank (10 repeat runs) plus 6 standard deviations (determined for these 10 runs), but when high confidence is required it may be better to use the mean reading for the blank plus 10 standard deviations. When specifying LLD values, it is important to note the type of sample to which they apply, because for many elements and compounds the LLD for a particular analytical technique will depend on the type of matrix to the analyte (i.e., what other elements or compounds are present and in what relative concentrations). A table comparing typical LLD values for a range of analytical techniques is given by Potts (1992).

Standards and Blanks

During the pretreatment stage of sample preparation (e.g., Chapter 5), thought should be given to the preparation of a blank and a series of standards. As a means to compensate for

any contamination (e.g., from fluxes, digestion reagents, grinding procedures, etc.), both standards and blanks should be prepared by exactly the same procedures as for the samples. The blank should contain all fluxes or digestion reagents that were used to prepare the samples, and they should be in the same proportions as were used in sample preparation. The only difference should be that the substance used should contain none of the analyte (e.g., spectrographically pure quartz for XRF studies).

International standard rock samples are the best standards for most analyses because they have the most accurately known compositions, but they are expensive and are not available in unlimited quantity. Hence, most laboratories develop their own internal standards that are calibrated against the international standards and are analyzed at several different laboratories; laboratories often exchange internal standard material and data as part of their normal quality-control program. For samples to be analyzed as liquid digests (e.g., in electrochemical, atomic absorption, or ICP analyses), artificially prepared standards are widely used. However, be-

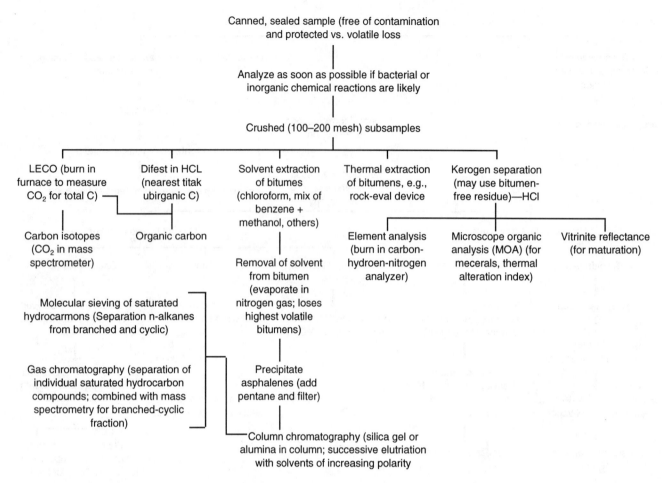

Figure 9-2.   Example of a flowchart showing the stages at which chemical compositional studies could be incorporated into a study of coal and bituminous sediments.

cause nonanalyte elements or compounds in the digest can result in suppression or enhancement of the value recorded for the analyte, the matrix for the synthetic standard must be as similar as possible to that of the sample. The best way to ensure that the matrix for the sample and the standard are as identical as possible is to standardize using *spike additions;* i.e., a very small and accurately known amount of the analyte is added to a subsample of the sample, then both the spiked sample and the sample are analyzed. Note that the method of spike addition is generally useful only if the calibration plot (the plot of instrument signal against analyte concentration) is linear over the range of concentrations that includes the sample and the spiked sample concentrations; the method can be used when the calibration plot follows a mathematically definable curve, but the calculations are no longer simple. When using spikes and the calibration plot is linear, the analyte concentration in the sample is given by

$$\text{concentration in sample} = \frac{\text{concentration in standard} \times Vb \times A}{(Va + Vb)\,(B - A)}$$

where A = the peak height for the sample

B = the peak height for the sample + spike mixture

Va = the volume of sample the spike was added to

Vb = the volume of spike added

LOSS ON IGNITION

Loss on ignition (LOI) is the mass of material lost by firing a sample, previously dried at 105°C, at a high temperature (usually to 1000°C for 1 hr). Determination of LOI is essential for X-ray fluorescence (XRF) analyses (see below) involving the use of fused discs (see Chapter 5) because the mass of analyzable sample material in the disc will have been reduced by the mass of material lost during the fusion process of forming the disc. LOI also can be used to obtain a rough guide to the amount of organic carbon in an organic matter–rich sample. However, it must be remembered that the LOI recorded actually results from the combined effects of loss of lattice water (H_2O^-), loss of any organic carbon, loss of CO_2 from carbonates, oxidization of any ferrous iron to ferric iron (a weight gain), possible loss of some sulfur, and the possible loss of other minor volatile components. Except as a necessary adjunct to XRF analyses of fused discs, we see little practical use for determinations of LOI because the contributory causes of the weight change cannot be quantified; if data on changes in the sample due to heating are required, it is better to consider using DTA or TGA methods (see Chapter 8).

ORGANIC AND INORGANIC CARBON

Total Carbon

The total carbon (both organic and carbonate carbon) in dried (105°C for 24 hr) rock and sediment samples is usually determined by combustion in an atmosphere of oxygen in either an induction or an equivalent high-temperature furnace, followed by determination of the amount of carbon dioxide evolved, either by gravimetric absorption (onto Ascarite™) or by infrared gas analysis (e.g., Fig. 9-3 and see Lee 1980). The Laboratory Equipment Corporation (LECO™) dominates the manufacturing of instruments for this work to the extent that the method is often referred to as LECO analysis. The equipment to carry out this work is not cheap, but it is efficient, provides good precision, and in our experience is easy to maintain and operate. Most instruments require only about 0.5 g of sample and many models are now automated.

Total Organic Carbon

Two common methods are used to determine the concentration of total organic carbon in sediment or soil samples (i.e., all carbon and carbonaceous matter present in the sediment, excluding carbon present as carbonate minerals). The first involves selective removal of inorganic (i.e., carbonate) carbon from the sample by digestion in 1 M acetic acid (dilute HCl can be used, but do not use an oxidizing acid such as nitric acid or some organic carbon may be lost), followed by determination of the carbon content of the remaining sediment by the oxidative combustion (LECO) method. Estimation of the organic carbon content by measuring the weight loss on firing the sample to temperatures above 550°C (most organic matter will combust between 450°C and 500°C) is not usually advisable for the reasons mentioned in the section on LOI.

The second method involves wet chemical oxidization of the organic carbon and back titration of the unused oxidant with a reducing agent. This method (Walkley and Black 1934), commonly referred to as the Walkley-Black method, provides results that are comparable with LECO analyses (e.g., Gaudette et al. 1974). A high calcium carbonate content in sediments may result in a slight overestimation of the organic carbon content (<2%; Gaudette et al. 1974). For marine sediments, the presence of chloride in salts remaining after drying also can cause an overestimation of the organic carbon content, but this effect can be overcome by either washing the salts from the sample prior to analysis (see Chapter 5) or by applying a chloride correction factor (Walkley 1947). The procedure for determining total organic carbon by wet chemical oxidation is described in Table 9-1.

The presence of oxidizable ferrous iron or manganese compounds in the sediment will result in an overestimation of the total organic carbon content in the sample, but if the sample has been oven-dried in air for 24–48 hr, most of the soluble reduced forms of Fe and Mn should have oxidized. The most likely problems when analyzing modern sediments

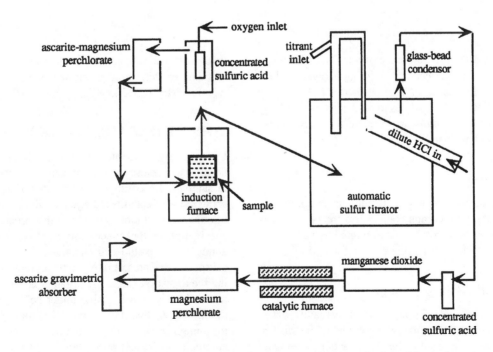

Figure 9-3.   Schematic diagram showing the principal components of instrument for determining total carbon and sulfur by inductive combustion with a titrimetric finish for sulfur and gravimetric determination of carbon.

Table 9-1.   Procedure for Determining Total Organic Carbon by Wet Chemical Oxidation

The analysis should be carried out in triplicate.

1. Grind about 2g of sample using a mortar and pestle so that the material will pass through a 0.5-mm sieve; weigh out between 0.1 g and 1.0 g of the fine material (use lower weights when you think the sample may have a high organic carbon content) and record the weights.

2. Place the weighed material into a 250-mL conical flask and add 10 mL of 0.1 M potassium dichromate solution; mix by gentle swirling for about 1 min. At this stage it is wise to prepare one or more blanks, which will contain all reagents except the sample, so that they will be ready to titrate at the same time as the test samples.

3. Add 20 mL of concentrated sulfuric acid and swirl gently. *Caution:* The reaction at this stage will generate considerable heat and splattering is possible, so ensure that the mouth of the flask is not pointed at yourself or anyone else. Carry out this operation in a fume cupboard.

4. Allow the mixture to stand on an asbestos mat for about 20–30 min, then carefully add c. 150 mL of distilled water, followed by 10 mL of 85% orthophosphoric acid.

5. Filter the mixture into a clean conical flask, and wash the filter with a small volume of distilled water.

6. Set up a titration unit and fill the burette with 0.1 M ferrous sulfate solution (this solution should be freshly prepared to ensure that no ferrous iron has already oxidized). Immediately before titrating, add 4 drops of ferroin indicator to each blank and sample, then titrate to the end point (the end point is indicated by a color change to red; the color change to red is preceded by the development of a bluish color); record the volume of titrant used for each titration. If the solution goes red as soon as the indicator is added, you have used too much sample and all the dichromate has been consumed so that there is none left with which to titrate the ferrous sulfate. *Be careful with the titration because the end point can be easily missed!* To help recognize the end point, it is wise to run a blank first—the equation in step 7 indicates that 10 mL of the 0.1-M dichromate will require 60 mL of the 0.1-M ferrous sulfate solution if none of the dichromate has reacted (i.e., there is no organic carbon in the sample).

7. Calculate the organic carbon content of the sample as

$$TOC\% = \frac{\left(mLCr_2O_7{}^{2-}\ used - [0.167 \times mL\ FeSO_4 \cdot 7H_2O]\right) \times 180}{Dry\ weight\ of\ sample\ (mg)}$$

If titration of the blank requires more or less than 60 mL of 0.1-M ferrous sulfate solution, either the dichromate or the ferrous sulfate solutions, or both, were not made up to exactly 0.1 M. In these circumstances, if you can be sure that the dichromate solution has the correct molarity, the calculation can be completed by replacing the value of 0.167 in the above equation by: (10)/(mL of ferrous sulfate used to reach the end point for the blank).

will be associated with the presence of metastable hydrated iron sulfides, which can decompose during analysis and release oxidizable ferrous iron. This problem can be largely overcome by allowing the sulfides to oxidize very slowly by storing the sample in a moist atmosphere for a few weeks before drying and analysis.

Inorganic Carbon

The inorganic carbon content of a sediment is commonly determined as the total carbon content less the organic carbon content, after both have been determined by LECO. Alternatively the procedure of Table 9-2 can be used.

TOTAL SULFUR

The total sulfur content of samples is most readily determined using the LECO oxidative combustion instrument coupled either to an infrared gas analyzer or to an automatic sulfur titrator using the iodide/iodate titration (see Lee 1980 and Fig. 9-3). This very efficient procedure has high sensitiv-

ity and precision if the equipment is available. An alternative method for determining oxidizable sulfur (primarily sulfides) is listed in Table 9-3.

TOTAL KJELDAHL NITROGEN

In many environmental studies involving sediments, it is necessary to determine the concentration of organic nitrogen in samples. Determination of organic nitrogen is achieved using the well-established Kjeldahl digestion, ammonia distillation, and titration procedure; this procedure permits determination of the concentration of organic nitrogen plus ammonia, but excludes nitrate, nitrite, and a few other chemical species that are not likely to be present in sediments. During the digestion stage, the organic nitrogen is converted to ammonium, which is stable under acid conditions. During the subsequent distillation stage, the ammonium is converted to ammonia by addition of excess sodium hydroxide, and the ammonia is distilled into a boric acid solution. In the final titration stage, the boric acid is titrated back to its starting pH with standardized sulfuric or hydrochloric acid to determine

Table 9-2.   Procedure for Laboratory Determination of the Inorganic Carbon Content of Samples

1. Crush predried sample material to <300 μm and weigh 1.0 g into a 250-mL Erlenmeyer flask. The analysis should be carried out in triplicate so weigh out three subsamples.

2. Add 50 mL of analytical grade water and 25 mL of standardized 1 M HCl to each flask.

3. Prepare a blank (water and acid only) and a pure (1 g) calcium carbonate reference.

4. Boil all flasks for 2 min, cool to room temperature, and add a few drops of phenolphthalein indicator.

5. Titrate the unused acid in the flasks with prestandardized 1 M sodium hydroxide solution. The blank should require 25 mL of sodium hydroxide solution and the calcium carbonate reference should require 5 mL; 1 g of pure calcite reacts with 20 mL of 1 M HCl.

6. Determine the volume of acid used as 25 mL – the volume of sodium hydroxide used.

7. Calculate the calcium carbonate equivalent and % carbonate CO_2 content of the sample as

$$\% \text{ CaCO}_3 \text{ equivalent} = 5 \times \text{volume of acid used (mL)}$$

and

$$\% \text{ carbonate CO}_2 = 2.2 \times \text{volume of acid used (mL)}$$

If greater sensitivity is required, more dilute sodium hydroxide can be used in the titration.

Table 9-3.   Procedure for Determining Oxidizable Sulfur (Mainly in Sulfide Form) in Samples

1. Shake a 5-g subsample of dried sample with 20 mL of analytical-grade water for 5 min, filter, then analyze the sulfate content of the water by ion chromatography (see below). The results of this step will enable the concentration of sulfate in the sample to be determined.

2. Wash the solids retained on the filter at step 1 into a beaker, add 40 mL of 10% hydrogen peroxide, and leave for 3–5 days to oxidize all remaining sulfides to sulfate.

3. Filter the mixture from step 2, make the liquid up to 50 mL, and analyze by ion chromatography.

4. Analyze a blank prepared using exactly the same amount of hydrogen peroxide as was used in step 2. This step is very important because commercial hydrogen peroxide is stabilized using a small amount of sulfuric acid.

5. Calculate the amount of sulfate formed by oxidation of oxidizable sulfur by subtracting the concentration in the blank from the concentration measured at step 3 and convert the sulfate concentration to a sulfide concentration by multiplying by 0.3338.

ment. Kjeldahl determinations can be made using standard laboratory equipment by the procedure set out in Table 9-4.

OILS AND GREASE

Total oils plus grease in a sediment is defined as that mass of organic material that can be extracted from the sediment by refluxing with chloroform under the analytical conditions specified in Table 9-5 (chlorofluorocarbon extraction has largely been abandoned because of the deleterious impact on the ozone layer). Determination of total oils and grease in both sediment and water is important in pollution monitor-

the amount of ammonia produced and thus the amount of nitrogen in the original sample. Kjeldahl digests can now be carried out using programmable digestion blocks, and a variety of instruments are on the market that automate the ammonia distillation and titration steps in the procedure; this automated equipment speeds up the whole process considerably, but it is an expensive approach to analyzing one ele-

Table 9-4.   Procedure for Determining the Total Kjeldahl Nitrogen Content of Sediments Using Standard Laboratory Equipment

1. Accurately weigh 1.0–5.0 g dried sample (crushed to pass through a 2-mm sieve) and place it in a 250-mL or larger Kjeldahl digestion flask (these are long-necked bulb-bottom flasks sold as Kjeldahl digestion flasks). Use less sample material when the organic nitrogen content is expected to be high.

2. Add a few glass beads, 10–15 mL of concentrated sulfuric acid, and either five commercial (1 g) Kjeldahl catalyst tablets or about 5 g of a catalyst mixture (prepared by thoroughly mixing 10 g selenium powder with 500 g potassium sulfate).

3. Heat the mixture in a fume hood for 15 min at 200–250°C, then for 1 hr at 380°C, then allow to cool. The digest should be colorless or have a clear pale color (usually greens or yellows) depending on trace components in the samples. Nitrogen is now present as ammonium sulfate.

4. Dilute the mixture by adding 100 mL of analytical-grade water (in case of splattering do not point the mouth of the flask toward yourself when making the dilution).

5. Set up a standard distillation unit, transfer the diluted digest to a round-bottom distillation flask, and immerse the condensate outlet in a 100-mL beaker containing about 30 mL of boric acid solution prepared by dissolving 20 g of analytical-grade boric acid in 1 L of analytical-grade water.

6. Place a few boiling chips to the distillation flask, then carefully but quickly (to avoid mixing and possible loss of ammonia) add about six times as much 40% (i.e., 40 g per 100 mL) sodium hydroxide solution as the volume of sulfuric acid added at step 2. Fit the flask to the distillation apparatus.

7. Measure the pH of the boric acid solution accurately and record the reading, then begin distillation and continue until about 50 mL of distillate have been collected. Remove the condensate outlet tube from the boric acid solution and turn off the distillation unit.

8. Place the beaker on a magnetic stirrer, add a stirring bar, insert a pH probe into the solution, turn on the stirrer, and titrate in prestandardized HCl (use 0.01 M HCl for low nitrogen samples and 0.1 M HCl for more nitrogen-rich samples) until the pH is restored to what it was before distillation commenced.

9. Repeat steps 1–8 for a blank such as pure quartz sand previously boiled in aqua regia and washed with analytical-grade water. *Caution!*

10. Calculate the concentration of nitrogen in the original sample as

$$N (\%) = \frac{1400 \times (\text{mL HCl used - mL HCl for the blank}) \times \text{molarity of HCl}}{\text{wt of digested sample (mg)}}$$

Table 9-5.   Procedure for Determining the Total Oils and Grease Content of Samples

1. Accurately weigh about 20 g of sediment into a round bottom flask, add 100 mL of analytical-grade water, and acidify to pH 0–1 with hydrochloric acid (a large volume of acid may be necessary for calcareous sediments).

2. Add 100 mL of chloroform, reflux for 30 min, and allow to cool.

3. When most of the sediment has settled, decant off the liquid and transfer it to a 500-mL separating funnel and allow the layers to separate.

4. Rinse the sediment in the round bottom flask with two 50-mL aliquots of chloroform and add this liquid to the separating funnel.

5. Filter the lower chloroform layer (leave the upper water layer in the separating funnel) in the separating funnel through a phase-separating filter paper (using diatomaceous earth as a filtering aid if necessary) into a preweighed quick-fit boiling flask.

6. Rinse the separating funnel and filter system with 50 mL of chloroform and add the liquid to the preweighed quick-fit boiling flask.

7. Recover as much of the chloroform as possible by distillation using a water bath at 70°C in a fume cupboard (a stream of warmed air will assist evaporation). *Caution:* Do not breathe the chloroform fumes that may be produced during this analysis.

8. Dry the preweighed quick-fit boiling flask in a drying oven for 4 hr at 110°C, then cool in a desiccator and reweigh accurately.

9. Calculate the concentration of total oils and grease in the sample as 100 times the weight gained by the boiling flask divided by the original sample weight.

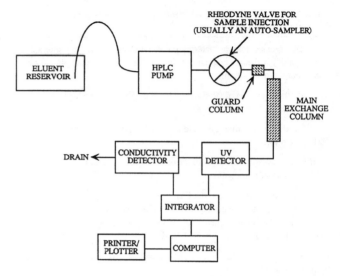

Figure 9-4.   Schematic diagram showing the principal components of an ion chromatograph.

ing, and most environmental protection bodies have set upper limits on acceptable oils and grease loadings.

ION CHROMATOGRAPHY

Principles

Ion chromatography, largely based on high-performance liquid chromatography (HPLC) but also using capillary electrophoresis, is effectively a highly controlled application of the same type of ion-exchange process with which sedimentologists working with clays and similar minerals will be familiar. In a basic HPLC system, a very stable high-pressure, low-volume pump drives an eluent through a column packed with an ion-exchange resin, on through a conductivity detector and/or a UV detector (an electrochemical detector is commonly used for transition metal ions) to a waste drain (e.g., Fig. 9-4). A small volume of sample solution (usually between 10 and 100 μL) is injected into the eluent stream just before the column through a rheodyne valve, which may (but need not) be part of an autosampler. The ions in the sample are adsorbed onto the exchange resin in the column and progressively removed by the eluent, with the smallest and lowest-charged ions coming off the column first (see Fig. 9-5). As the ions come off the column in the eluent stream, they are detected and the detector signal is passed either directly to a chart recorder, or to an integrator and on to a computer for further analysis. The time required for each type of ion to

come off the column will depend on the type of ion, the rate of eluent pumping, the composition and concentration of the eluent (particularly the pH), and the type of column. For a particular eluent and pumping rate, standards are used to indicate which ion comes off the column after what time interval since sample injection, and to provide the peak-height or area data that will be used to determine the concentrations of each ion type; both peak height and area increase in proportion to the amount of the ion in the sample.

The choice of chromatography column and eluent will depend on whether anions, groups 1 and 2 cations, or metal ions and rare earth elements are being analyzed; the concentration and composition of the eluent also will depend on the specifications of the column used and how much separation of the

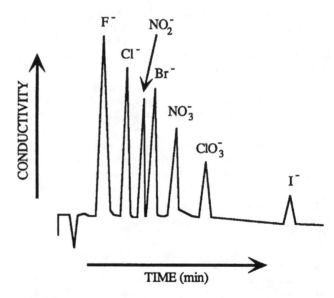

Figure 9-5.   Sketch of a chromatogram for anions determined by ion chromatography.

ions on the chromatogram is required (e.g., see papers in Jandik and Cassidy 1989). Column manufacturers normally provide information on eluent concentrations and compositions suited to their columns; this information provides a starting point for adjustments to suit the needs of a particular analytical task; books on ion chromatography such as Heftmann (1983a, 1983b), Lindsey (1987), Smith (1988), and Jandik and Cassidy (1989) provide further suggestions, and most publications involving ion chromatography routinely provide information on the type of column and eluent used.

A basic ion chromatograph is inexpensive relative to most analytical instruments; even elaborate systems can be established for under U.S. $30,000. Sophisticated instruments are fitted with chemical-suppressor systems that improve the resolution of the chromatogram and dual pumps that facilitate gradient chromatography (i.e., where the concentration or composition of the eluent is progressively changed to adjust the time at which different ions come off the column). Most modern instruments are fitted with computers for storing and analyzing chromatograms for samples and standards, and particularly when using an autosampler, computerization makes the whole analytical procedure very efficient. With an appropriate change of columns and eluents, an HPLC used for ion chromatography can be quickly adapted to analyze organic materials in sediments, but most users find their equipment is fully committed analyzing ions.

Ion chromatography requires very small sample volumes and the technique has an excellent sensitivity (most ions can be detected at part per billion concentrations). Hence, solutions of many geological materials (e.g., dissolved evaporite minerals) will require several dilutions before analysis, and care is required to ensure that these dilutions do not become a serious source of error.

Applications

Because wet chemical analysis of anions is such a tedious process, many analysts have tried to minimize or avoid anion analysis. However, with the ability of ion chromatography, which can analyze most anions simultaneously in about 15 min per sample, these analyses are being undertaken more commonly. In environmental sudies, analysis of nutrients (nitrates and phosphates) adsorbed onto soils and sediments is commonly achieved by ion chromatography, and in sedimentology the method is being increasingly applied to the determination of anions in evaporite minerals, brines, and groundwaters. Ion chromatography is rapidly becoming the dominant technique for anion analysis.

Ion chromatography can be used to analyze cations and metals, but these are more commonly analyzed by other techniques. However, in combination with electrochemical detection, it is being applied increasingly to rare-earth element analyses.

The main advantages of ion chromatography are its capacity to analyze many ions simultaneously in a short time; its high sensitivity for most ions; the reasonably low cost of

the equipment; and the fact that for anions it provides an alternative to tedious wet-chemical techniques. The main disadvantages are columns are not cheap and have to be replaced regularly; pumps seem to require a lot of maintenance; it is often difficult to analyze two ions that elute at similar times if one is much more concentrated than the other (i.e., the peak for the more concentrated ion may mask the peak for the other ion whether or not dilutions are used); and columns need to be flushed for a long time (ideally at least 12 hr) whenever the system has been shut down for a while, or the composition of the eluent has been changed.

ELECTROCHEMICAL METHODS

Principles

Electrochemical methods include Eh and pH determinations (discussed in Chapter 3), anodic-stripping voltammetry, cathodic-stripping voltammetry, square-wave voltammetry, and differential pulse polarography. These latter techniques are extremely sensitive (detection limits below 1 part in 10^{10} are possible for some elements), analyses are rapid, and in some configurations several elements can be analyzed simultaneously. Because it is not possible to examine all electrochemical techniques and component configurations here, the following discussion will focus on anodic-stripping voltammetry. This analytical method is finding rapidly increasing application in geology and environmental geochemistry because of the excellent sensitivity of the technique and because of the recent development of field-portable instruments (e.g., Mann and Lintern 1984a, 1984b) such as the PDV2000™ and TEA3000™ manufactured by Chemtronics™ in Western Australia. Discussion of other electrochemical techniques and operating procedures can be found in Florence (1986), Breyer and Gilbert (1987), Li and van den Berg (1989), Harman and Baranski (1990), Aydin and Tan (1991) and Van Den Berg (1991).

Anodic-stripping voltammetry (ASV; e.g., Vydra, Stulik, and Julakova 1976; Wang 1982; Mann and Lintern 1984b) essentially involves the concentration of metals onto a small electrode by electrodeposition (as in electroplating). Deposition spans a predetermined period of time, during which the working electrode is held at a negative potential relative to the reference electrode (the positively charged metal ions in solution are attracted to the electrode, where they accept electrons and deposit as metals; Fig 9-6). An electrolyte in the cell of the instrument provides a conductive medium between the electrodes, a buffer against extremes in pH and chloride ion concentration, a means of deoxygenating the cell, and a means of inhibiting changes in the oxidation state of metal ions in solution. Following the electrodeposition stage, the polarity of the electrodes is reversed and the voltage is progressively raised to sequentially strip the metals from the electrode. The voltage at which each element is stripped identifies the element, while the current flowing at that voltage is directly proportional to the amount of the ele-

PLATING

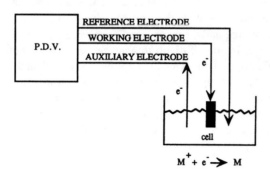

STRIPPING

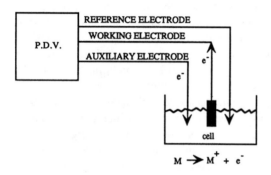

Figure 9-6.   Diagrammatic representation of the current flow directions and metal deposition/dissolution reactions during the plating and stripping cycles in anodic stripping voltammetry.

ment on the electrode, hence to the amount of element in the sample solution (e.g., Fig. 9-7); for most elements, calibration curves are nearly linear over about four orders of magnitude. The highly stable and precise voltage control required by the method is achieved by using a Ag/AgCl reference

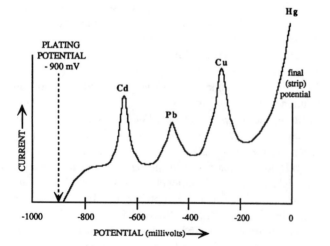

Figure 9-7.   Sketch of a typical voltammagram for copper, lead, and cadmium analyzed by anodic stripping voltammetry using a mercury-coated glassy carbon electrode.

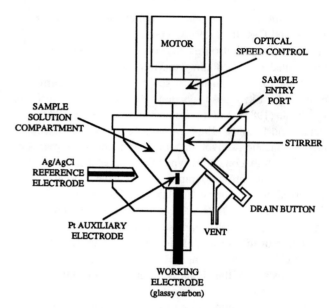

Figure 9-8.   Schematic diagram of the analytical cell in the Chemtronics™ (Western Australia) field-portable instrument for metal analysis by anodic stripping voltammetry.

electrode built into the analytical cell (e.g., for the field-portable instrument see Fig. 9-8).

The method has high precision and excellent sensitivity (e.g., Mann and Lintern 1984b; McConchie et al. 1988), but the number of metals that can be analyzed is limited. When hanging-mercury drop or mercury-coated glassy carbon working electrodes are used, metals that can be analyzed are restricted to those that form an amalgam with mercury and behave in a reversible manner. Other metals such as Au, Ag, and Hg (which have high reduction potentials) can also be analyzed, but they need to be plated directly onto an inert electrode (e.g., glassy carbon), and there is some loss in sensitivity. Common metals that can be analyzed by ASV include Ag, As, Au, Bi, Cd, Cu, Fe, Hg, Pb, Sb, Sn, Tl, and Zn; other metals can be analyzed by other similar electrochemical procedures.

It is possible to calibrate electrochemical instruments by analyzing a few artificial reference standards, but because the response of the working electrode will change slowly over time (particularly if an electrode with a thin mercury-film coating is being used), we have found that it is better to calibrate using the method of spike addition described at the beginning of this chapter. The use of spike-addition methods also compensates for many of the interferences that can affect some electrochemical analyses (e.g., the suppression of the zinc signal due to the formation of brass when copper is present).

Because electrochemical techniques are so sensitive, it is essential that all precautions be taken to avoid contamination of the cell of the instrument; in our experience, when poor results are obtained the problem can be traced to contamination, a general lack of cleanliness around the cell, a poorly maintained reference cell, or corrosion of electrical contacts caused by chemical spills.

Applications

Anodic-stripping voltammetry is applied to the determination of very low (to ultra-trace) concentrations of metals in sample digests and waters, and it offers the significant additional advantage over alternative analytical techniques of making it possible for the analyses to be carried out in the field. The capacity to undertake metal analyses in the field makes it much easier (and more cost effective) to trace sources of metals, whether they be from a pollutant discharge or from a natural geochemical anomaly such as an ore body. The equipment is reasonably priced, with a field-portable instrument costing about U.S. $15,000; it is easy to operate, but good results do demand considerable care during operation.

The major disadvantage of electrochemical methods is that they are limited in the range of elements they can analyze, although continuing research on electrodes, electrolytes, and chemical pretreatments is continuing to expand the range of both organic and inorganic analytes. Another limitation that can be annoying is the interference in analyses caused by some nonanalyte elements or chelating agents in sample solutions (e.g., Mann and Lintern 1984b); for many samples this interference can be overcome by calibrating using the method of spike addition, but where chelating agents (and some other organic compounds) are present, they usually need to be destroyed by acid decomposition before analysis.

X-RAY FLUORESCENCE

Principles

X-ray fluorescence (XRF) is probably the most widely used instrumental analytical technique for the chemical composition of rocks and minerals, primarily because it can analyze most elements of interest to acceptable limits of detection and because sample preparation (see Chapter 5) is much easier than for procedures requiring samples to be in solution. Figure 9-9 provides a schematic diagram of the basic parts of an XRF instrument.

The principles of wavelength-dispersive XRF (the most widely used method) are similar in many respects to those of XRD (see Chapter 8). In XRF instruments, crystals with known d spacings are used to determine the wavelengths of fluorescent X-rays emitted by elements in the sample; i.e., in Bragg's law n and d are known, the angle θ is measured, and λ is calculated. Common high-efficiency crystals for analyzing wavelengths include lithium fluoride 200 ($d = 4.028$Å); lithium fluoride 220 ($d = 2.848$Å); pentaerythritol ($d = 8.742$Å); ethylene diamine d-tartrate ($d = 8.804$Å); rubidium acid phthalate ($d = 26.1$Å); and thallium acid phthalate ($d = 25.75$Å). The most appropriate crystal for each element is selected on the basis of its d spacing and efficiency. Each element in a sample emits its own characteristic set of X-ray wavelengths, and the intensity of emission at each wavelength is proportional to the concentration of the emitting element in the sample (after correction for suppression or enhancement caused by the other elements in the sample and other atoms of the same element).

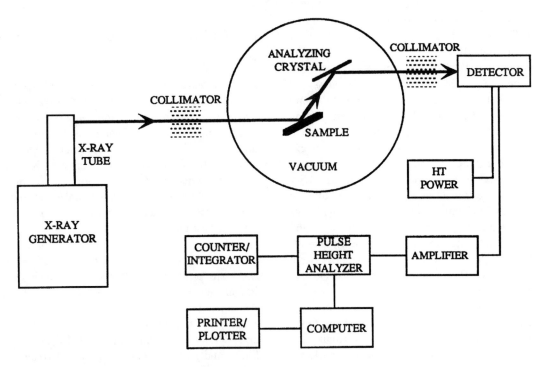

Figure 9-9.   Schematic diagram of the principal components in an X-ray fluorescence spectrometer.

The emission of the fluorescent (or secondary) X-rays from elements in the sample is induced by bombarding the sample with primary X-rays from an X-ray tube; the most widely used tubes have Ag, Au, Rh, Cr, or Mo anodes and are operated at 50–60 kV and 2.5–3.0 kW. Primary X-rays are used that have sufficient energy to knock electrons out of inner shells in the target elements; when electrons move to fill the vacancy in the inner shell, they lose energy by emitting secondary X-rays—the wavelength of which is determined to identify the element. The fluorescent X-rays are detected with an appropriate counter, usually either a scintillation counter (unsuitable for lighter elements that emit longer-wavelength X-rays) or a gas-flow proportional counter. The signal pulses from the counter are amplified, processed by a pulse-height analyzer (to exclude pulses not due to the X-rays being examined), a single-channel analyzer, then passed to a chart recorder and pulse counter (or, more commonly, to a computer for further processing). Almost all XRF instruments are now fitted for computer data processing, and most are equipped with an autosample changer.

Processing of X-ray fluorescence data is quite complex; complete processing requires a computer, because for major elements a correction matrix of suppression or enhancement factors must be developed to correct count rates (for each element) for the influence of the concentration of every other element present (including itself). For trace elements, mass-absorption corrections are usually determined using the Compton scattering technique (e.g., Nesbitt et al. 1976; Lee and McConchie 1982). For some elements, the analyzed wavelength coincides with a wavelength corresponding to another element, and correction for these overlap effects also is necessary. Fortunately most of the work is done by the computer once the appropriate correction factors have been determined by experimentation or from the literature (e.g., Lachance and Traill 1966; Rasberry and Heinrich 1974; Nesbitt 1976; Leroux and Thinh 1977; Lee and McConchie 1982). Calibration is usually based on a blank and a series of reference standards (e.g., USGS standard rock powders).

For most elements, XRF techniques have a minimum limit of detection around 10 ppm or less, but on most existing machines the method cannot analyze elements with atomic weights below fluorine, and there are some problems in examining many rare earth elements. The latest XRF instruments, with end window tubes and small high-sensitivity detectors, can analyze elements with atomic weights down to beryllium. Additional information on wavelength-dispersive XRF instrumentation and operation procedures can be found in many publications (e.g., Jenkins 1976; Bertin 1978; Herglotz and Birks 1978; Nisbet, Dietrich, and Esenwein 1979; Schroeder et al. 1980; Jenkins, Gould, and Gedcke 1981; Williams 1987; Potts 1992).

Energy-dispersive detection methods (e.g., Aiginger, Wobrauschek, and Brauner 1974; Giauque, Garrett, and Goda 1977; Gedcke, Elad, and Denee 1977; Furnas, Kuntz, and Furnas 1982; Johnson 1984; Potts, Webb, and Watson 1984; Potts 1992) are also being successfully applied in XRF

analyses. This approach permits the simultaneous determination of several elements, but it is less sensitive than wavelength-dispersive procedures (detection limits for most elements are above 100 ppm). An interesting extension of energy-dispersive XRF instruments is their combination with radioactive sources (e.g., Americium) to construct field-portable XRF spectrometers; although these instruments are less sensitive than laboratory energy-dispersive spectrometers, they are a highly desirable field tool.

Applications

XRF represents a quantitative tool for compositional analysis particularly suited to whole-rock studies and analyses of mineral separates; in modern geological laboratories, almost all whole-rock analyses are carried out by XRF. With the advent of fully automated equipment and computerized data processing, the analysis of samples by XRF has become almost routine; sample preparation (see Chapter 5) also is relatively simple. Modern instruments seldom break down, provided they have a stable power supply and are given the benefit of regular maintenance and cleaning.

The main limitation to XRF instruments is their very high capital cost (over U.S. $200,000 for a basic instrument). Instruments in general use are also limited by their inability to analyze elements with an atomic weight less than fluorine, but the latest instruments largely overcome this problem. There is some difficulty in analyzing some higher-molecular-weight trace elements, but overall the limitations on XRF are minor when compared with the range of components that can be analyzed and the rapid rate at which the analyses can be performed.

ELECTRON MICROPROBE AND EDAX

Principles

Electron microprobe (EPMA) equipment basically involves a combination of an electron microscope and an X-ray fluorescence spectrometer (see Figs. 9-9, 9-10). In these instruments, generation of the fluorescent X-rays is caused by the high-energy electron beam of the microscope striking atoms in the sample. Detection of the fluorescent X-rays is either by wavelength dispersion (a standard electron microprobe) or by energy dispersion (the EDS or EDAX systems). Processing of X-ray count data involves similar procedures to those for XRF instruments.

A large volume of literature exists on electron microbeam techniques involving X-ray analysis; good introductions are provided by Bence and Albee (1968), Boyd, Finger, and Chayes (1969), Sweatman and Long (1969), Andersen (1973), Gould and Healey (1975), Reed and Ware (1975), Dunham and Wilkinson (1978), Robinson and Nickel (1979), Hall and Lloyd (1981), Heinrich (1981), Smith and Reed (1981), Tovey and Krinsley (1991), and Potts (1992).

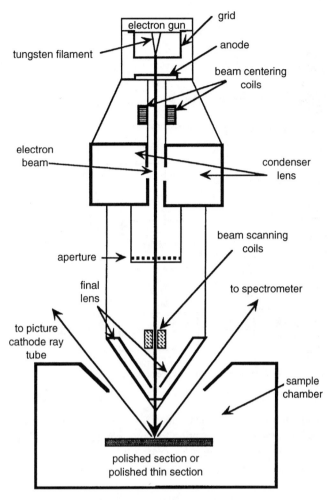

electron gun

grid

tungsten filament

anode

beam centering coils

electron beam

condenser lens

beam scanning coils

aperture

final lens

to spectrometer

to picture cathode ray tube

sample chamber

polished section or polished thin section

Figure 9-10. Schematic diagram of the principal components in an electron microprobe. The arrow marked "to spectrometer" refers to either a wavelength-dispersive X-ray fluorescence spectrometer (i.e., the arrow would go to the analyzing crystal in Fig. 9-9) or an energy-dispersive X-ray fluorescence spectrometer.

Applications

EPMA techniques are a powerful tool for mineralogical analysis, and because only a small area of the sample is emitting X-rays at any one time, they can be used to analyze the composition of inclusions in minerals, exsolution lamellae or zoning within single crystals, etc. In sedimentology, the small spot size is very useful for examining microfracture fillings, diagenetic overgrowths, and cements. If the beam is moved in scanning mode across the sample, it is also possible to make an element distribution map for selected areas of polished sections and polished thin sections. The principal advantage of these procedures is their ability to provide compositional analyses of very small areas of sample, and to allow the analyzed areas to be positioned accurately on an electron optical image. Features seen on samples studied by electron microscopy can be compositionally characterized,

and semi-quantitative analyses can be performed on parts of grains and on other nonpolished samples studied by scanning electron microscopy.

The small spot size analyzed also constitutes a limitation on the EPMA technique because it is difficult to obtain analyses of large areas and virtually impossible to obtain whole-rock analyses without analyzing an unrealistically large number of points. A further limitation on microbeam analyses is that results can be affected by inhomogeneities in the sample (e.g., minute fractures or inclusions, or a thin layer of one mineral over another such as might occur near crystal boundaries) that may not be observed by the analyst. Other advantages and limitations that apply to XRF instruments and procedures also apply to microbeam techniques involving X-ray analysis.

ATOMIC ABSORPTION

Principles

Until the recent development of efficient inductively coupled plasma (ICP) instruments, the atomic absorption (AA) technique was the most widely used technique for analyzing most elements in water or digest samples. In atomic absorption analyses, a lamp with a cathode that is made of (or contains a significant quantity of) the element being investigated is used as a light source. The sample to be studied, which must be in solution, is sucked up by an aspirator and fired through a burner (mixed with the fuel C_3H_6 or C_2H_2 and oxidant N_2O or air) such that the light from the source lamp passes through the flame (e.g., Fig. 9-11). A monochromator allows a single spectral line from the source to be examined. The settings (wavelength, lamp current, slit width, fuel and oxidant type, and flame condition—oxidizing or reducing) that should be used for each element are usually provided by the instrument manufacturer. The amount by which the intensity of the selected spectral line from the lamp element is reduced when the sample is fed into the burner is proportional to the concentration of the element in the sample.

The concentration of the element being analyzed is determined from a calibration curve prepared using a series of standards and blanks; the use of spiked samples is also a useful way to compensate for absorption or enhancement effects due to the composition of the solution containing the sample. Because the calibration curve for many elements is not linear over a very wide range of concentrations, it is necessary to use dilutions to adjust the concentration of both standards and samples so that they fall in or near the optimum working range of the instrument. Most modern AA spectrophotometers are fully automated, with microprocessor control that permits rapid processing of samples and internal calculation of element concentrations.

The AA instrument has a good lower limit of detection (sub-ppm concentrations are detectable for most elements), is quite stable, and can be used to analyze most elements in the periodic table quickly and reliably. If a graphite furnace is

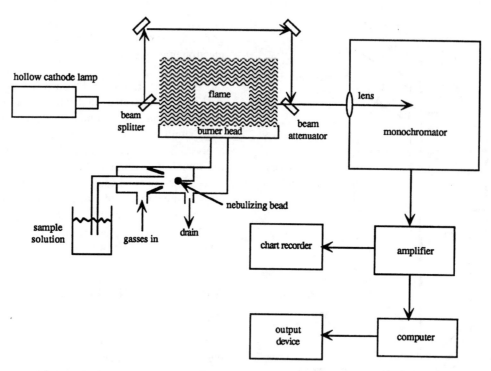

Figure 9-11.   Schematic diagram showing the principal components in an atomic absorption spectrometer.

used in the instrument instead of a flame, detection limits down to ppb levels (sub-ppb for many elements) can be readily achieved. For elements such as Hg (flameless atomic absorption) and As, which form hydrides, low detection limits can be achieved using a hydride generator. Additional information on AA equipment and procedures can be found in Angino and Billings (1967), Oslen (1975), Salvin (1978), Aklemade and Herrman (1979), Price (1979), Van Loon (1980), and Varma (1984).

Applications

AA methods are still widely used for the analysis of water samples and for total and selective digests (see Chapter 5) of geological and biological material; the equipment is much less expensive than ICP equipment and much less costly to operate. Atomic absorption instruments are easy to operate, and once the instrument is operational (lamp warmed up, etc.) it is possible to analyze over fifty samples per hour by hand (even more with an autosampler). Lower limits of detection are good for most elements (there are some elements that cannot be analyzed by AA and others that can be detected only at high concentration), particularly if a graphite furnace is used and instrumental stability is high.

The main disadvantages of AA methods are that a separate lamp is needed for each element analyzed (a few combination lamps are available) and the sample must be in solution. Although the need for samples to be dissolved is a disadvantage in some respects, it is an advantage if selective extractions are being used to obtain element speciation data. Matrix effects, due to other elements in the sample solutions,

can interfere with the absorption response for some elements as can nonatomic absorption for some elements such as lead; interference can cause a lower or a higher absorption to be recorded depending on the cause.

FLAME PHOTOMETRY

Principles

Flame photometry is one of the simplest analytical procedures used in earth sciences; basic instruments (see Fig. 9-12) are easy to operate and cheap (even for instruments with autosamplers and microprocessor control). In flame photometry, the intensity of the spectral emissions from the sample is examined directly, as opposed to AA methods; the spectral lines are examined using filters, a prism, or a diffraction grating. Sample solutions are aspirated through a flame (usually an air/propane mix), as they are in AA analysis, and atoms excited by the flame emit light, from which the appropriate wavelengths pass through the filter (or are selected by the prism or diffraction grating); those wavelengths then pass to a detector and intensity recording and display device. Because many elements emit a very large number of spectral lines, spectral overlap is the biggest problem affecting flame photometry; computerized iterative application of appropriate correction factors is necessary to enable analysis of most elements (except those of groups I and II on the periodic table).

Calibration is achieved using a series of standards and a blank for each element analyzed, and to check for unexpected interferences and spectral line overlaps it is wise to

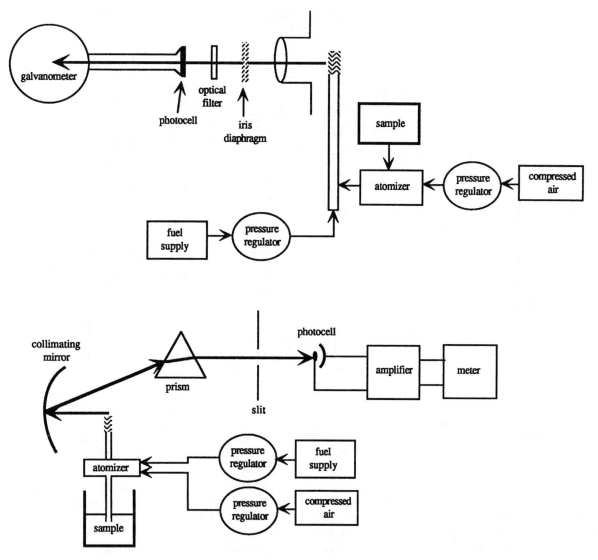

Figure 9-12.   Schematic diagram showing the principal components in a filter flame photometer (upper figure) and a prism flame photometer (lower figure).

run several spiked samples; element concentrations are determined from the calibration plot.

Applications

Flame photometry is usually used in geology only to analyze Li, Na, Ca, K, Sr, and Rb. For these elements it is chosen in preference to flame AA because it is more sensitive and stable and the calibration plot is more linear over a greater range of concentration. Sophisticated flame photometers can be used to analyze other elements, but full computerization is generally necessary to correct the readings for all possible spectral line overlaps. Flame photometry is susceptible to more inter-element interference and spectral line overlap than AA methods; hence there is a greater need for verification with spiked samples. In our experience, if flame photometers are kept clean they require little maintenance and seldom break down.

Because samples must be in solution, flame photometry has similar advantages and disadvantages to AA methods.

ICP AND ICP-MS

Principles

There are two variations on the inductively coupled plasma (ICP) theme: (1) ICP-optical emission spectrometry (ICP-AES) and (2) ICP-mass spectrometry (ICP-MS; Fig. 9-13). In both approaches, samples (which must be in solution) are aspirated in an argon stream through high-energy radio-frequency coils to form a plasma with a temperature about 10,000 K. In ICP-AES the wavelengths emitted by sample elements excited in the plasma are split up using a diffraction grating and detected by a photomultiplier tube; the pulses from the photomultiplier are analyzed by computer. Com-

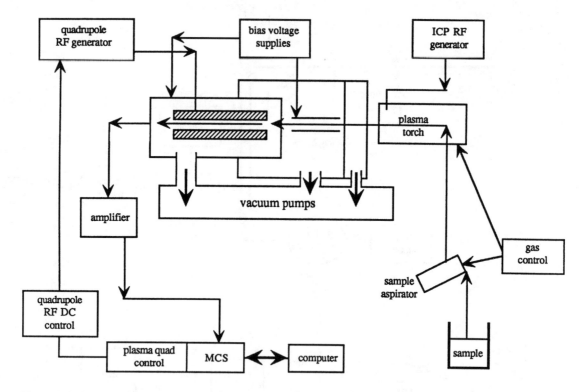

Figure 9-13.   Schematic diagram showing the principal components in an ICP-MS instrument.

puter analysis is essential because so many emission lines are produced that complex mathematical corrections for spectral line overlaps are required. Careful selection of which wavelengths to analyze for each element can help reduce the problems caused by spectral overlaps. ICP-AES has good detection limits (to sub-ppm concentrations for most elements) and for most elements (particularly the more refractory elements) it is better than AA procedures, but for some elements such as Na and K better detection limits can still be obtained by AA.

In ICP-MS the ions in the plasma are analyzed with a quadrupole mass spectrometer instead of their emission wavelengths being analyzed as in ICP-AES. Because very low detection limits (sub-ppb concentrations for most elements) can be achieved using ICP-MS, and multielement analyses can be performed rapidly, it is likely that ICP-MS will become a standard instrument for environmental and exploration geochemistry in the near future. However, with the significant increase in sensitivity provided by ICP-MS there is a need, as there is with electrochemical techniques, for extra caution to avoid sample contamination during preparation and analysis. For further details of principles, equipment, and applications, see Date and Gray (1988), Thompson and Walsh (1988), and Jarvis, Gray, and Houk (1992).

Applications

ICP is rapidly becoming the most widely used technique for analyzing elements in solution; it provides good detection limits for most elements and excellent detection limits if ICP-

mass spectrometry is used. The advantage of ICP methods over AA methods (which are the main rival) include increased sensitivity for most elements, faster analyses, the lack of a need to change lamps between elements, and the ability to analyze some elements that cannot be analyzed (or are difficult to analyze) by AA. On the negative side, the price of an ICP instrument is about three to four times as much as a reasonable AA instrument; ICP-MS instruments cost about 50% more than ICP-AES instruments, but the price of both types has fallen substantially over the past few years. ICP instruments also are expensive to run because they have a high argon consumption and require more comprehensive routine maintenance than most other analytical instruments.

SPECTROPHOTOMETRY

Principles

The spectrophotometer uses a prism or a diffraction grating to select a particular wavelength of light from a source lamp, then measures the amount of light absorbed by a sample solution placed in the light beam. The absorption is measured relative to the absorption for a blank placed in a reference beam (double-beam spectrophotometers [Fig. 9-14]), or relative to zero absorbance set using a blank in place of a sample (single-beam spectrophotometers).

Applications

In sedimentology, the spectrophotometer is essentially used only to analyze nitrate, nitrite, and phosphate (see Table 9-6

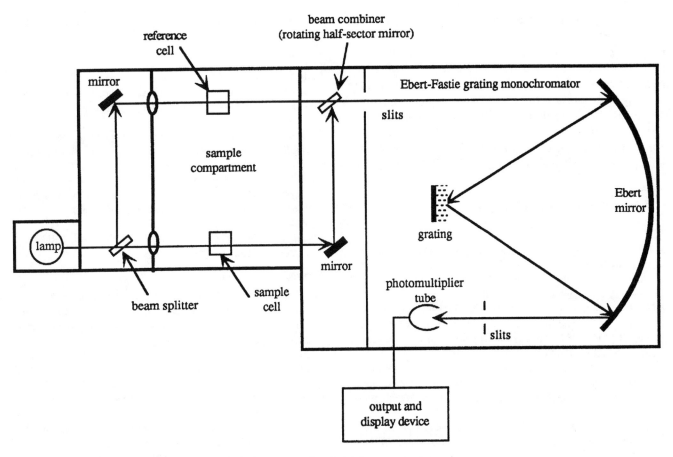

Figure 9-14.   Schematic diagram showing the principal components in a double-beam spectrophotometer.

Table 9-6.   Procedure for the Spectrophotometric Determination of Phosphate as the Phosphomolybdate Blue Complex

Phosphate concentrations can be determined spectrophotometrically at 882 nm by forming the phosphomolybdate blue complex. The exact chemistry of this complex is not yet known, but it is known that it can be used to analyze phosphate concentrations with high sensitivity, precision, and accuracy relatively easily.

For 100 mL of the phosphate reagent. Mix the following:

1. 15 mL of ammonium molybdate solution (4 g of ammonium molydate in 100 mL)

2. 50 mL of 10 M H_2SO_4 (70 mL of concentrated H_2SO_4 in 500 mL)

3. 30 mL of ascorbic acid solution (1.76 g ascorbic acid in 100 mL—this goes off rapidly if stored in light or for long periods; if it has gone yellow, throw it away and make a new lot)

4. 5 mL of potassium antimony tartrate solution (1.3715 g potassium antimony tartrate in 500 mL)

Procedure

1. Filter 20 mL of the sample through a membrane filter into a test tube. Add 1 drop phenophthalein indicator. If a red color develops, add dilute H_2SO_4 by drops until the color just disappears.

2. Set up an ultra-pure water blank and a range of standards (10–1000 µg/L). Add 20 mL of blank and each of the standards to separate test tubes.

3. Add 4 mL of phosphate reagent to each test tube (samples, standards, and blank) and mix thoroughly.

4. Analyze spectrophotometrically at 882 nm after 15 min, but before 1 hr. Use 1- or 5-cm sample cells depending on the concentration expected. Zero the spectrophotometer using the blank.

5. Construct a calibration plot using the standards and determine the phosphate concentration in the samples; samples with a high phosphate content may need to be diluted to fall in range.

for the phosphate procedure, and Clesceri, Greeberg, and Tressell 1989 for nitrate and nitrite procedures). These compounds are measured in water samples and as selective extracts from sediments where other ions in high concentration make ion chromatography difficult. The equipment is simple and easy to operate, and spectrophotometers suitable for sedimentological work are inexpensive; however, sample preparation can be time consuming (particularly for nitrate). For nitrate, nitrite, and phosphate it is possible to detect concentrations around the low ppb range using a 5-cm sample cell.

SELECTED BIBLIOGRAPHY

General

Abbey, S., 1983, Studies in "standard samples" of silicate rocks and minerals 1969–1982. *Geological Survey of Canada Paper 83-15*.

Albaiges, J., (ed.), 1982, *Analytical Techniques in Environmental Chemistry*. Pergamon Press, Oxford.

Bauer, H. H., G. D. Christian, and J. E. O'Reilly, 1978, *Instrumental Analysis*. Allyn and Bacon, Inc., Boston, 832p.

Braun, R. D., 1987, *Introduction to Instrumental Analysis*. McGraw-Hill, New York.

Clesceri, L. S., A. E. Greeberg, and R. R. Tressell (eds.), 1989, *Standard Methods for the Examination of Water and Wastewater*, 17th ed. American Public Health Association, Washington, D.C., 1466p.

Flanagan, F. J., 1973, 1972 values for international geochemical reference samples. *Geochimica et Cosmochimica Acta* 37:1189–1200.

Fripiat, J. J., (ed.), 1982, *Advanced Techniques for Clay Mineral Analysis*. Developments in Sedimentology 34, Elsevier, Amsterdam, 234p.

Fritz, J. S., and G. H. Schenk, 1979, *Quantitative Analytical Chemistry*, 4th ed. Allyn and Bacon, Boston, 661p.

Gasshoff, K., M. Ehrhardt, and K. Kremling, 1983, *Methods of Seawater Analysis*, 2d ed. Verlag Chemie, Germany, 419p.

Gaudette, H. E., W. R. Flight, L. Toner, and D. W. Folger, 1974, An inexpensive titration method for the determination of organic carbon in recent sediments. *Journal of Sedimentary Petrology* 44:249–53.

Krom, M. D., and R. A. Berner, 1983, A rapid method for the determination of organic and carbonate carbon in geological samples. *Journal of Sedimentary Petrology* 53:660–3.

Lee, R. F., 1980, Simultaneous determination of carbon and sulphur in geological material, using inductive combustion. *Chemical Geology* 31:145–51.

Harris, D. C., 1987, *Quantitative Chemical Analysis*. Freeman & Co., New York, 818p.

Olsen, E. G., 1975, *Modern Optical Methods of Analysis*. McGraw-Hill, New York.

Potts, P. J., 1992, *A Handbook of Silicate Rock Analysis*. Blackie & Son Ltd., Glasgow, 622p.

Van Loon, J. C., 1982, *Chemical Analysis of Inorganic Constituents of Water*, vol. 1. CRC Press, Boca Raton, FL, 248p.

Walkley, A., 1947, A critical examination of a rapid method for determining organic carbon in soils—Effect of variations in digestion conditions and of inorganic soil constituents. *Soil Science* 63:251–64.

Walkley, A., and I. A. Black, 1934, An examination of the Degtjareff method for determining soil organic matter and a proposed modification of the chromic acid titration method. *Soil Science* 37:29–38.

Ion Chromatography

Heftmann, E. (ed.), 1983a, *Chromatography: Fundamentals and Applications of Chromatographic and Electrophoretic Methods. Part A: Fundamentals and Techniques*. Elsevier, Amsterdam, 388p.

Heftmann, E. (ed.), 1983b, *Chromatography: Fundamentals and Applications of Chromatographic and Electrophoretic Methods. Part B: Applications*. Elsevier, Amsterdam, 564p.

Jandik, P., and R. M. Cassidy (eds.), 1989, *Advances in Ion Chromatography, vol. 1*. Century International Inc., Franklin, Mass. 523p.

Lindsey, S., 1987, *High Performance Liquid Chromatography*. Wiley, New York, 244p.

Smith, R., 1988, *Ion Chromatography Applications*. CRC Press Inc., Boca Raton, FL, 177p.

Electrochemical Methods

Adeloju, S. B., and K. A. Brown, 1987, Determination of ultra-trace amounts of cadmium in natural waters by the combination of a solvent extraction procedure and anodic stripping voltammetry. *Analyst* 112:221–6.

Aydin, H., and G. H. Tan, 1991, Differential-pulse polarographic behavior of selenium in the presence of copper, cadmium, and lead. *Analyst* 116:941–5.

Breyer, P., and B. P. Gilbert, 1987, Determination of very low levels of selenium(4) in sea water by differential-pulse cathodic stripping voltammetry after extraction of the 3,3-diaminobenzidine piazselenol. *Analytica Chimica Acta* 201:33–41.

Florence, T. M., 1970, Anodic stripping voltammetry with a glassy carbon electrode mercury plated in situ. *Journal of Electroanalytical Chemistry* 27:273–81.

Florence, T. M., 1986, Electrochemical approaches to trace element speciation in waters: A review. *Analyst* 111:489–505.

Harman, A. R., and A. S. Baranski, 1990, Fast cathodic stripping analysis with ultramicroelectrodes. *Analytica Chimica Acta* 239:35–44.

Li, H., and C. M. G. van den Berg, 1989, Determination of titanium in sea water using adsorptive cathodic stripping voltammetry. *Analytica Chimica Acta* 221:269–77.

Mann, A. W., and M. Lintern, 1984a, Field analysis of heavy metals by portable digital voltammeter. *Journal of Geochemical Exploration* 22:333–48.

Mann, A. W., and M. Lintern, 1984b, Portable digital voltammeter for field analysis of trace metals. *Australian Water Resources Council Technical Paper 83*, 80p.

McConchie, D. M., A. W. Mann, M. J. Lintern, D. Longman, V. Talbot, A. J. Gabelish, and M. J. Gabelish, 1988, Heavy metals in marine biota, sediments, and waters from the Shark Bay area, Western Australia. *Journal of Coastal Research* 4:51–72.

Van Den Berg, C. M. G., 1991, Potentials and potentialities of cathodic stripping voltammetry of trace elements in natural waters. *Analytica Chimica Acta* 250:265–76.

Van Den Berg, C. M. G., and G. S. Jacinto, 1988, The determination of platinum in sea water by adsorptive cathodic stripping voltammetry. *Analytica Chimica Acta* 211:129–39.

Vydra, F., K. Stulik, and E. Julakova, 1976, *Electrochemical Stripping Analysis*. Ellis Horwood Ltd., Sussex, 283p.

Wang, J., 1982, Anodic stripping voltammetry as an analytical tool. *Environmental Science and Technology* 16:104–9.

Warner, M., 1987, Trace analysis. *Analytical Chemistry* 59:1311–4.

X-Ray Fluorescence

Aiginger, H., P. Wobrauschek, and C. Brauner, 1974, Energy-dispersive fluorescence analysis using Bragg-reflected polarized X-rays. *Nuclear Instrumental Methods* 120:541–2.

Bertin, E. P, 1978, *Introduction to X-ray Spectrometric Analysis.* Plenum Press, New York, 485p.

Furnas, T. C., G. S. Kuntz, and R. E. Furnas, 1982, Toroidal monochromators in hybrid XRF system improve effectiveness of ED-XRF tenfold. *Advances in X-ray Analysis* 25:59–62.

Gedcke, D. A., E. Elad, and P. B. Denee, 1977, An intercomparison of trace element excitation methods for energy dispersive fluorescence analyzers. *X-ray Spectrometry* 6:2–29.

Giauque, R. D., R. B. Garrett, and L. Y. Goda, 1977, Energy dispersive x-ray fluorescence spectrometry for determination of twenty-six trace and two major elements in geochemical specimens. *Analytical Chemistry* 49:62–7.

Herglotz, H. K., and L. S. Birks (eds.), 1978, *X-ray Spectrometry.* Marcel Dekker, New York, 518p.

Jenkins, R., 1976, *An Introduction to X-ray Spectrometry.* Heyden, London.

Jenkins, R., R. W. Gould, and D. Gedcke, 1981, *Quantitative X-ray Spectrometry.* Marcel Dekker, New York.

Johnson, R. G., 1984, Trace element analysis of silicates by means of energy-dispersive X-ray spectrometry. *X-ray Spectrometry* 13:64–8.

Lachance, G. R., and R. J. Traill, 1966, Practical solution to the matrix problem in X-ray analysis. *Canadian Spectroscopy* 11:43–8.

Lee, R. F., and D. M. McConchie, 1982, Comprehensive major and trace element analysis of geological material by X-ray flourescence using low dilution fusion. *X-ray Spectrometry* 11:55–63.

Leroux, J., and T. P. Thinh, 1977, *Revised Tables of X-ray Mass Attenuation Coefficients.* Corporation Scientifique Claisse, Quebec.

Nesbitt, R. W., H. Mastins, G. W. Stolz, and D. R. Bruce, 1976, Matrix corrections in trace element analysis by X-ray fluorescence: An extension of the Compton scattering technique to long wavelengths. *Chemical Geology* 18:203–13.

Nisbet, E. G., V. J. Dietrich, and A. Esenwein, 1979, Routine trace element determination in silicate minerals and rocks by X-ray fluorescence. *Fortschr. Miner.* 57:264–79.

Potts, P. J., P. C. Webb, and J. S. Watson, 1984, Energy dispersive X-ray fluorescence analysis of silicate rocks for major and trace elements. *X-ray Spectrometry* 13:2–15.

Potts, P. J., P. C. Webb, and J. S. Watson, 1985, Energy dispersive X-ray fluorescence analysis of silicate rocks: comparisons with wavelength-dispersive performance. *Analyst* 110:507–13.

Rasberry, S. D., and K. F. J. Heinrich, 1974, Calibration for interelement effects in X-ray fluorescence analysis. *Analytical Chemistry* 46:81–9.

Schroeder, B., G. Thompson, M. Sulanowska, and J. N. Ludden, 1980, Analysis of geological materials using an automated X-ray fluorescence system. *X-ray Spectrometry* 9:198–205.

Williams, K. L., 1987, *An Introduction to X-ray Spectrometry.* Allen & Unwin, London.

Electron Microprobe

Andersen, C. A. (ed.), 1973, *Microprobe Analysis.* John Wiley and Sons, New York.

Bence, A. E., and A. L. Albee, 1968, Empirical correction factors for the electron microanalysis of silicates and oxides. *Journal of Geology* 76:382–403.

Boyd, F. R., L. W. Finger, and F. Chayes, 1969, Computer reduction of electron probe data. *Carnegie Institute of Washington Yearbook* 67:210–5.

Dunham, A. C., and F. C. F. Wilkinson, 1978, Accuracy, precision and detection limits of energy dispersive electron microprobe analysis of silicates. *X-ray Spectrometry* 7:50–6.

Gould, R. W., and J. T. Healey, 1975, Secondary fluorescent excitation in the scanning electron microscope: Improved sensitivity of energy dispersive analysis. *Review of Scientific Instruments* 46:1427–8.

Hall, M. G., and G. E. Lloyd, 1981, The SEM examination of geological samples with a semiconductor backscatter electron detector. *American Mineralogist* 66:362–8.

Heinrich, K. F. J., 1981, *Electron Beam X-ray Microanalysis..* Van Nostrand Reinhold, New York.

Reed, S. J. B., and N. G. Ware, 1975, Quantitative electron microprobe analysis of silicates using energy dispersive X-ray spectrometry. *Journal of Petrology* 16:499–519.

Robinson, B. W., and E. H. Nickel, 1979, A useful new technique for mineralogy: The backscattered-electron/low vacuum mode of SEM operation. *American Mineralogist* 64:1322–8.

Smith, D. G. W., and S. J. B. Reed, 1981, The calculation of background in wavelength dispersive electron microprobe analysis. *X-ray Spectrometry* 10:198–202.

Sweatman, T. R., and J. V. P. Long, 1969, Quantitative electron probe microanalysis of rock forming minerals. *Journal of Petrology* 10:332–379.

Tovey, N. K., and D. H. Krinsley, 1991, Mineralogical mapping of scanning electron micrographs. *Sedimentary Geology* 75:109–23.

Atomic Absorption and Flame Photometry

Aklemade, C. T. J., and R. Herrman, 1979, *Fundamentals of Analytical Flame Spectroscopy.* Adam Hilger Ltd., Bristol, U.K., 441p.

Angino, F. F., and G. K. Billings, 1967, *Atomic Absorption Spectrometry in Geology.* Elsevier, Amsterdam.

Oslen, E. G., 1975, *Modern Optical Methods of Analysis.* McGraw-Hill, New York.

Price, W. J., 1979, *Spectrochemical Analysis by Atomic Absorption.* Heyden and Son, London, 392p.

Salvin, W., 1978, *Atomic Absorption Spectrometry,* 2d ed. Wiley, New York.

Van Loon, J. C., 1980, *Analytical Atomic Absorption Spectroscopy, Selected Methods.* Academic Press, New York.

Varma, A., 1984, *C.R.C. Handbook of Atomic Absorption Analysis,* vols. 1 and 2. Chemical Rubber Co., Florida.

ICP, ICP-MS

Date, A. R., and A. L. Gray, 1988. *Applications of ICP-MS.* Blackie, New York.

Jarvis, K. E., A. L. Gray, and R. S. Houk, 1992. *Handbook of Inductively Coupled Plasma Mass Spectrometry.* Blackie, New York, 380p.

Thompson, M., and J. N. Walsh, 1988, *Handbook of Inductively Coupled Plasma Spectrometry,* 2d ed. Blackie, New York, 316p.

10
Borehole Sedimentology

Boreholes provide most primary data on sediments below the surface. Mastering borehole methodology and interpreting borehole information are vital in applied fields such as petroleum exploration, mining, geothermal energy, hydrogeology, and environmental science. Logs generated from geophysical instruments lowered by wires down into boreholes have been traditionally underutilized by sedimentologists because of a lack of education in the techniques and applications, as well as by costs (shallow drilling costs double if reasonable logging is carried out); they are misconceived as principally an engineering tool. This chapter provides an elementary understanding of the collection and interpretation of basic borehole geologic information from cuttings, cores, and geophysical (wire-line) logs.

THE BOREHOLE ENVIRONMENT

Despite differences in scale, petroleum, mineral, and water drilling all have similar components: a bit, a mechanism for dropping or turning the bit, and a system to clean away the cuttings (chips of rock cut by the grinding action of the drill bit), cool the bit, and prevent formation fluids (and gas) from invading the borehole and either causing collapse or blowouts (Fig. 10-1). For deep drilling, the most common drilling system is a rotary rig (the bit and the drill string or stem are mechanically turned by a motor). Drilling "mud" serves these purposes—it is circulated to clean the borehole by carrying cuttings to the surface, to cool and lubricate the bit, and to contain the formation fluids; it is composed of swelling clays (smectites), barite, various lighteners and viscosity changers, plus water (or oil) with a chemistry that is varied to optimize the borehole environment with drilling speed and costs. For shallow holes, the use of auger and percussion drilling rigs is still widespread; these technologies utilize simple, water-based (no additive) muds or air because of their low cost and the shallow depths that are penetrated.

For shallow boreholes (seeking water and proving most mineral ores and coal), rigs are small and commonly mounted on truck or tractor (helicopters may locate the rig in steep terrain). Water wells seldom are cored (due to the cost) and cuttings comprise the primary geologic information. Mineral exploration wells frequently will cut core continuously from the surface to total depth. Petroleum rigs are large and have elaborate mud systems; cores are expensive and therefore only taken over intervals of interest, and most of the direct geologic information comes from cuttings or wire-line logs. Even where core is cut, core recovery is seldom 100% and there may be appreciable gaps in the core record. Table 10-1 summarizes the common drilling technologies, their appropriate depths, and hole diameters. Hole diameter is an important factor influencing which wire-line logs can be run and interpretation of the data.

Mud-rock interaction is one of the most critical aspects of drilling technology. During drilling, the mud is circulated through the borehole by pumping it down the drill string and letting it flow up through the space between the drill string and the side of the hole (this zone is called the annulus). As the mud circulates, fluid filtered from it will invade porous and permeable formations; the filtrate displaces the formation fluids to create the *invaded zone*. The depth of invasion is a function of the rock permeability—a quantitative measure of the ease with which fluids can pass under a potential gradient; this gradient has a feedback relationship with mudcake (a buildup of mud from which the fluid has filtered into the rock; once created, this dictates the *rate* of mud filtrate flow into the formation—i.e., it acts as a seal to fluid migration), the mud pressure, mud viscosity and density, and the fluid pressure in the formation. Thus, around the well

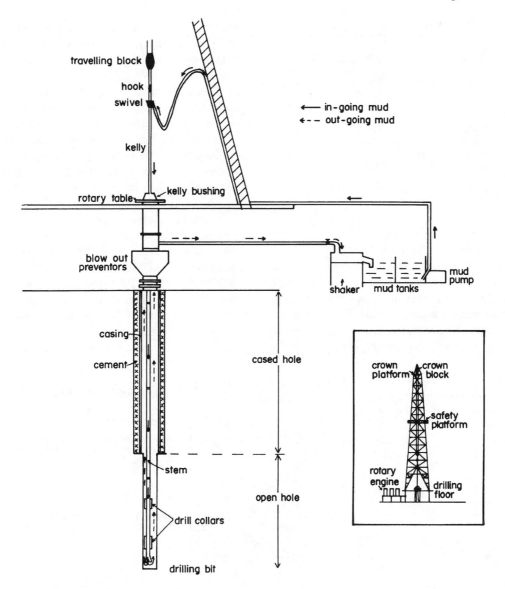

Figure 10-1.   Major features of a drilling rig.

bore there will be a zone (also called the *flushed zone*) where the mud fluids have displaced the original pore fluids. Generally this zone is only a few centimetres thick, but whatever the thickness, the effect has important implications for geophysical log interpretations, particularly because of the effect of the mud and its salinity on electrical properties of the system.

LITHOLOGICAL DRILLING LOGS

A lithological drilling log (commonly called the *mud log* because lithology is only one part of the log) comprises the record of well cuttings, drill rate, and various other specialized measurements such as calcimetry or gas detection (e.g., Fig. 10-2). It provides a description of the lithology, age, and thickness of units relative to depth. For any kind of exploration, these data are the foundation upon which subsurface mapping

and modeling rests, as well as geologic and economic interpretation. To construct a drilling log, considerable care must be taken to ensure both the cuttings and their corresponding depths are accurately recorded. Discussion here concentrates on the use of well cuttings in preparing the log.

Sample Gathering and Preparation of Well Cuttings

Selection of a well-cutting sample interval depends on the rate of (drilling) penetration and objectives of the well. In all cases, the logger must coordinate with the driller to calculate the lag between the time the rock is cut and the time it arrives at the surface in order to accurately log lithology with depth. As the well gets deeper, this lag period will increase. Because cuttings are frequently contaminated with cavings (pieces of rock from collapsing walls of the borehole well above the bit), geologic

Table 10-1.   Drilling Technology

Industry	Drilling Method	Borehole Diameter (cm)	Maximum Depth (m)
Petroleum	Rotary	90–10	7,000
	Turbo	90–20	11,000
	Cable	50–10	2,000
Hydrogeology	Cable	60–10	450
	Rotary	50–10	450
	Rotary/percussion	30–50	600
	Auger	90–15	25
Mining	Rotary	50–0	3,000

resolution does not improve greatly by sampling more frequently than every 1.5 metres. At fast penetration rates it may be possible to collect samples only every 5 metres. In areas where substantial lithologic variation occurs over thin intervals, as is common in coal measures, coring of the well is necessary for precise mapping of facies. In thick homogeneous lithologies, coring is generally unnecessary (exceptions exist, such as where the unit is mineralized or is a hydrocarbon reservoir); in any situation the very expensive coring technique of sample collection depends on the balance between potential value in information gained versus finance available.

Well cuttings are frequently contaminated not only with cavings from the hole but with casing cement or metal fragments from the surface casing or drilling equipment (also recirculated rock fragments and/or microfossils present in the drilling mud!). To reduce the number of contaminants, all cuttings should be washed through a coarse sieve, which will selectively remove most of them (but not all of the casing and bit fragments). Whereas in deep drilling rigs the drilling mud is separated from the well cuttings on a shaker table, in shallow rotary and auger drilling the mud normally remains mixed with the cuttings (air-drilled samples will contain rock dust). Such samples must be washed *gently* to remove the drilling mud or dust; note that when dealing with unconsolidated or weakly indurated sediment this procedure will remove the clay and silt from the cuttings and bias the sample. In addition, sample constituents may alter during the treatment or storage (e.g., sulfides can decompose, with the for-

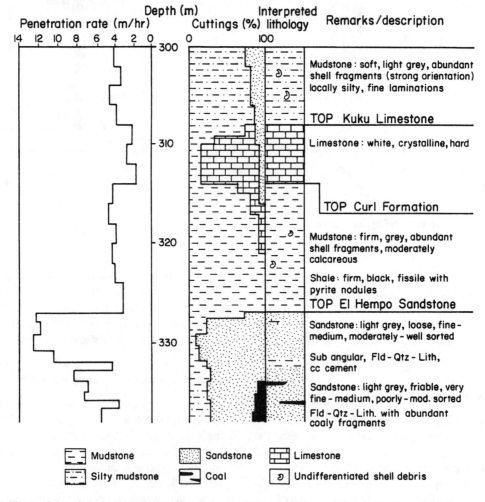

Figure 10-2.   Example of part of a well mud log, showing penetration rate, depth, and lithology. Complete mud logs also provide information on hole diameter, bit type used, and other data.

mation of sulfate compounds; clays may change their properties during washing or drying). In most circumstances, samples destined for micropaleontological examination should not be washed at the drill site but should be sent directly to the laboratory.

Each washed cutting sample should be placed in a separate small dish that is numbered by depth. Many geologists prefer to keep the samples wet for easier description. Assemble or purchase a reference manual containing the charts, methods, classifications, and other reference material necessary for the descriptions (and use it!); the specific composition of the collection will generally be dictated by the company for which you are working.

Examination and Description

Use a high-quality binocular microscope: your eyes and your geology will be better for it! Under the microscope at low magnifications, actual cuttings tend to be small and angular, whereas cavings will commonly be larger and more rounded. When describing cuttings, arrange the samples in a continuous series so that you can more easily detect changes in texture, lithology, and color, as well as detecting caving contamination.

Since time is precious at the well site, use symbols and shorthand annotations for lithologies and fossils whenever possible; ensure that all available information is on the logs. Be sure to make a key to all symbols on the cuttings description. Do not leave anything to memory. In the laboratory, as in the field, describe the samples using a consistent classification system (commonly specified by the employer) to ensure complete and systematic descriptions.

When describing arenites, there is a common tendency to overestimate the quartz content, partly because quartz grains are the easiest components to identify, but also because feldspar and rock fragments may be selectively crushed by the bit because they are mechanically less competent. Some experience—and even better, a reasonable reference sample—is necessary to permit compensation. Carbonates can be difficult to describe from cuttings because fabrics and fossils are commonly obscure; because of its simplicity, the Dunham classification (**PS** Chapter 8) is far easier to use with cuttings than other alternatives. Misidentification of basement has occurred when dealing with conglomerates because samples can contain a high proportion of crystalline basement clasts (the variety of basement lithotypes and the proportion of fines in the cuttings should help to differentiate between the lithologies).

During the description of the well cuttings (or core), samples must be selected for detailed laboratory analysis, such as paleontological, mineralogical, or geochemical. The choice of samples and the type of analysis to be performed is initially based on the experience of the geologist, within the constraints (cost, purpose) set by the employer. Reasons for the selection of each such sample should be noted.

Lithologic Log Preparation and Interpretation

Keep observation separate from interpretation. Prepare lithologic logs by plotting lithologies with descriptions against depth at a scale that provides maximum information. Compare the geologist's description (which is the % of observed cuttings in the sample for the interval drilled) with the driller's penetration log (Fig. 10-2); the penetration rate is a good indicator of lithologic uniformity or variability and will assist in picking formation boundaries (the rate is faster through less indurated rock such as mudstone than through arenites or rudites, although penetration rate also depends on bit type and hole size). Interpret from the lithological log you have prepared.

GEOPHYSICAL LOGS

Geophysical or wire-line logs depict information derived from torpedo-shaped instruments (*tools* or *sondes*) that measure rock and fluid characteristics such as electrical properties, radioactivity, temperature, and density. (Downhole cameras are being developed and may rapidly find widespread application but are not geophysical tools.) The tools are lowered down an open borehole and measure formation properties as they are raised to the surface (a few, as they are lowered); modern tools make readings every 15 cm (some read continuously). Results graphically display a continuous record of rock properties relative to depth on long strips of paper—the geophysical logs. By convention, lithology in the borehole is presented at the same scale as the logs and is represented graphically in the center, with lithological logs on the left and other measurements on the right (Fig. 10-3). Correlation of lithology and geophysical properties is made between wells by comparing these logs. Although the petroleum industry was the first to develop and use wire-line logs, they now are commonly used in hydrogeology and mineral exploration (Table 10-2).

Because of the expense, operators do not run wire-line logs on many water or chemical observation wells. Logs are run routinely on most deeper-water wells (>100 m) and all petroleum wells. Wire-line logs are particularly useful because they are recorded digitally and can be manipulated and displayed by computer, facilitating computer mapping and modeling. Logs also have the advantage of being measurements of in situ rock properties; laboratory measurements requiring correction for temperature, pressure, and different chemistries are not necessary.

The Caliper Log

A caliper tool is a mechanical device that uses either hinged arms or bow springs to measure the average borehole diameter; the variation in diameter with depth is recorded. The caliper provides information critical to the interpretation of other logs. It is important to recognize intervals where the

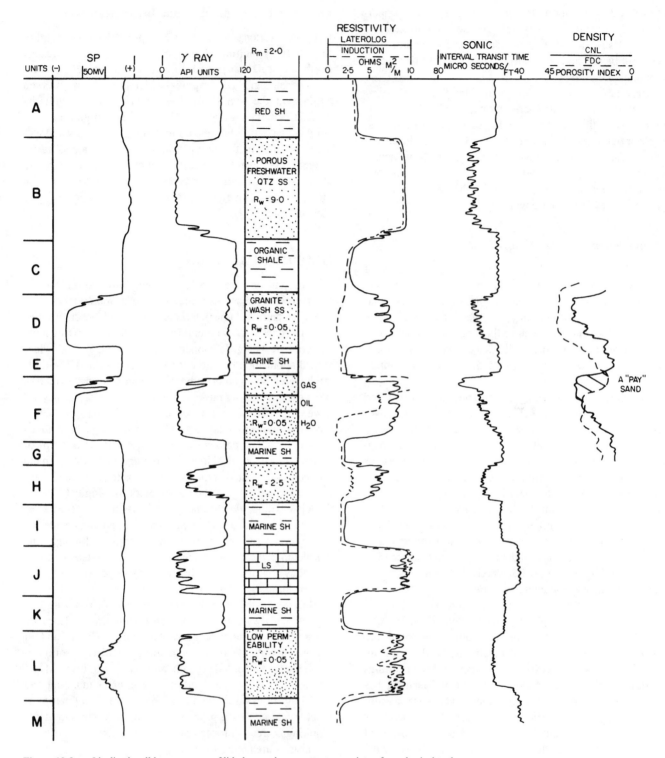

Figure 10-3.   Idealized well log responses of lithology and pore water to a variety of geophysical tools.

Table 10-2.   Wire-Line Log Uses

Log Application	Log Tools Used
Petroleum Applications	
lithology identification	GR SP Cal FD ND
fault detection	Dipmeter
fracture detection	Sonic
porosity calculations	Sonic FD ND
hydrocarbon detection	Restv FD ND
subsurface fluid identification	Restv
depositional environments	SP GR Dipmeter
subsurface mapping, both structural and stratigraphic	SP GR CP Dipmeter
Coal Exploration	
coal identification	GR FD ND Restv
Young's modulus calculations	FD
coal seam mapping	GR ND Restv Cal
Hydrogeology	
water table surface	ND
lithology identification	SP GR Cal
water salinity	Restv
aquifer identification	GR Temp
porosity	Sonic FD ND
specific yield	ND

SP: spontaneous potential log	Cal: caliper log
GR: gamma-ray log	FD or BD: formation or bulk density log
ND: neutron density log	Sonic: sonic log
Temp: temperature log	Restv: resistivity log

hole is considerably larger than the bit diameter (called a washout): over these zones the porosity tools may fail to make contact with the side of the hole and anomalous readings result. Similarly, resistivity and density logs must be corrected for bore diameter before many calculations (a task easily accomplished by computer).

The caliper also provides information on lithology and fractures. Evaporites, friable sands, and coals are prone to wash out, mudstones are generally slightly overgage, and well-indurated formations such as limestone and firm sandstones will be at gage (*gage* is the diameter cut by the bit). A log that shows rapid and substantial variation in diameter is an indication of abundant natural fractures.

The caliper can also indicate permeability through the presence of mudcake. As the bit penetrates a permeable layer, some of the drilling mud (which is commonly at a higher hydrostatic pressure than the formation fluids) will penetrate into the formation, leaving a buildup of mud on the inside of the hole. This mud is called *mudcake* and its presence indicates some permeability (see Fig. 10-4). This mud loss may also show up on the driller's log.

The Spontaneous Potential (SP) Log

SP logs measure the natural electrochemical differences between lithologies in a well. The tool functions by measuring the (milli)voltage difference between a reference electrode at the surface and an electrode in the tool. The voltage difference is a function of the salinity of the drilling mud, the salin-

ity of the formation fluid, and the lithology. The SP log is particularly valuable in telling geologists the salinity of the formation waters relative to the drilling mud. If the salinity of the drilling mud and filtrate is known, there are a number of methods for calculating the salinity of the formation water. Qualitatively, if the SP response in a clean formation is a shift to the right, the salinity of the formation water is fresher than the salinity of the drilling mud.

The SP log will respond a number of ways to different lithologies (Figs. 10-3, 10-5). In theory, the electrochemical effect is different for clay-rich vs. clay-poor (clean) formations because of the high cation-exchange capabilities of clay minerals. A relatively negative response (a shift to the left on a log) is indicative of a clean formation (sandstone, carbonate, or salt), whereas a positive response (shift to the right) indicates clay-rich strata (mudstones, siltstones, marls). Introducing clay into a clean formation (e.g., by drilling mud infiltration) decreases the SP deflection in proportion to the amount of clay. Well-indurated, impervious carbonates may have little or no SP response. However, the SP log response depends on factors other than lithology, and the gamma-ray log (see below) has replaced the SP log as the primary lithology indicator.

The Gamma-Ray Log

The gamma-ray log depicts natural gamma radioactivity of strata in the borehole. For most sediments, the primary source of radioactivity is potassium, with relatively minor contributions by uranium and thorium; hence highs tend to reflect the abundance of illitic clays, mica, and/or organic matter (other than most coals) that adsorbs radioactive ele-

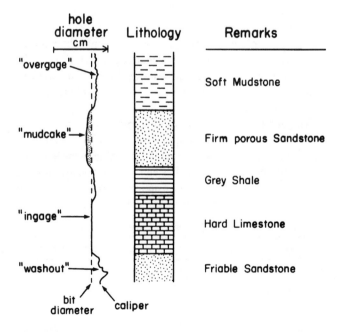

Figure 10-4.   Example of caliper log, showing presence or absence of mudcake.

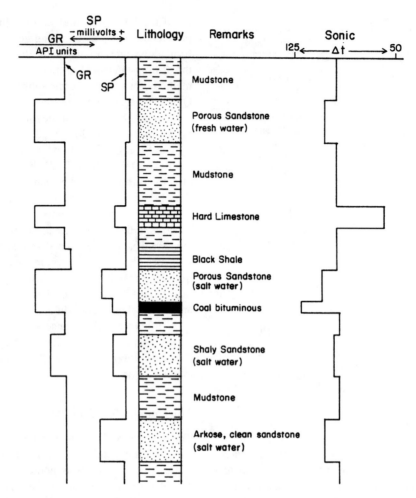

Figure 10-5. Example of geophysical log showing SP (self potential), GR (gamma ray), and sonic log profiles.

ments (Table 10-3; Fig. 10-3). Gamma rays have high energy; thus, the gamma-ray log can be run inside casing. Due to absorption effects, gamma rays penetrate only about 30 cm of rock; rocks immediately adjacent to the well bore contribute most to the readings. Gamma-ray logs are reported in standardized API (American Petroleum Institute) units, commonly from 1 to 150.

The gamma log is a popular lithological indicator. However, geologists should supplement logs with cuttings (or core information whenever possible) because intensity readings for carbonates and arenites and some coals are similar; in addition, common attempts to delineate grain size trends in

sandstones from the logs are beset by major problems (e.g., Rider in Hurst, Lovell, and Morton 1990). In addition, whereas the most common potassium-rich sediments are shales with abundant illitic clays, micas, and disseminated organic matter, alkali feldspars and potash-rich evaporites can also provide strong responses. Consequently, clean feldspathic or glauconitic sandstones may have high gamma readings and be confused for dirty, clay-rich sediments. Coal normally has low radiation, but there are rare exceptions. The gamma log is particularly useful in picking contacts between formations, detecting hydrocarbon source rocks, and mapping net aquifer, reservoir, or coal thicknesses. Comparison of the gamma log with the SP and resistivity logs permits maximum lithologic resolution.

Resistivity Logs

Resistivity logs depict the resistance of formations to an applied electrical current. Resistivity tools vary in the method by which they emit an electrical current and in the spacing of electrodes in the tool. *Induction* resistivity tools emit an electric current by inducing a magnetic field into the zone

Table 10-3. Radioactive Elements in Lithologies (ppm)

Lithology	Uranium	Thorium	Potassium
Shale	3.7	5.4	15.0
Sandstone	1.0	1.4	4.8
Limestone	1.3	0.5	0.6
Granite	4.0	5.9	18.3

Source: Schlumberger Ltd. 1972.

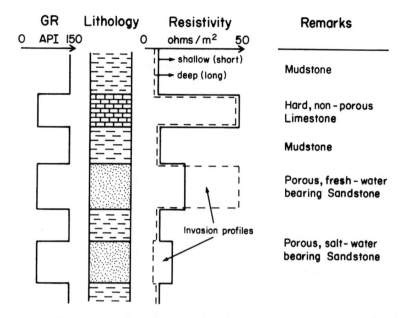

Figure 10-6. Example of log showing relationship of shallow and deep resistivity responses to lithology and pore waters.

around the borehole. *Laterolog* tools emit a current directly between two electrodes. Generally, the closer the electrodes in the tool, the closer to the borehole the resistance is measured. Induction logs perform better in holes filled with air or freshwater-based muds. Laterologs are normally used in saltwater-based muds. On all resistivity logs, resistivity measurements are displayed to the right of the borehole and are reported in ohm metres squared (Figs. 10-3, 10-6).

Most petroleum and hydrological logs contain at least two resistivity readings: deep and shallow. The shallow reading conveys information about the zone invaded by drilling fluids and can provide information on bed boundaries and rock characteristics (Fig. 10-6). The shallow resistivity is thus commonly used for correlation and sedimentological interpretations. The deep resistivity is designed to "look" beyond the zone contaminated by drilling mud, and reflects the porosity and fluid content of the formation. Consequently, deep resistivity is used mainly for evaluating groundwater and hydrocarbons. The two resistivity readings are displayed together to reveal the presence of porous formations. Generally the resistivity devices will have similar readings in impervious strata (such as mudstones) but differences when measuring the resistivity of a porous formation (such as sandstone). The separation between the readings is called an *invasion profile* (Fig. 10-6). All else being equal, in a porous formation the deep resistivity will increase as the water salinity decreases. (*Note:* For any salinity, the resistivity will decrease as temperature increases.)

In petroleum applications, the resistivity readings are used along with porosity values to calculate the relative amount of hydrocarbons within a reservoir (a value called water saturation or SW for short). High SW values indicate low amounts

of hydrocarbon and vice versa. Archie in 1942 (see Archie 1950) empirically derived the equation for SW as follows:

$$SW^n = aR_w / \phi^m R_t$$

where

SW = water saturation
n = the saturation exponent (= 2 for basic interpretations)
a = an empirical constant
m = the cementation factor
ϕ = porosity
R_t = the formation resistivity
R_w = the resistivity of the formation water

Essentially, this equation states that the water saturation is a function of the formation's porosity and the ratio between the resistivity of the formation water and the measured resistivity.

Sonic Logs

Sonic logs measure the velocity of acoustic waves through the rock adjacent to the borehole. Sonic tools transmit sound waves from a point on the tool and measure the time the wave takes to reach two or more receivers. The time of wave emission and the distance between receivers are known, and the interval transit time between the two receivers is automatically calculated. The sonic log is normally plotted to the right of the borehole and is reported in either microseconds per foot per foot or microseconds per metre per metre.

For most petroleum wells the sonic log is compensated for borehole size. In water wells, compensation is less common, and it is important to compare noncompensated sonic logs with caliper logs to detect spurious readings. A primary function of sonic logs is the calculation of porosity. Theoretically, the faster the transit time, the more dense the formation. The general formula for sonic porosity is:

$$\theta = (dT_{log} - dT_{ma}) / (dT_f - dT_{ma})$$

where

θ = porosity
dT_{log} = the velocity measured by the log
dT_{ma} = the velocity of the rock matrix (a constant)
dT_f = the velocity of the formation fluids (a constant)

Essentially, the equation states that the porosity is equal to the ratio of the difference between the log and the matrix velocities and the fluid and the matrix velocities. In logging nomenclature, *matrix* refers to all the mineral components of a rock. Usually the density of the framework grains (in sandstones) is used as a proxy for the total matrix density; some common constants for rock matrix are given in Table 10-4 . In more advanced logging applications, geologists can actually model the matrix density for different components.

In terms of lithological identifications, most lithologies do not have a characteristic sonic velocity (Fig. 10-3; Table 10-4). However, anhydrite normally has an anomalously high velocity, and coal and halite have unusually low velocities, and in some cases sonic resolution is good for other lithologies (e.g., Fig. 10-5). Sonic logs are also used to identify specific horizons or intervals that, because of velocity contrasts between two lithologies, act as seismic reflectors in seismic profiles. This application of the logs is more important in coal and petroleum work than in hydrogeology. Other important geologic uses of sonic logs include identifying and correlating marker horizons such as thin hard limestones, tuffs, and shales; describing the compaction profile and thermal maturation in a basin; assessing the significance of unconformities; determining the ash content of coals.

The Formation Density (FD) Log

Formation or bulk density logs measure the electron density of the rock adjacent to the well bore. The tool bombards the surrounding rock with gamma rays and then calculates the

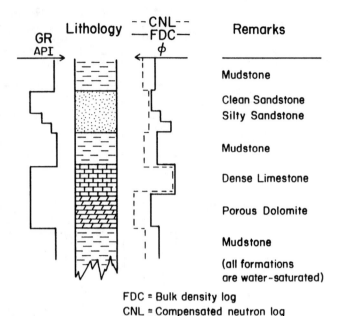

Figure 10-7.   Example of log showing relationship of neutron and bulk density logs to lithology.

rock density based on the scattering of electrons. The gamma-ray source is mounted on a skid and must be in contact with the side of the borehole for the tool to function properly. For this reason, rough or oversized holes (see caliper logs above) will cause incorrect density measurements.

Like the sonic log, the bulk density log is presented to the right of the well bore on logs (e.g., Fig. 10-7). The measurements are recorded in grams per cubic centimetre. In petroleum applications, the bulk density log is run and presented with a neutron density log.

The density log is also used to calculate porosity. Bulk density porosity calculations reveal both the primary and secondary porosity but theoretically do not detect fracture porosity (exceptions exist). The equation is similar to acoustic log porosity:

$$\theta = (P_m - P_b)/(P_{ma} - P_f)$$

where

θ = porosity
P_{ma} = the density of the rock matrix (a constant)
P_b = the density measured by the log
P_f = the density of the formation fluid

Essentially, the equation states that the porosity is equal to the ratio of the difference between the log and the matrix densities and the difference between the fluid and matrix densities. Some common constants for matrix densities are given in Table 10-5.

The bulk density log can be used in conjunction with other logs to identify lithology by plotting density against the neutron or sonic log (particularly useful in carbonate facies dis-

Table 10-4.   Common Rock Matrix Values (in μs/ft/ft)

Anhydrite	50
Halite	67
Calcite	47.5
Dolomite	43.5
Quartz	51.5
Shale	70–150
Coal	120–170

Table 10-5.   Common Constants for Rock Matrix Densities (in g/cm³)

Anhydrite	2.97
Halite	2.03
Quartz	2.65
Calcite	2.71
Dolomite	2.87
Freshwater	1.0
Saltwater	1.14
Coal	1.2–1.8

crimination; e.g., Fig. 10-7). A plot of porosity using a lime-stone matrix displayed alongside the neutron (limestone) porosity can reveal the degree of dolomitization. In coal exploration the bulk density log is used to calculate Young's modulus of formations and evaluate their competence relative to mining design, and to calculate the ash content of coal in situ.

Neutron Logs

Neutron logs measure the gamma radiation emitted when a formation is bombarded by neutrons. The tools emit neutrons from a radioactive source and measure the resultant gamma radiation as the neutrons collide with nuclei in the adjacent rocks. Maximum energy loss occurs when neutrons collide with hydrogen; hence the intensity of gamma recordings is a measure of hydrogen concentration (Fig. 10-7). Like the other porosity tools, neutron log response depends on borehole conditions. Some tools (which produce compensated neutron logs) are designed with two receivers to correct for borehole effects; others (usually slim-hole tools) are not compensated and must be interpreted differently.

In petroleum applications, neutron logs display their measurements as apparent porosity units relative to a reference matrix (usually sandstone, limestone, or dolomite). In hydrological applications, readings may be in percent moisture content or a hydrogen index (quantity of hydrogen per unit rock volume).

Since hydrogen is most abundant in water or hydrocarbons, neutron logs are used for interpreting porosity and the associated pore fluids. Hydrogeologists use the neutron log to identify and characterize the undersaturated zone, the phreatic surface, and aquifer storage. Petroleum geologists use the neutron log in identifying hydrocarbons and lithology.

In lithologic interpretation, geologists use the neutron log in conjunction with the sonic and bulk density logs to construct a geophysical signature for rock facies. This technique is especially useful in interpreting carbonates. In both qualitative and quantitative interpretations, the user must remember the bias of the neutron tool's calibration (i.e., sandstone, limestone, or dolomite).

CORRELATION USING GEOPHYSICAL LOGS

The first step in interpretation is to compile all relevant information from core and cutting examination plus geochemical and paleontological data from laboratory analyses and record them on the wire-line log. Care should be taken to reconcile cutting and core information with the wire-line logs: depths recorded for each seldom match. With computer log software packages, the geologist can select the tools and scales that best suit interpretation purposes and establish separate fields for each kind of datum. Without the computer, geologic information must be manually annotated in the center of the log.

Selection of Units

The initial phase of interpretation requires identification of units (stratigraphic and/or genetic) with which to work. When dealing with large-scale basins, fill tends to be episodic and gives rise to thick packages of strata, differing in age and genetic relationships. When dealing with smaller areas, as is common with coal exploration and exploitation, units may be highly discontinuous and facies relationships highly complex (and well spacing must be much shorter to permit stratigraphic correlation). For most subsurface work, practical resolvable units must be recognizable (mappable) over the entire study area; hence they are comparable to the groups, formations, and members used in field mapping. On the smaller scale, strata can be organized into genetically related units based on the hierarchy of contacts within larger units. Among the lowest order are individual cross-bed sets, whereas at the upper level are sequences bounded by regional unconformities.

Generally speaking, few procedures are more valuable than visiting the nearest outcrop or using a geologic map and trying to "tie" the lithologies into what has been interpreted in the well. Particularly in water wells or waste-disposal wells, many misidentifications of aquifers have been made because the geologist did not visit the outcrop (in some cases only 1 km away). When working with subsurface data, stratigraphic units in the hierarchy are identified by:

1. Marker beds, which should be consistently recognizable, relatively continuous lithologies, and the character and lateral continuity of major bounding surfaces. Modern wire-line logs, particularly the short resistivity and gamma-ray logs, have sufficient resolution for determining whether contacts are sharp or gradational.

Some units that are particularly distinctive in the subsurface are volcanic ash (high-gamma, low-resistivity log response); thin carbonate beds (high-resistivity, high-sonic response); coal beds (low-density, low-sonic response); black shales (high-gamma, low-sonic response). Genetic intervals may manifest themselves as rhythmic repetitions or cycles of lithologic units.

Unconformities make particularly good boundary surfaces for large-scale units. Unconformities may be identified by correlating known unconformities from outcrop to the well, by dramatic changes in the characteristics of the sediment sequence above and below a surface, or by other criteria such as those listed in **PS** Table 4-4.

2. The vertical profile and geometry of the strata between the bounding surfaces. Wire-line logs and lithologic profile columns are vital in comparisons between sites. Redisplay of wire-line logs at a larger scale permits the geologist to "see the forest through the trees" and recognize sedimentary packages based on log response.

3. The ages of the strata between bounding surfaces. Insofar as the bounding surfaces as well as the lithologic units commonly cross time planes, care must be taken when utilizing and interpreting this information.

Stratigraphic resolution must be kept compatible with available geologic detail and the purpose of the current project. There is no point in making microstratigraphic delineations without substantial, unambiguous evidence, or more importantly, if the problem does not require it. Be particularly careful when using global concepts such as eustatic sea-level changes to interpret local stratigraphy. Be sure that the correlations on a small or basin scale are correct before imposing an outside model.

Correlating Logs

A central part of all mapping and interpretation is the three-dimensional variation of the unit under study. To assess and map a unit's lateral changes it is necessary to correlate selected units or intervals between wells and outcrops. Most subsurface correlation proceeds from the postulate that wire-line log responses are consistent enough to allow matching log signatures between wells. Correlation is thus easiest and most reliable within lithologies that are widespread (principally mudstones and evaporites). When comparing two wells, missing sections can be due to either facies changes, erosion, strike-slip or normal faulting; repeated sections are a result of thrust faulting. There are basically two types of correlation procedures: by lithology or lithofacies (Fig. 10-8) and by time using either dates or with seismic profiling (Fig. 10-9). General rules for correlating are as follows:

1. Begin correlating on a large scale and work progressively down to finer detail.

2. Make a grid of cross sections that are relatively close in space and wherein correlations between wells (and outcrop if possible) can be made with precision (e.g., a fence diagram, see Chapter 2). Early establishment of such a grid will force correlations to be consistent among those data points and will provide a firm stratigraphy that can more easily be extended to more distant data points. Initially construct a series of cross-sections that pass through the same points, and build outward in a radial pattern. Select distinctive boundaries between lithologic units to draw between the borehole control points. Correlate the easiest intervals first (e.g., mudstones), then use the pattern of these units to help unravel the more difficult sections. The general picture of onlap, erosion, thickening, and thinning probably will be indicated by the initial pattern. (Once the general pattern is known, cross-sections constructed parallel and perpendicular to the facies changes can depict the geology most effectively.) Only when convinced of reliable correlation should you proceed to the next step.

3. Select "marker" horizons that are continuous and easily recognized, such as unconformities, to act as the baseline on which isopach maps can be hung. Then construct a variety of

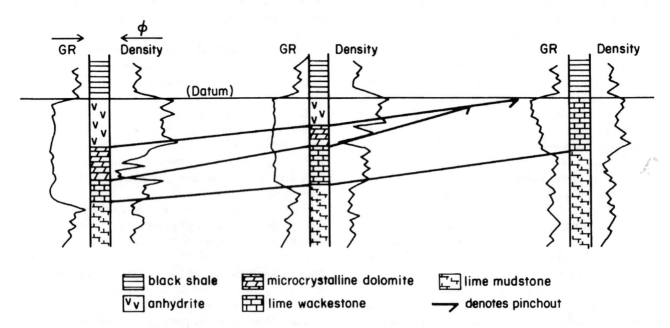

Figure 10-8.   Example of correlation between logs based on lithology.

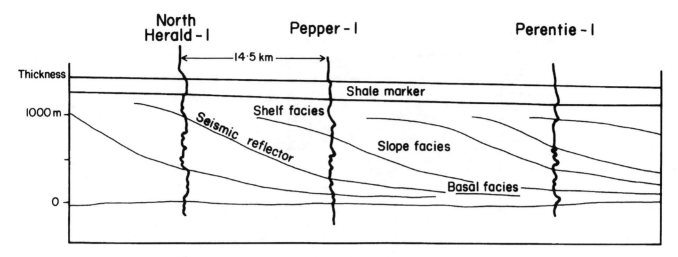

Figure 10-9.   Example of correlation based on interpretation of seismic patterns.   (After Boots and Kirk 1989.)

maps (e.g., isopach, net sand, percentage carbonate or dolomite, log character, or lithofacies maps; see Chapter 2) for correlative intervals use a series of different lower marker beds and see if they make geologic sense.

Forcing of lithologic correlations should be avoided. When sections become difficult to correlate, there usually is an important geologic cause and these areas should be noted on the regional map. For example, shelf channels pose a common correlation problem because the channel fill both cuts out

widespread marker beds and superimposes localized and environmentally out-of-place units into the shelf succession (Fig. 10-10). Some submarine fan and terrestrial environments may be extremely difficult to correlate because sections lack marker beds and/or fossil control. In these cases the geologist must make extensive use of seismic profiles (if available).

All available data must be used when correlating units, particularly over long distances. Although seismic principles and interpretations are beyond the scope of this book, information from the sound waves reflected from lithological boundaries is

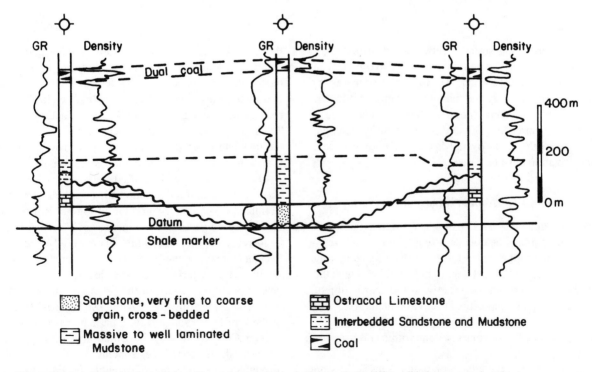

Figure 10-10.   Example of correlation of logs when a probable shelf channel with different lithology appears in only one well.   (After Wood and Hopkins 1989.)

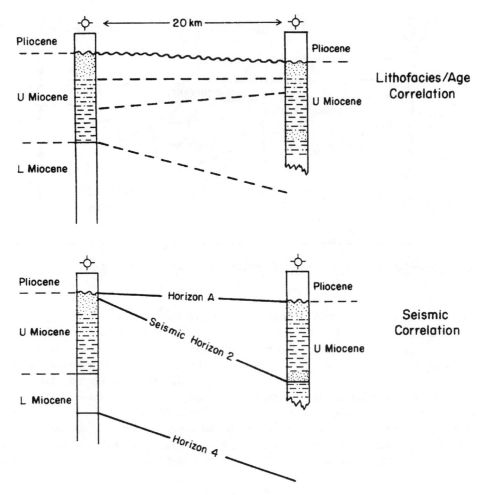

Figure 10-11.   Example of the differences possible in correlating between wells when using different kinds of geophysical data.

important in correlating surfaces between distant localities. Fig. 10-11 shows how two entirely different correlations can be made between distant sections. The top sketch shows a correlation based on lithology and microfossil dating; the lower, based on the seismic pattern, shows that the two sections are in fact genetically unrelated. Both correlations make geologic sense; it is the *quality* of the overall data available on which the final interpretation must be based.

Interpretation of Depositional Environment

Because logs generally lack sedimentary detail (although some information may be obtained—e.g., Table 10-6), sedimentological interpretation in the subsurface tends to rely heavily on concept. The most important is Walther's law: the vertical distribution of lithofacies reflects the distribution of laterally continuous depositional environments (see **PS** Chapter 2). Remember that vertical profile analysis (an application of Walther's law) must be done within conformable stratigraphic packages.

Subsurface environmental interpretation relies heavily on

the log (vertical) profile, three-dimensional geometry (including thickness and lateral extent) of each unit, paleontology, and the character and inferred environment of the underlying and overlying units. The first step in all cases is to recheck lithological correlations and ensure they are correct. The second step is to examine vertical and lateral patterns in the logs and place them in a three-dimensional framework (i.e., with maps of intersecting cross-sections).

Once the units are identified, their internal geophysical characteristics and interrelationships can be interpreted. For example, a common assumption in log vertical profiling is that decreased gamma-ray or increased SP deflection reflects an increase in grain size, which is commonly inferred to reflect an increase in energy of the depositional processes. Basically, three types of log profiles for intervals can be determined by the shape of the SP or GR curves: fining upward, coarsening upward, and blocky (Fig. 10-12). The scale of the profile can range from a particular bed to a thick sequence. Contrary to what is written in some books, none of the basic shapes constitutes conclusive evidence for any environment, but in conjunction with the three-dimensional

Table 10-6.   Paleoenvironmental Information That May Be Inferred from Well Cuttings and Cores

Color: Whereas color may be imparted both before, during, and after deposition, in conjunction with other factors it may prove to be useful evidence of depositional environmental conditions (e.g., black suggests anoxic conditions; mottling may indicate bioturbation).

Structures: The dipmeter can provide useful information on primary structures down to the scale of cross-bedding (e.g., Serra 1985). Well cuttings generally are too small for recognition of stratification, although fine lamination (mainly indicative of still waters) and bioturbation (indicative of oxygenated bottom waters) may be apparent. Small-scale structures (e.g., ripple cross-lamination indicative of the lower flow regime) are well displayed in core, and can provide important diagnostic information (see **PS** Chapter 4). Fracture analysis is generally performed from sonic logs.

Texture: The degree of sorting is generally a good indicator of the relative energy of entrainment deposits (see **PS** Chapter 5), although grain history and provenance (particularly multiple cycles of sedimentation) are also involved. Grain size is less useful because it is a function of provenance, weathering, and other nondepositional factors; similarly, roundness is a function of the grain history and may not be related to the environment. In all cases, one must remember that the sample may contain a mix of environments and processes because bioturbation may have mixed initially discrete layers. Analysis of thin sections of cuttings or cores can provide further information.

Mineralogy: The presence of certain minerals can give environmental clues. The presence of glauconite suggests a marine environment, and if concentrated, an unconformity immediately below or nearby. Phosphate nodules also tend to indicate marine conditions and prominent unconformities. Pyrite commonly occurs in mudstones and generally indicates a reducing environment. Analysis of thin sections of cuttings or cores can provide further information (e.g., **PS** Chapter 6).

Fossils: The most common fossils in well cuttings are foraminifera, pollen and spores, and shell fragments that cannot be identified to species level but can commonly show morphological characteristics or can be identified to class (or even genus level) and may be generally indicative of environment (e.g., **PS** Chapter 2). In both cases, reworking in the depositional environment must be carefully evaluated before reaching paleoecological conclusions. Microfossils can be plentiful in mudstones samples; a paleontologist will render an environmental interpretation. Without a paleontologist, a geologist can record the ratio of planktic to benthic forms and construct a crude, relative depositional depth profile, but exceptions to the generalized trends may exist in particular basins because of their hydrodynamic situation or varied substrate character.

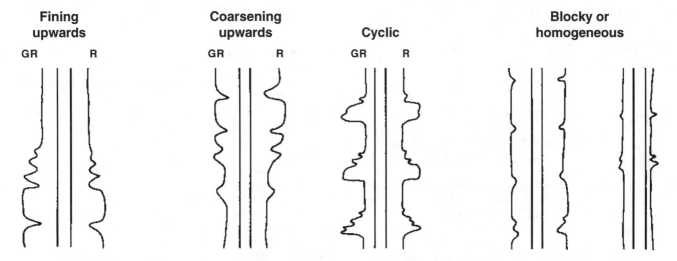

Figure 10-12.   Idealized electic log profiles for commonly recognized lithological sequences.

Table 10-7.   Sample Tabulation Entries to Summarize Major Recognition Criteria for Subsurface Units

Interval	Log Pattern	Geometry	Color/Texture	Associated Facies
Zeus shale	High gamma	Blanket	Black	Overlain by Athena chalk
Thor sandstone	Fining up	Linear strike	Moderate–poorly sorted	Rests on unconformity overlain by Zeus shale

geometry of the units they may assist environmental interpretation.

The traditional analysis procedure is crudely sequential: (1) recognition and description of units, then of distinctive intercals (each of which may comprise one unit or a genetically associated mix of units). Since subsurface interpretation relies on an assemblage of nondefinitive factors, it may be helpful to create a table summarizing the evidence for each unit or interval as illustrated in Table 10-7. The procedure continues with (2) a comparison of the genetic sequence to idealized models (see **PS** Chapter 2) and (3) a comparison of vertical and lateral facies changes seen in wells to available outcrops to determine the degree of variability.

The most common interpretative pitfall is the misapplication of vertical profile models. Situations where idealized profiles can be misapplied include:

Superposed environments (e.g., stacked barrier bars).

Compound environments with rapid facies changes (e.g., a fluvial channel incised into a shelf environment). Coal measure sequences generally show much greater facies variability than in many other settings.

Local spurious effects induced by borehole size changes, cave-in and other contamination, or by local fracturing and/or diagenetic modifications.

From these brief comments on interpreting subsurface data it is clear that several to many possible interpretations may be equally viable for a given situation; the geologist should clearly indicate alternatives and point out means of making interpretations more certain.

SELECTED BIBLIOGRAPHY

Archie, G. E., 1950, Introduction to petrophysics of reservoir rocks. *American Association of Petroleum Geologists Bulletin* 34:943–61.

Argall, G. O., Jr., 1979, *Coal Exploration, 2nd International Coal Exploration Symposium Proceedings* vol. 2. Denver, Colo., October 1978, Miller-Freeman, San Francisco, 560p.

Asquith, G., and C. Gibson, 1982, *Basic Well Log Analysis for Geologists, Methods in Exploration.* American Association of Petroleum Geologists/Society for Sedimentary Geology, Tulsa, Okla., 216p.

Boote, D. R. D., and R. B. Kirk, 1989, Depositional wedge cycles on evolving plate margin, western and northwestern Australia. *American Association of Petroleum Geologists Bulletin* 73:216–43.

Busch, D. A., 1974, Stratigraphic traps in sandstones—Exploration techniques. *American Association of Petroleum Geologists Memoir 21,* 174p.

Cripps, J. C., F. G. Bell, and M. G. Culshaw, 1986, Groundwater in engineering geology. *Geological Society, Engineering Geology Special Publication No. 3,* Geological Society of London, London, U.K.

Fons, L., 1969, *Geological Applications of Well Logs.* Society of Professional Well Log Analysts 10th Logging Symposium, Houston, Texas, pp.1–41.

Galloway, W. E., 1978, *Exploration for Stratigraphic Traps in Terrigenous Clastic Depositional Systems.* American Association of Petroleum Geologists Short Course No. 3, lecture notes, Houston Geological Society.

Hobson, G. D., and W. Pohl (eds.), 1973, *Modern Petroleum Technology.* Applied Science Publishers, Barking, Essex, 996p.

Hurst, A., M. A. Lovell, and A. C. Morton (eds.), 1990, Geological applications of wireline logs. *Geological Society of London Special Publication 18,* Geological Society of London, London, U.K.

Jaegler, A. H., and D. R. Matuszak, 1972, Use of well logs and dipmeter in stratigraphic-trap exploration. In R. E. King (ed.), *Stratigraphic Oil and Gas Fields,* American Association of Petroleum Geologists Memoir 16, pp. 107–35,

LeRoy, L. W., D. O. LeRoy, and J. W. Raese, (eds.), 1977, *Subsurface Geology: Petroleum Mining Construction,* 4th ed. Colorado School of Mines, Golden, Colo., 941p.

LeRoy, L. W., D. O. LeRoy, S. D. Schwochow, and J. W. Raese, 1987, *Subsurface Geology, Petroleum, Mining, Construction.* Colorado School of Mines, Golden, Colo.

Low, J. W., 1951, Examination of well cuttings. *Quarterly of Colorado School of Mines 46,* 47p.

Lynch, E. J., 1962, *Formation Evaluation.* Harper & Row, New York, 422p.

Merkel, R. H., 1981, *Well Log Formation Evaluation.* American Association of Petroleum Geologists Continuing Education Series No. 14, American Association of Petroleum Geologists, Tulsa, Okla., 82p.

Moore, C. A., 1963, *Handbook of Subsurface Geology.* Harper & Row, New York, 235p.

Neidell, N. S., 1980, *Stratigraphic Modeling and Interpretation: Geophysical Principles and Techniques.* AAPG Continuing Education Series No. 13, American Association of Petroleum Geologists, Tulsa, Okla. 145p.

Payton, C. E. (ed.), 1977, Seismic stratigraphic applications to hydrocarbon exploration. *American Association of Petroleum Geologists Memoir 26,* 516p.